MICROBIOLOGY RESEARCH ADVANCES

OYSTERS AND CLAMS

CULTIVATION, HABITAT THREATS AND ECOLOGICAL IMPACT

MICROBIOLOGY RESEARCH ADVANCES

Additional books in this series can be found on Nova's website
under the Series tab.

Additional e-books in this series can be found on Nova's website
under the eBooks tab.

MICROBIOLOGY RESEARCH ADVANCES

OYSTERS AND CLAMS

CULTIVATION, HABITAT THREATS AND ECOLOGICAL IMPACT

JESÚS L. ROMALDE
EDITOR

nova publishers
New York

Library of Congress Cataloging-in-Publication Data

Names: Romalde, Jesús L., editor.
Title: Oysters and clams : cultivation, habitat threats, and ecological impact / editor, Jesús L. Romalde (Professor of Microbiology, Departmento de Microbiologbia y Parasitologbia, Facultad de Biologbia and Institute of Aquaculture, Universidad de Santiago de Compostela, Spain).
Description: Hauppauge, New York : Nova Science Publishers, 2016. | Series: Microbiology research advances | Includes index.
Identifiers: LCCN 2016019042 (print) | LCCN 2016024448 (ebook) | ISBN 9781634850742 (hardcover) | ISBN 9781634850964 ()
Subjects: LCSH: Oyster culture. | Clam culture.
Classification: LCC SH371 .O967 2016 (print) | LCC SH371 (ebook) | DDC 639/.41--dc23
LC record available at https://lccn.loc.gov/2016019042

Published by Nova Science Publishers, Inc. † New York

To Mado and Totó

CONTENTS

PREFACE

Bivalve molluscs, and particularly oysters and clams, are scientifically and economically important invertebrates from a global distribution and aquaculture production standpoint. In recent years, due to the advances in farming technology, environmental monitoring and disease control, together with the diversification of cultured species, new findings and outcomes have emerged. The purpose of this volume, which presents a compilation of the latest research on oyster and clam culture, is to gather together information on aspects of ecological distribution and management, disease, physiological response to environmental stressors, or host-pathogen interactions, in most cases from a multidisciplinary approach.

The distribution, habitat and ecology of the "disco" or "electric" clam *Ctenoides ales* (*Limidae*) is described by Lindsey Dougherty in the first chapter. This clam species is the only known bivalve with a vivid flashing display, result of specialized tissue that scatters light. Three locations in Indonesia and one in Australia were explored, identifying four main habitats including coral reef walls, coral reef bommies, caves, and karst islands, where the clams were often found in clustered populations. The clam size distribution, indicating a sex ratio favoring males, as well as issues associated with *C. ales* cultivation and habitat threats are also discussed in the chapter.

The second chapter entitled *Culture of wedge-shell clam Donax trunculus. New developments*, describes the advances in the intensive farming of this valuable molluscan species. Since populations in natural beds are currently in regression, the development of hatchery culture is a clear benefit to shellfish industry, both as a strategy for the obtention of seed for restocking and to increase production. All stages of cultivation are considered in this chapter by Andrea Louzán et al., from stabling or conditioning of broodstocks at hatchery facilities to on-growing in the natural beds.

In the third chapter, Luci Jesus and coworkers present the results of a study on the possibility of maintaining already depurated clams for longer, up to 3-4 weeks' periods in a parallel system to the depuration unit. Such possibility can be considered interesting from a commercial point of view, having in mind the importance of food safety related to the consumption of shellfish and taking into account the pertinent legislation, since it would allow circumventing periods of ban on harvesting and still ensuring the condition and safety of bivalves for further marketing and human consumption. The authors utilize an individual bioenergetics model based in the scope for growth concept to predict the condition and growth of clams maintained in a system similar to that used for depuration at different temperatures and concentrations of microalgae.

The study of the microbiota associated to clams and oysters is essential not only to know the sanitary status of mollusc populations but also to design new strategies to prevent and control disease outbreaks. The fourth chapter, authored by Sabela Balboa and coworkers, aims to give an overview of the state of art on the composition of the microbiota associated to clam and oyster cultures, and its putative role on the success of their culture. They reviewed not only the bacterial pathogens, which are the main cause of the repetitive mortality episodes that seriously decimate the mollusc production worldwide both in hatcheries and natural beds, but also bacteria with antimicrobial activities (probiotics) which are becoming essential to control pathogenic microorganisms and therefore improve larvae survival. Finally, the application of next generation sequencing techniques, which had provided a window into the complex world of microbial communities, to the study of the interactions microbe/mollusc and and their usefullness in the improvement of shellfish production are also discussed.

The fith chapter describes the microbiological results obtained within a multidisciplinary study performed by Juan L. Barja and coworkers trying to explain the significant reduction of the wedge-shell clam (*Donax trunculus*) populations in Galicia (NW Spain) during the last ten years. The results obtained seem to indicate that there is no apparent infectious cause to explain the regression of this bivalve species in the area, and further analysis of the physicochemical, toxicological, and physiological parameters are needed in order to understand the problem and find apprpriate solutions.

In the next chapter, *Spatial and temporal distribution of Vibrio vulnificus in the different molluscs cultivated in the Ebro delta*, Carmen Lopez-Joven and coworkers try to clarify aspects related to the prevalence and characteristics of *Vibrio vulnificus* in the environment. To better understand the ecology of *V. vulnificus*, the distribution and prevalence of this bacterium was investigated over four campaigns in two bays of the Ebro delta, in four species of commercially important bivalve molluscs, taking into account environmental factors such as temperature and salinity. The authors generated a significant logistic regression model, using multivariate analysis, to explain the presence/absence of *V. vulnificus* in bivalve molluscs, which allowed to establish that the presence of *V. vulnificus* depends on molluscs, year of harvest, and temperature. Results presented in this chapter may contribute to the production of data body to be used in quantitative risk analysis for this important human and fish pathogen.

Christine Paillard, in *An ecological approach to understanding host-pathogen-environment interactions: the case of brown ring disease in clams*, describes an ecological approach to understand host-pathogen interactions using brown ring disease (BRD), caused by the bacterial microparasite *Vibrio tapetis*, as a case study. The phenology and dynamics of BRD are described by coupling epidemiological, ecological and physiological data. Spatial distribution of BRD prevalence in natural clam populations and *V. tapetis* load in the sediment allowed the investigation of the transmission mode of this pathogen, and experimental testing of different transmission modes has demonstrated the predominant role of feces and pseudofeces in *V. tapetis* transmission and diffusion processes. The authors present evidences that cross-species transmission could potentially play a major role in the dispersion of the pathogen during its life cycle, as well as that environmental factors such as temperature, salinity and nutrients strongly modulate the coevolution between *Vibrio* and bivalve.

In recent decades a substantial intensification of environmental stressors has been generated by a combination of anthropogenic and natural causes. In the coastal marine environments that harbour mollusc populations the effects of this problematic are reflected in the development of diseases capable of producing massive mortalities in both wild and cultivated organisms, as well as alterations at different levels of biological organization. Boucard and coworkers evaluate in chapter 8 the effects of environmental stressors in order to better understand the response and adaptive mechanisms of individual organisms and populations exposed to stressful conditions. In addition, the authors examine the responses produced by pathogens-hosts interactions at the transcriptional level, as well as immunogenic responses in resistant and susceptible species of molluscs caused by pathogens capable of causing massive mortality.

In chapter 9, entitled *Heavy metals and platinum group metals (PGMs) determination in oysters, mussels and clams as bio-monitors of the aquatic system pollution*, Locatelli and Melucci present a study on these dangerous pollutants of aquatic systems owing to their bioaccumulation and toxicity. The detection of these metals at trace and ultra-trace level is essential to establish reasonable water quality criteria. On the other hand, molluscs, and other aquatic organisms, sequestrate and concentrate several metals from their aqueous environment, possibly becoming dangerous to human health in consequence of their consumption, and therefore they require particular attention and inspections before being sold on the market. The authors discuss the different analytical methodologies for the voltammetric and spectroscopic determination of heavy metals in oysters, mussels and clams.

In the last chapter, *Oyster summer mortality: the example of a multifactorial syndrome*, is reviewed one of the most threatening problems for the oyster culture in the last decades, the summer mortality syndrome. Since 2008, successive disease outbreaks associated with high temperatures have been reported in Pacific oyster (*Crassosstrea gigas*) populations mainly from France but also from other countries, resulting in huge losses in production which jeopardized the oyster-farming industry in some areas. Several hypothesis about the etiology of these summer mortalities have been proposed including pathogenic vibrios (*Vibrio aestuarianus* and *V. splendidus*) and viruses (oyster herpesvirus). In addition, the influence of environmental (dissolved oxygen, harmful algae, pollutants, etc) and host (physiological condition, reproductive status, glycogen metabolism, etc) factors on the appearance of the disease has also been studied. The chapter summarizes the most recent studies that seem to indicate that the syndrome is the result of a complex relationship among all the mentioned parameters, where co-infection *Vibrio*-herpesvirus is needed to induce the full expression of the disease and where envionmental factors may have an impact on the host, the pathogen and the host/pathogen interactions. The need of multidisciplinary studies in the future to fully understand the disease is also highlighted and discussed.

I would like to thank all the authors of the different chapters for their contribution, the reviewers for their invaluable assistance in enhancing the quality of this book, and the team at Nova Publishers, especially Carra, for their understanding and patience.

Jesús L. Romalde
December 2015

In: Oysters and Clams
Editor: Jesús L. Romalde
ISBN: 978-1-63485-074-2

Chapter 1

DISTRIBUTION, HABITAT, AND ECOLOGY OF THE "DISCO" CLAM, *CTENOIDES ALES*

Lindsey F. Dougherty[*]
University of California, Berkeley, CA, US

ABSTRACT

The "disco" or "electric" clam *Ctenoides ales* (*Limidae*) is the only known bivalve with a vivid flashing display, despite living as deep as 50 m underwater and inside small crevices, where ambient light is dim and wavelength-restricted. The flashing is the result of specialized tissue that scatters light using silica nanospheres. This tissue is repeatedly exposed then concealed to create the flashing appearance. The behavioral purpose of the flashing is still under investigation. The distribution, habitat and ecology of the clams is described herein. Three locations in Indonesia and one location in Australia were explored and 210 individual clams were located at depths of 3-50 m. The properties of the reflective silica spheres did not change with the depth at which the clam was collected. There were four main habitats identified, including coral reef walls, coral reef bommies, caves, and karst islands. These habitats shared characteristics of low water flow, sedimentation, and numerous holes or crevices. The clams were often found in clustered populations at these sites. *C. ales* is thought to be a protandrous hermaphrodite, and based on shell height at the time of change in the congener *C. scaber*, the clam size distribution indicated a sex ratio favoring males in *C. ales*. Clam movement over a 2-hour period did not differ based on sex. Clams were found in conspecific groupings (≥2 clams) in more than half of *in situ* observations (n = 109/210), and the average difference in shell height between organisms was 10.2 ± 7.2 mm (n = 36). Issues associated with *C. ales* cultivation and habitat threats are also discussed.

Keywords: clam, *Ctenoides*, habitat, ecology, distribution, *Limidae*

[*] Corresponding Author address: 3040 Valley Life Sciencies Building #3140, Berkeley, CA 94720-3140, Email: lindseydougherty@berkeley.edu.

INTRODUCTION

File clams (family *Limidae*) are a cosmopolitan group of bivalves that possess mantle tentacles, are capable of swimming, and attach to substrates. Their habitat, morphology and ecology differ drastically from the more well-known burying clams that are harvested commercially as a food source. File clams are popular in aquaria, and similar to most ornamental aquarium species, are collected and imported from the wild [1]. Many, such as *Ctenoides scaber*, are known for their vivid red coloration, which is due to carotenoids [2]. There are few studies of file clam aquaculture [3], but there is no large-scale commercial harvesting of file clams, and therefore there is limited information on cultivation compared to the more commercially-relevant bivalves such as oysters, scallops and mussels. Little is known about the distribution, habitat, ecology, or life history of many species of file clam. For those same species, there is also no documentation of growth rates and sex ratios, which influence population composition and spawning.

One of the best studied file clams, which is also very popular in aquaria, is *C. scaber*, which a Western-Atlantic file clam that ranges from North Carolina to Brazil [4]. It is a congener to *C. ales*, but does not have the characteristic flashing display of *C. ales* [5]. *C. scaber*, along with several other species of file clams, are protandrous hermaphrodites, maturing from small males to large females as they grow [3, 6, 7, 8]. One reason for protandrous hermaphroditism in marine file clams is the permission of rapid somatic growth due to lower energetic costs of spermatozoa compared to ovules, but in *C. scaber*, there were similar slopes for male and female regressions of somatic tissues to shell height, suggesting the costs of producing male or female gametes were similar [8]. In *C. scaber*, the switch from male to female occurs around 40 mm shell height [9], and clams starting at 25 mm shell height were found to be reproductively active [8]. Although shell growth rate has not been measured in many file clams, including *C. ales*, some species of scallops can grow up to 10 mm per month [10, 11]. The growth rate of file clams is essential to understanding sex ratios in populations, and to predicting how changes in environmental conditions may affect growth and reproduction.

Many file clams are filter feeders, so their growth and reproduction are linked to plankton production [12–14]. However, reproductive strategy is not dependent solely on feeding method, as the deep-sea dwelling file clam *Acesta* sp., which feeds on tube worm eggs, has a similar reproductive strategy to its ecologically similar congener *Acesta excavata*, which is a filter feeder [7]. Filter feeders such as *C. ales* and *C. scaber* obtain all of their nutritional requirements by filtering, as they do not contain any photosynthetic symbionts. The habitat in which they live must therefore provide a continuous supply of plankton. Filter feeding occurs in the gills of file clams, which are also used for respiration. Bivalves exclude particles of particular sizes and compositions when they feed, and continually filter particles of a specific size range [15–19]. When plankton concentrations are high, even after filtering, more than 70% of captured particles can be rejected as pseudofeces prior to ingestion in mussels and oysters [16]. Sedimentation can also affect filter-feeding bivalves, such as the hard clam *Mercenaria mercenaria,* which exhibited a decline in algal ingestion rate with increasing sediment loads [20]. Habitat and feeding are linked, especially in coastal areas that are prone to sedimentation when undergoing development.

The size of the particles that bivalves ingest is important. The oyster *Crassostrea virginica* was found to reject particles >100 μm, while ingesting particles of <40 μm, or equivalent to the size of phytoplankton [18]. However, taste or odor is important as well, as chemical feeding stimulants were used to induce the oysters to ingest larger particles [18]. It is therefore not only the size, but the taste, that affect preferential ingestion. Because of their complicated feeding preferences, file clams are difficult to maintain in aquaria without extensive care and individual feeding, and are vulnerable to changes in sedimentation levels *in situ*. Some bivalves are known to consume conspecific larvae, which also has implications for successful distribution of larvae after spawning events [18].

Ctenoides ales is a filter feeder, and is found throughout the Indo-Pacific, where recent coral cover loss has been greater than expected [21] due to a variety of factors but including sedimentation and storms, which can influence feeding. *C. ales* is thought to be a protandrous hermaphrodite, lives in small crevices, and has been found as shallow as 3 m and as deep as 50 m. Its photic environment changes as depth increases, as long wavelengths attenuate rapidly in the first 50 m of sea water, leaving the majority of wavelengths in the blue-green range (400-500 nm) [22]. Even at shallow depths, most illuminating light inside the small crevices where *C. ales* are found is from the horizontal direction, and therefore also dominated by blue-green wavelengths [5]. *C. ales* is the only bivalve known to have a flashing light display, and the ultrastructure of the tissue and its rapid movement, which are unique to this species, suggest it may serve as a signaling function [5]. Little is known about the life history or behavior of *C. ales*, and an obvious question is why the display occurs. Possible functions of the flashing display currently being tested include a signal facilitating the recruitment of conspecifics, prey luring, or predator deterrence. *C. ales* are often found clustered in high density in relatively small areas. Additionally, more than one clam is often found inside the same crevice within the reef. The geographical, depth, and spatial distribution of *C. ales* were unstudied.

In this chapter, ecological findings from 72 SCUBA dives in the habitats of *C. ales* are discussed, as well as what can be learned from their size distribution, and hence, their sex ratios and associated movement. The nature of their grouping in the wild is examined, including size differentials between individuals grouped together. Finally, habitat threats and cultivation issues are discussed.

METHODS

Field Observations

Ctenoides ales were located, observed, photographed and filmed over the course of 72 SCUBA dives on coral reefs in three locations in Indonesia; Lembeh, Sulawesi (1°27'N, 125°14'E), Pemuteran, Bali (8°8'S, 114°39'E) and Kri Island, Raja Ampat (0°34'S, 130°40'E), and one location in Australia; Lizard Island (14°38'S, 145°27'E). *C. ales* were located through exploratory dives in areas whose coral reef habitat was deemed suitable for occupation. Characteristics defining a suitable habitat were identified. Observations included measurements of clam depth, clam size, crevice size and distance to neighboring clams. Depth was determined using an Aeris Atmos© dive computer, and clam and crevice size were measured by placing a metal pointer with an attached ruler into the crevice as close to the clam as possible and taking a photograph. Shell width was measured, as shell height is not measurable *in situ* without removing the clam from its crevice. Shell width was then converted to shell height by measuring 55 single valves of dead shells found while SCUBA diving at the field site in Pemuteran, Indonesia, and calculating a simple linear regression value. Shell width was measured using the Adobe Photoshop© ruler function. Shell area was analyzed using ImageJ© software. The distance to nearest clam was measured using transect tape. Photographs were taken using a Canon 60D DSLR in Ikelite Housing or a GoPro Hero 3© camera. Video was taken using three GoPro Hero 3© cameras. Cameras were attached to 2kg weights and were positioned to film the clam(s), the larger crevice, or the coral formation in which the clam lived. Depth of video setup varied from 3 m to 22 m. Light and Motion© underwater lights (SolaDive 1200 and Gobe) were left with the cameras when filming at night (n = 3) or in caves where low ambient light made camera visibility impossible to analyze (n = 2). Cameras were set up on one dive and recorded until the battery died or the memory card was filled. Cameras were retrieved on the next consecutive dive. When necessary, cameras and lights were left in place overnight. Dive sites were chosen for video setup based on reef topography and feasibility of camera positioning. Sites were also chosen based on the ability to leave the video setup unattended for prolonged periods of time. The 30 videos captured ranged in length from less than one minute to over one hundred minutes. Total video footage time was 16 hours and 57 minutes. Film was reviewed at 4x speed using MovieMaker© software and film clips were made of significant events.

Laboratory Experiments

Ctenoides ales were purchased through BlueZoo© Aquariums and housed in a 100-gallon aquarium with Fluval© filters in laboratory tanks at the University of California, Berkeley. Two *C. ales* were purchased through CairnsMarine© in Cairns, Australia, which were collected from approximately 50 m depth. Clam movement was measured by placing individual clams in 20 gallon tanks for 2 hours and photographing their position at 0, 1, and 2 h. The silica spheres found inside the reflective portion of the clam tissue were measured using the measurement tool in Adobe Photoshop©. The photographs showing the spheres were obtained through tissue prepared as outlined in Dougherty et al. [5].

RESULTS

Geographical, Depth and Spatial Distribution

Three locations in Indonesia were explored to determine the number and density of *C. ales* present; Lembeh, Sulawesi (2 dives sites, n = 3 clams), Pemuteran, Bali (9 dive sites, n = 83 clams) and Kri Island, Raja Ampat (4 dive sites, n = 117 clams). One location in northeast Australia was explored; Lizard Island (2 dive sites, n = 7 clams). Clams were found between depths of 3 m – 50 m (collected by Cairns Marine). The diving for this study was done at ≤24 m due to scientific diving regulations through the American Association of Underwater Scientists (AAUS), but *C. ales* has been found as deep as 50 m, and could possibly live deeper. Photographs of tissue from clams collected from three different depths (10 m, 50 m, and unknown) were analyzed to determine differences in sphere size. Results indicated that the sphere sizes were significantly different, but that the difference was optically negligible because the variation did not result in a change in the light-scattering properties of the spheres (Figure 1).

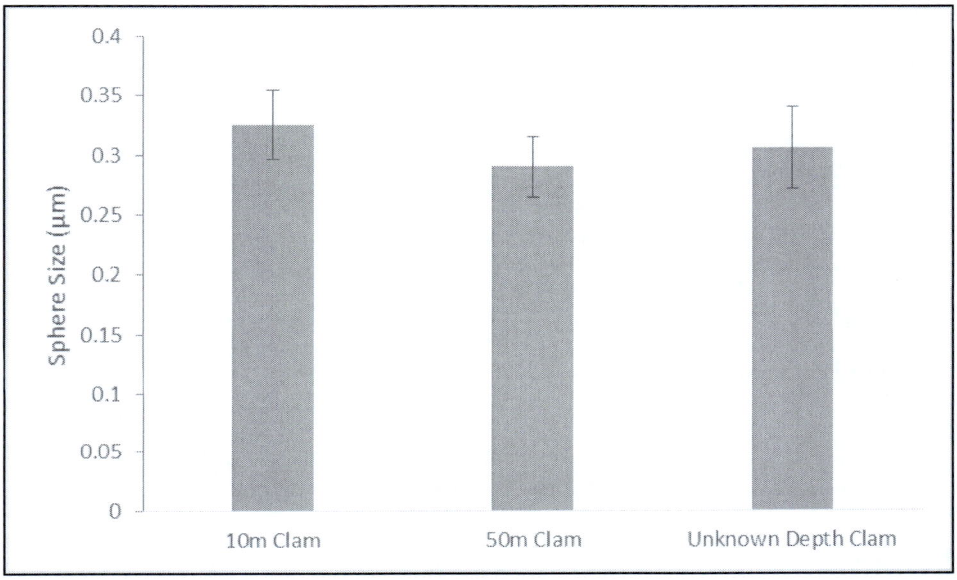

Figure 1. Sphere size (µm) based on depth (m) at which clam was collected (n = 150 spheres/clam).

C. ales can be found along a single depth gradient, and they are often in dense groupings on coral reefs. In Australia, the clams were found an average of 3.3 ± 1.9 m apart from one another (n = 7). The clams are distributed throughout the Indo-Pacific region, as indicated by the GBIF (Global Biodiversity Information Facility), which has 47 specimen listings in 11 countries, including Indonesia, the Philippines, Australia (and Christmas Island), Japan, Fiji, Papua New Guinea, New Caldeonia, the Solomon Islands, Samoa, Timor-Leste, and Palau.

Habitat Characteristics

Four main habitats were identified that were suitable for *C. ales* colonization; coral reef walls, coral reef bommies, caves, and karst islands. Small crevices were the primary requirement for occupation. These crevices generally ranged from 5-20 cm (n = 7). In addition to small crevices, holes made by boring clams were often found occupied by *C. ales*. Areas with heavy sedimentation, caves with little to no water flow, and areas of reef wall with limited competition from other corals, sponges and sea fans were the areas in which *C. ales* were found in the highest densities.

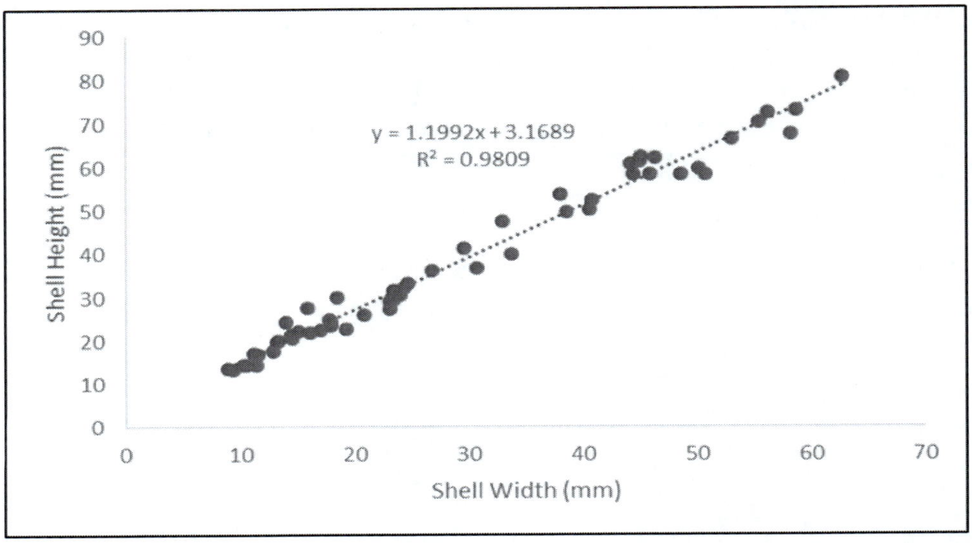

Figure 2. Shell width (mm) measured against shell height (mm) to obtain a simple linear regression equation.

Size, Associated Sex Ratios and Movement

The regression analysis of shell width and shell height gave a linear equation of y = 1.1992x + 0.3169 with an R^2 value of 0.9809, suggesting 98.09% of shell height can be explained by shell width and indicating that the line is a good fit (Figure 2).

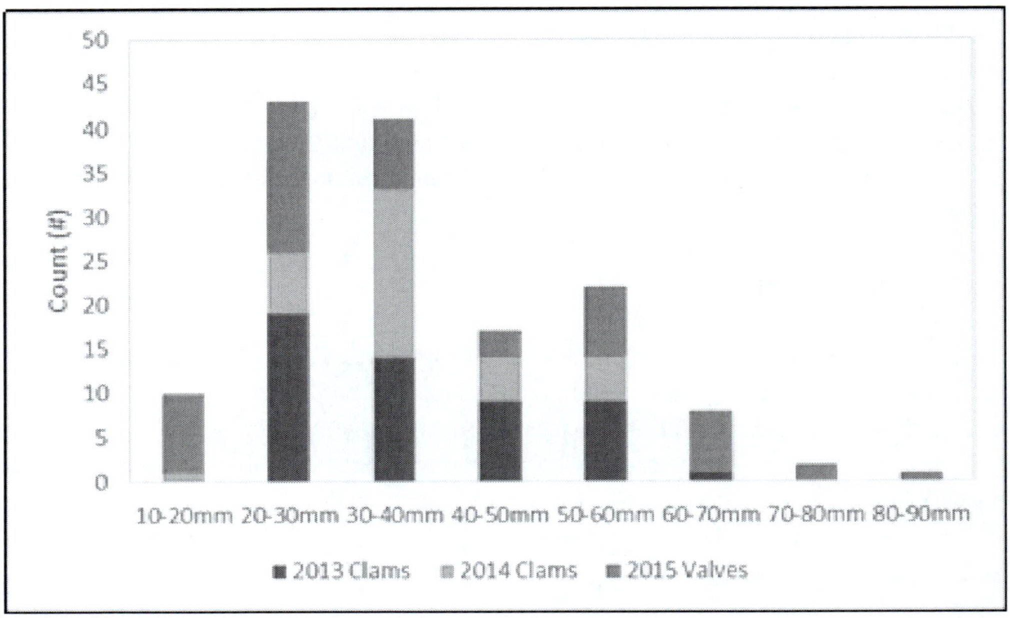

Figure 3. Shell height (mm) based on live clams photographed in 2013 and 2014, and valves collected in 2015.

All measured shell widths were converted to shell height using this equation, and the resulting clam shell heights ranged from 13.3 mm to 80.2 mm, with an average height of 37.4 ± 14.3 mm (Figure 3).

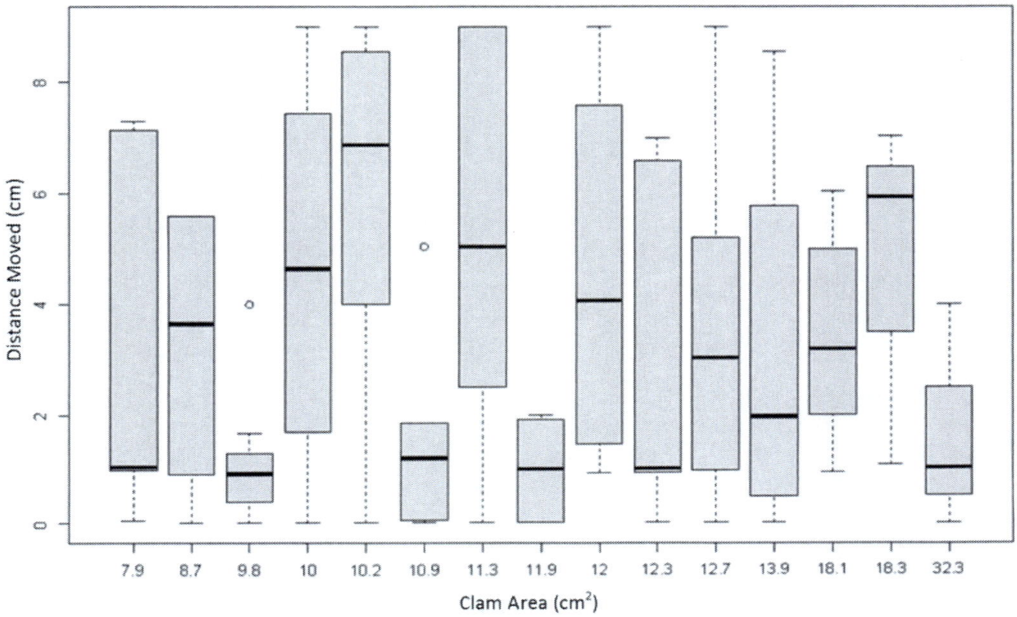

Figure 4. Clam shell area (cm^2) and associated movement (cm) during a 2-hour trial.

Two recent papers have examined sex determinations based on shell height in the protandrous hermaphrodite *C. scaber*. Lodeiros and Himmelman [8] found the following proportions to be female: 21% of clams 15-30 mm, 22% of clams 30-40 mm, 55% of clams 40-50 mm, 59% of clams 50-60 mm, 70% of clams 60-70 mm and 78% of clams 70-80 mm. Dukeman et al. [9] found 17% of clams <40 mm and 71% of clams ≥40 mm to be female. Applying the ratios used by Lodeiros and Himmelman [8] to *C. ales*, with the implied caveat that there may be a differential in shell size at the time of sex change between these two species, approximately 34% of the live clams measured in Raja Ampat in 2013 were female (n = 52), 31% of the live clams measured in the same area in Raja Ampat in 2014 were female (n = 37), and 37% of the dead shells found while diving in Bali in 2015 were female (n = 55). The same three groups were 36%, 31% and 37% female when applying the ratios used by Dukeman et al. [9], respectively. The sex ratio of these three populations of *C. ales* would lie somewhere between 1M:0.5F and 1M:0.6F, which varies from the 0.6M:1F reported by Dukeman et al. [9] and the 1M:0.84F reported by Gomez et al. [3]. In the 2013 measurement, the average live clam shell height was 28.4 ± 9.8 mm (n = 52). In the 2014 measurement, the average live clam shell height was 28.0 ± 7.3 mm (n = 37). In the 2015 measurement of dead shells collected while diving, the average clam shell height was 38.1 ± 19.0 mm (n = 55). *C. scaber* may be slightly larger than *C. ales*, so *C. ales* may transition to female at smaller shell heights, which would increase the number of observed specimens that were female. Given these approximate size ratios in relationship to sex, movement over a 2-hour period was monitored based on clam size to determine if males or females were more active, but there was no relationship between the sex of the clam and distance moved (Figure 4).

Although specimens kept in the laboratory were likely mixed-sex, spawning was never witnessed in any of the groups of ≤20 *C. ales* during the ≤3 months they were kept in aquaria. Spawning was also never seen in clams exposed to stressors which could induce spawning such as movement between tanks or exposure to potential predators.

Conspecific Grouping and Size Variation

In field observations, 210 clams were found throughout four geographic locations (Figure 5).

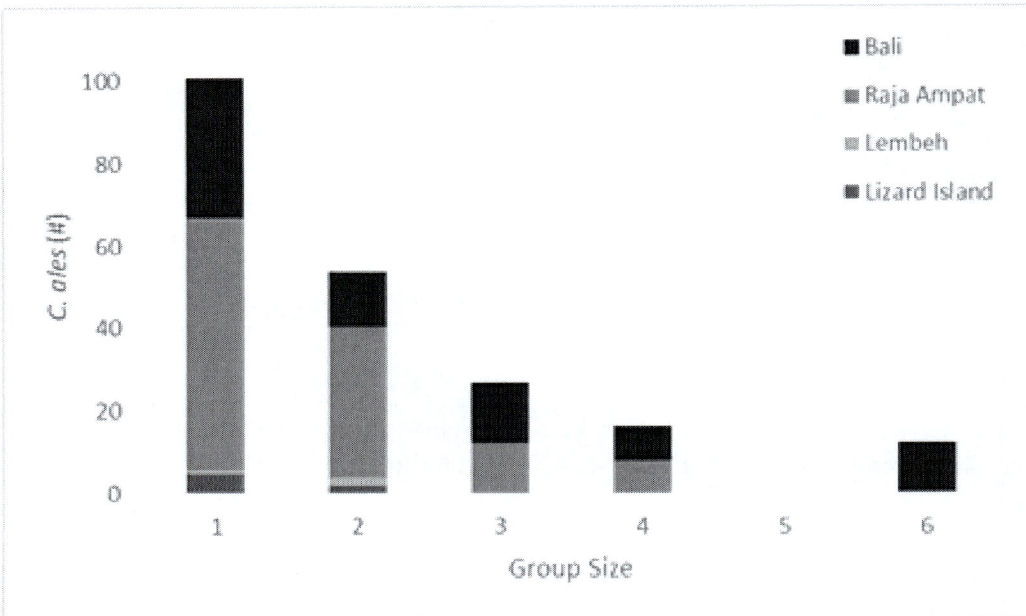

Figure 5. Clam groupings (1-6 individuals) based on location.

Observations indicated 48% (n = 101) of the clams were alone, while 52% were found co-inhabiting crevices (approximately ≤25 cm^2) in groups of ≥2 (n = 109). Of the clams that were in groups, 49% of clams were in pairs of two (n = 54), 25% were in groups of three (n = 27), 15% were in groups of four (n = 16), and 11% were in groups of six (n = 12). The average difference in shell height of clam pairs photographed together was 10.2 ± 7.2 mm (n = 36). When comparing the sizes of the two (or more) clams inside a given crevice, the smaller clam was an average of 22.9 ± 8.4 mm (n = 18), and the larger clam was an average of 33.1 ± 9.6 mm (n = 18), suggesting a <20% chance that the two clams were different sexes according to sex determination based on size in *C. scaber* as measured by Lodeiros and Himmelman [8] and a <17% chance as measured by Dukeman et al. [9]. When the pairs are examined individually, in 13 of the 18 pairs, both clams had a shell height of <40 mm, suggesting an 83% chance they were both male based on Dukeman et al. [9]. In 1 of the 18 pairs, both clams had a shell height of >40 mm, suggesting a 71% chance they were both female. In the remaining 4 of the 18 pairs, one clam had a shell height of <40 mm and one clam had a shell height of >40 mm, with an average difference of 15.4 ± 8 mm.

CONCLUSION

Geographical, Depth and Spatial Distribution

Ctenoides ales are broadly distributed throughout the Indo-Pacific region and can range as deep as 50 m. The depth at which they are located does not affect the optical properties of silica spheres in their tissue, which produce their distinctive flashing display. The spheres are statistically different in size, but still scatter the same wavelengths, resulting in the broadband display of white light. In observations of depth along a vertical gradient, as well as observations of width along a horizontal gradient, clams were found in high densities within relatively small (≤15 m) patches of habitat. Of the 16 total dive sites explored throughout Indonesia and Australia, which had 210 clams total, 6 sites had ≤2 clams. The other 10 sites had an average of 26 clams each, suggesting clustered population densities in small patches of habitat. These concentrated population densities could be a byproduct of currents or some other physical force, as *C. ales* are thought to be broadcast spawners. There may be some form of conspecific chemosensory attraction, such as in larvae from the oyster *Crassostrea virginica* which show attraction to cues from both live adult oysters and bacterial films growing on shell surfaces. The larvae respond by swimming vertically down in the water column, slowing horizontal swimming and then contacting the bottom and attaching with their foot, indicating settlement [23]. It is unknown what cues affect settlement in *C. ales*.

Habitat Characteristics

The habitats in which high densities of *C. ales* were found shared common characteristics; relatively weak currents, high sedimentation, and numerous crevices or holes. Caves also consistently housed large numbers of clams in all three Indonesian locations. The number of clams found on a given reef did not change noticeably with depth, although the exploration of deeper habitats was limited by scientific diving restrictions. Dive sites with higher numbers of clams generally seemed to have less biodiversity of marine plants and animals, perhaps resulting in higher clam numbers due to decreased competition for space or fewer predation threats. The lack of biodiversity in addition to high sedimentation meant that the sites where the most clams were found were rarely used for SCUBA tourism.

Size and Associated Sex Ratios and Movement

If the size at which *C. ales* changes from male to female is similar to the congener *C. scaber*, approximately one-third of the specimens measured in the field were female (sex ratio between 1M:0.5F and 1M:0.6F). This varies significantly from the 0.6M:1F in *C. scaber* reported by Dukeman et al. [9] but was closer to the 1M:0.84F reported in *C. scaber* by

Gomez et al. [3]. Differences in this sex ratio are difficult to explain using sampling bias, as larger clams are easier to find while SCUBA diving. This sex ratio may be affected by the collection of organisms for sale in aquaria, depending on what size specimens are targeted. There was a broad size range of both living and dead (based on discarded valves) *C. ales*, suggesting predators target various sizes of individual. These observations may be similar to previous findings of juvenile *C. scaber* being predated upon by crabs [9] and parrotfish [6]. There was no trend between clam size and how much they moved in the laboratory, suggesting that male clams do not move more than female clams. It was hypothesized that males may move more than females in order to settle near females for spawning, as males are smaller and less likely to be permanently settled on the reef. However, it is unknown at what size the clams become sessile, or permanently attached in their reef crevice habitats using byssal threads. The clams may not move at all after they have grown to a certain size, which could explain the lack of difference in movement based on shell height between males and females. Spawning was never witnessed either in situ or in the laboratory while studying *C. ales*. The congener *C. scaber* only spawns twice a year [8], once in June/July, and once in November/December. It is possible that spawning in captive *C. ales* either did not occur due to the time period kept in the laboratory being too short, the absence of cues that would occur *in situ*, such as temperature of plankton levels, or that spawning did occur but was not observed.

Conspecific Grouping and Size Variation

In field observations, 52% of specimens were found co-inhabiting holes in pairs of ≥2 (n = 109). The average difference in shell height between 18 measured pairs was 10.2 ± 7.2 mm (n = 36). Depending on growth rate, which has not been studied in *C. ales*, clams may have settled in the same crevice after the same spawning event based on this average size difference. Of the 18 pairs, 9 pairs had a size difference of less than 10 mm, 7 pairs had a size difference of 10-20 mm, and 2 pairs had a size difference of 20-30 mm. Without knowledge of the growth rate of *C. ales*, it is difficult to determine whether these pairs had asynchronous settlement, but a size difference nearing 30 mm would suggest it is possible for some of the pairs. As many species of bivalves possess light-sensitive eyes in the peidiveligeir stage, during which settlement occurs [24] it was originally thought that the flashing display might act as a settlement cue for conspecifics. Specifically, perhaps small males were attracted to the flashing of large females in order to settle nearby. However, given this analysis of size similarities of clams in the field, it is possible that the groupings are instead a result of currents, some other physical force, or perhaps a chemosensory attraction between conspecifics, such as in *Crassostrea virginica* [23]. Furthermore, examination of *C. ales* eyes using transmission electron microscopy revealed that the size and structural features of the eye tissues do not support the hypothesis that *C. ales* are able to detect the flashing displays of neighboring conspecifics. The analysis also determined that the eyes are not capable of image formation [Dougherty, unpublished data].

Cultivation and Habitat Threats

To my knowledge, there have been no studies attempting laboratory cultivation of file clams. Organisms which have a pelagic larval stage possess many complex needs in terms of temperature, salinity and food [25], and are therefore difficult to cultivate in aquaria. Aquarists report that survivorship of adult file clams is often only a matter of months, making captive spawning unlikely in species such as *C. scaber*, which only has two major spawning events per year [8]. There are no studies assessing the feasibility of farming file clams, but such studies have been conducted on other bivalves such as scallops, mussels, and oysters, which are of a greater commercial value. Collecting wild file clams for cultivation or for study is problematic due to their fragile coral reef habitats. Additionally, many file clams such as *C. ales* use byssal threads to attach themselves inside of small crevices within the coral [5], which makes it difficult to collect them without causing damage to the surrounding reef. If file clams are successfully collected and moved to aquaria, there remain challenges associated with their feeding, growth rate, and reproduction. Many file clams are broadcast spawners, and according to field findings based on the sizes of the congener *C. scaber*, only 31-37% of the *C. ales* seen were female. It is unknown whether collection for aquaria has influenced these numbers. Spawning was never witnessed in captive clams during this three-year study. Like many organisms, file clams are threatened by changes in water temperature, which is often a catalyst for spawning. Some bivalves spawn when temperatures increase [10, 13, 14], some bivalves spawn when temperatures decrease [8, 26], and some bivalves have no correlation between temperature and spawning [11, 27]. In addition to feeding, spawning and reproductive challenges, file clams are considered by many to be ill-suited for commercial exploitation due to both the difficulty associated with their collection as well as the potential damage that harvesting would cause to surrounding coral communities [8]. These complications, combined with the clams' sporadic distribution on reefs, suggest that further study is needed to understand their distribution, life cycle and other threats to their habitat.

ACKNOWLEDGMENTS

The author would like to thank Dr. Roy L. Caldwell for experimental feedback and guidance, undergraduate research assistants Alexandria Niebergall and Katherine Meyer for their work in both the lab and the field, Jim Hayward for dive support and planning assistance, Karin Van Beeck and staff from the Bali Dive Academy for dive and scouting assistance, Christy Furman for dive assistance, Dita Cahyani and Aji Wahyu from the Indonesian Biodiversity Research Center for permit and research support, Kerri Bingham and Hergen Spalink from GotMuck for logistical and dive support, Tertius Kammeyer from the Raja Ampat Research and Conservation Centre for dive assistance, Danny Charlton and Dimpy Jacobs from Lembeh Resort for housing and dive assistance, and the Lizard Island Research Station.

REFERENCES

[1] Chapman, F. A., Fitz-Coy, S. A. Thunberg, E. M., Adams, C. M. 1997. United States of America trade in ornamental fish. *J. World Aquacult. Soc.* 28: 1-10.

[2] Lin, A. L., Pompa, L. A. 1977. Carotenoids of the red clam *Lima scabra*. *Bol. Inst. Oceanogr. Univ. Oriente* 16, 83-86.

[3] Gomez, J.A., Prieto, A., Lodeiros, C. 1990. Relaciones biometricas y biomasa especifica en el bivalvo *Lima scabra tenera* (Sowerby 1843). *Scientia* (Panama) 5, 13–17.

[4] Abbott, R. T. 1974. American seashells. New York: Van Nostrand Reinhold Company. pp. 471–482.

[5] Dougherty, L. F., Johnsen, S., Caldwell, R. L., Marshall, N.J. 2014. A dynamic broadband reflector built from microscopic silica spheres in the 'disco' clam *Ctenoides ales*. *J. R. Soc. Interface* 11, 20140407. http:// dx.doi.org/10.1098/rsif.2014.0407.

[6] Gomez, J. A., Linero, I., Fermin, J. 1995. Estudios ecologicos sobre *Lima scabra* (Born 1778) (*Pelecypoda:Limidae*) en el Golfo de Cariaco, Venezuela. I. Censo y relaciones morfometricas. *Bol. Inst. Oceanogr. Univ. Oriente* 34, 109–119.

[7] Järnegren, J., Rapp, H. T., Young, C. M. 2007. Similar reproductive cycles and life-history traits in congeneric limid bivalves with different modes of nutrition. *Mar. Ecol.* 28, 183–192.

[8] Lodeiros, C. J., Himmelman, J. H. 1999. Reproductive cycle of the bivalve *Lima scabra* (*Pterioida: Limidae*) and its association with environmental conditions. *Rev. Biol. Trop.* 47, 411-418.

[9] Dukeman, A. K., Blake, N. J., Arnold, W. S. 2005. The reproductive cycle of the flame scallop, *Ctenoides scaber* (Born 1778), from the Lower Florida Keys and its relationship with environmental conditions. *J. Shellfish Res.* 24, 341–351.

[10] Lodeiros, C. J., Himmelman, J. H. 1994. Relations among environmental conditions and growth in the tropical scallop *Euvola* (*Pecten*) *ziczac* (L.) in suspended culture in the Golfo de Cariaco, Venezuela. *Aquaculture* 119, 345-358.

[11] Lodeiros, C. J., Rengel, J. J., Freites, L., Morales, F., Himmelman, J. H. 1998. Growth and survival of the tropical scallop *Lyropecten* (*Nodipecten*) *nodosus* maintained in suspended culture at three depths. *Aquaculture* 165, 41-50.

[12] Barber, B. J., Blake, N. J. 1991. Reproductive physiology. Pp. 377-428 in S. E. Shumway, (ed.) *Scallops: biology, ecology and aquaculture*. Developments in Aquaculture and Fisheries Science. Elsevier Science, New York, NY.

[13] Giese, A. C., Kanatani, H. 1987. Maturation and spawning. Pp. 251-329 in A. C. Giese, J. S. Pearse, V. Pearse (eds.) *Reproduction of marine invertebrates*. Blackwell Scientific, San Diego, CA.

[14] Sastry, A. N. 1979. Pelecipoda (excluding Ostreidae). pp. 113-292 in A. C. Giese, J. S. Pearse (eds.) *Reproduction of marine invertebrates*. Academic Press, New York, NY.

[15] Beninger, P. G., Veniot, A. 1999. The oyster proves the rule: Mechanisms of pseudofeces transport and rejection on the mantle of *Crassostrea virginica* and *C. gigas*. *Mar. Ecol. Prog. Ser.* 190, 179-188.

[16] Hawkins, A. J. S., Smith, R. F. M. Tan, S. H., Yasin, Z. B. 1998. Suspension-feeding behaviour in tropical bivalve molluscs: *Perna viridis*, *Crassostrea belcheri*,

Crassostrea iradelei, Saccostrea cucculata and *Pinctada margarifera. Mar. Ecol. Prog. Ser.* 166, 173-185.

[17] Shimeta, J., Koehl, M. A. R. 1997. Mechanisms of particle selection by tentaculate suspension feeders during encounter, retention, and handling. *J. Exp. Mar. Biol. Ecol.* 209, 47-73.

[18] Tamburri, M. N., Zimmer-Faust, R. K. 1996. Suspension feeding: basic mechanisms controlling recognition and ingestion of larvae. *Limnol. Oceanogr.* 41, 1188-1197.

[19] Ward, J. E., Levinton, J. S., Shumway, S. E., Cucci, T. 1998. Particle sorting in bivalves: In vivo determination of the pallial organs of selection. *Mar. Biol.* 131, 283-292.

[20] Bricelj, V. M., Malouf, R. E. 1984. Influence of algal and suspended sediment concentrations on the feeding physiology of the hard clam *Mercenaria mercenaria. Mar. Biol.* 84, 155-165.

[21] Bruno J. F., Selig, E. R. 2007. Regional decline of coral cover in the Indo-Pacific: timing, extent, and subregional comparisons. *PLoS ONE* 2, e711. doi:10.1371/journal.pone.0000711.

[22] Jerlov, N. G. 1976. *Marine Optics.* Amsterdam: Elsevier.

[23] Tamburri, M. N., Zimmer-Faust, R. K., Tamplin, M. L. 1992. Natural sources and properties of chemical inducers mediating settlement of oyster larvae: A re-examination. *Biol. Bull.* 183, 327-338.

[24] Carriker, M. R. 1990. The Bivalvia: Proceedings of a Memorial Symposium in Honour of Sir Charles Maurice Yonge. Vol. 1, Functional Significance of the Peidiveligeir in Bivalve Development.

[25] Thorson, G. 1950. Reproductive and larval ecology of marine bottom invertebrates. *Biol. Rev.* 25, 1-45.

[26] Velez, A., Epifanio, C. E. 1981. Effects of temperature and ration on gametogenesis and growth in the tropical mussel *Perna perna* L. *Aquaculture* 22, 21-26.

[27] Marquez, B. 1996. Variación estacional de la fijación de la ostra negra *Pteria colymbus* (Röding, 1798) (*Bivalvia: Pteriidae*) a diferentes profundidades en la localidad de Turpialito, Golfo de Cariaco, Edo. Sucre, Venezuela. M. Sc. thesis, Departamento de Biología, Universidad de Oriente, Curnaná, Venezuela. 69 p.

In: Oysters and Clams
Editor: Jesús L. Romalde
ISBN: 978-1-63485-074-2

Chapter 2

CULTURE OF THE WEDGE-SHELL CLAM *DONAX TRUNCULUS*: NEW DEVELOPMENTS

Andrea Louzán[1], Susana Nóvoa[1], Justa Ojea[1], Fiz Da Costa[2] and Dorotea Martínez-Patiño[1,*]
[1]Centro de Cultivos Mariños (CIMA), Consellería do Mar,
Xunta de Galicia, Ribadeo (Lugo), Spain
[2]Novostrea Bretagne, Route du Vieux Passage,
Banastère, Sarzeau, France

ABSTRACT

The wedge-clam *Donax trunculus*, is an important shellfish resource in the Iberian Península that it is in regression in Galician natural beds. The development of hatchery culture, as a strategy for the obtention of seed for restocking and to increase production is a clear benefit to shellfish industry. Intensive farming of this species is addressed in this chapter, including all stages of cultivation from stabling or conditioning of broodstocks at hatchery facilities, obtention of gametes after induction to spawning, fertilization and embryo development, the D larval rearing from larvae D to metamorphosis, the post-larvae and spat culture, intermediate culture or pre-fattening stage and on-growing in the natural beds. With all the results obtained in the different phases, it is demonstrated that there is a real possibility of obtaining seed in hatcheries *D. trunculus* and sow it in the natural beds to reach commercial size.

Keywords: wedge-shell clam, *Donax trunculus*, larval rearing, spat culture

INTRODUCTION

The bivalve *Donax trunculus* (Linnaeus 1758) belongs to the family *Donacidae*. Its geographical distribution extends from the French Atlantic coast to the Senegal and around

---

* E-mail: dorotea.martinez.patino@xunta.es.

the Mediterranean Sea [3]. This species lives buried in sandy banks, preferably located in the intertidal zone, a few centimeters from the surface where the hydrodynamics and strong waves not allow excessive deposition of organic matter. On the other hand, this bivalve is suspension feeder [19], so the surf zones ensure a constant renewal of the existing phytoplankton. Its distribution may be between 0 and 2 meters depth on the Mediterranean coast and between 0 and 6 meters on the Atlantic coast. It has been observed a relation between the size of the individual and its spatial distribution in the bank, where small individuals have a shallower distribution in comparison with the large individuals [12]. This can be interpreted as a strategy to overcome intraspecific competition between adults and juveniles. This spatial distribution was also observed by Guillou and Le Moal [13] on the French coast.

Aquaculture is an important activity that allows not only an increase in the production of molluscs, but also a recovery in those banks affected by phenomena of overexploitation, disease or other reasons that lead to a regression.

Wedge-shell clam is a species of bivalve mollusc which constitutes an important economic resource for shellfish gatherers in Galicia (NW Spain). Natural beds are currently under regression and, consequentely, the production is being drastically affected. The development of hatchery culture as a strategy for obtaining seed for restocking to increase production, constitute a clear benefit to the shellfish industry.

In this chapter, all stages of cultivation of the species *D. trunculus* are addressed, from stabling or conditioning of broodstocks at hatchery facilities to on-growing in the natural beds, inclusing the obtention of gametes after induction to spawning, fertilization and embryo development, the D larval rearing from larvae D to metamorphosis, the post-larvae and spat culture, and intermediate culture or pre-fattening stage.

BROODSTOCK CONDITIONING

Conditioning is the procedure by which hatcheries are able to extend their production season over the period when adults are bearing mature gametes in the sea. Adult specimens of *D. trunculus* were collected from natural stocks and immediately transferred to our facilities at the Centro de Cultivos Mariños (CIMA).

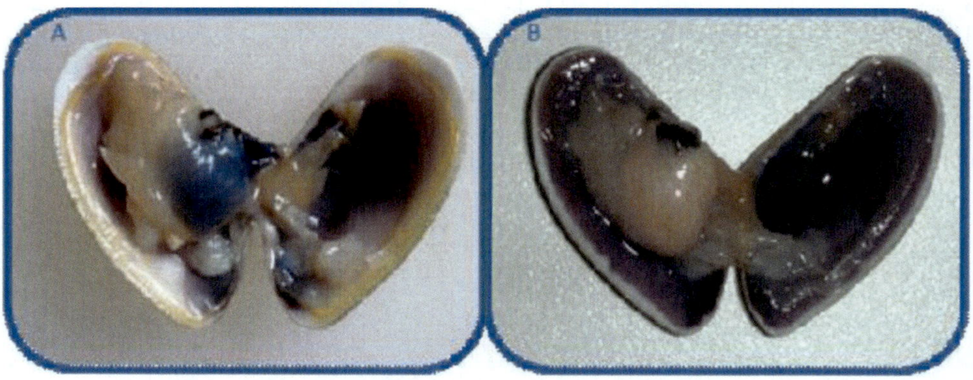

Figure 1. *Donax trunculus* broodstock. A) Ripe female with violet gonad. B) Ripe male with white gonad.

Figure 2. Broodstock conditioning system.

Depending on the maturation of the specimens, two strategies could be followed. Ripe adults (Figure 1) were maintained for a short period until the spawning.

In the case of immature individuals, they were conditioned. Adults, in number of 500, were disposed in 200-L tanks, in open circuit with 30-40 L/h of sand-filtered seawater flow (Figure 2). Wedge clams were fed with a mixed algal diet, including including *Isochrysis galbana* Parke 1949, *Isochrysis aff. galbana* (T-iso), *Diacronema lutheri* (Droop) Bendif and Véron 2011, *Chaetoceros calcitrans* (Paulsen) Takano 1968, and *Tetraselmis suecica* (Kylin) Butcher 1959. The daily dry weight of algae supplied was established at 4% of the mean dry flesh weight of clams.

Maturity may be accelerated by holding the wedge clams at elevated temperatures. The increase of temperature during conditioning enhances the vitellogenesis processes in most of the bivalves [22]. Ojea et al. [21] observed a higher gonad development in *R. decussatus* subjected to the highest temperature during conditioning at CIMA. In *D. trunculus*, the seawater at 21°C accelerated the maturation and a high number of eggs were obtained, but the survival rate of these adults was low, only 28%. The maturation of broodstock held in seawater at 17°C was slower, but they reached a higher survival rate, 44%.

The spawning period of *D. trunculus* in natural beds comprises five months, from March to July. In hatchery this period could be at present extended at least in one month by conditioning.

Spawning Induction

Spawning is the hatchery procedure by which ripe bivalves are induced to liberate their mature gametes using artificial stimuli. One of the best spawning methods for *D. trunculus* is to keep them dry at room temperature (14±1ºC) for 10 h, and then to introduce them in filtered seawater at 18ºC.

If spawning induction was not obtained by this method, ripe adults may be subjected to thermal shock stimulation with seawater temperatures between 13 and 25ºC. Effectiveness of this stimulus have been reported for several bivalve species, such as *Venerupis corrugata* [5], *R. decussatus* [21], *Solen marginatus* [8]. Martínez-Patiño [17] compiled the optimal conditions for every species, and on the basis of these data it could be concluded that, regardless of the differences on time or number of cycles, thermal shocks should be produced by cold seawater at 10-14ºC and warm seawater at 18-27ºC. In many species, including *D. trunculus*, thermal simulation is combined with the addition of gametes stripped from ripe individuals, thus improving spawning yields. Other authors described successful results inducing the spawning in members of family Donacidae by injection of ammonium hydroxide in gonad of *Donax variabilis* [6] and *D. trunculus* [18]. Once the males release the spermatozoids and the females release the oocytes, external fertilization initialize the embryo development.

Larval Rearing

After fertilization embryos develop by feeding the reserves stored in oocyte. Two days after fertilization, they reach the D-stage veliger larvae, beginning the larval culture, with successive stages until the settlement and transformation in post-larvae. The length of larval culture depends on the bivalve species and the culture conditions.

Embryonic and larval cultures of wedge clam at CIMA were performed in 500-L fiberglass conical tanks filled with filtered and UV-treated seawater at 21±1ºC.

Initial egg density was 20 eggs/mL and the larvae were maintained at 5 larvae/mL. Larvae were usually reared in closed circuit.

The seawater was renewed three times each week, after sieving the larvae through a serial mesh screens according to the size of the larvae into the tank, with a tray in the bottom to keep them moist (Figure 3). Food was delivered once per day to each tank, composed by a multispecies algal diet of *T. suecica*, *I. galbana*, T-iso, *D. lutheri*, y *Chaetoceros calcitrans*. The initial ration of 20 cell/µl for D-shaped larvae was increased according to the requirements until a final maximum ration of 40 cell/µl.

At the end of larval phase, a newly developed foot was observed between the shell valves, indicating immediacy of settlement and metamorphosis. When at least 50% of larvae reached the pediveliger stage, sieves (mesh screen 150 µm) with a thin layer of sand (particle size > 170 µm) were placed in the tanks. Table 1 summarizes the time of each embryonic and larval stage of *D. trunculus*. The time for transformation into the first stages (2, 4, 8 and 16 cells) were similar to those observed in other bivalves, as *Ensis magnus* and *E. siliqua* [7, 8], *R. decussatus* [1] and *V. corrugata* [5]. Blastula in *D. trunculus* developed at 7 hours 20 minutes post-fertilization, slightly later than the species above cited.

Figure 3. Larval rearing tanks.

Table 1. Stages in the embryonic and larval development of *D. trunculus*

Stage	Time post-fertilization*	Size
2-cell	1 h 15 min	
4-cell	2 h 15 min	
8-cell	3 h 15 min	
16-cell	4 h 20 min	
Blastula	7 h 20 min	
Gastrula	9 h	
Trochophore	15 h	
D-shaped larvae	41 h	105 - 110 μm
Umbonate	8 d	170 μm
Pediveliger	20-25 d	200 μm
Metamorphosis	30 d	280 μm

*h: hour, min: minute, d: day.

And the D-shaped larvae developed slower in comparison with *R. decussatus* (26 hours) [1] and *V. corrugata* (30-32 hours) [5].

The size of *Donax trunculus* eggs ranged from 75 to 85 μm in diameter. D-shaped larvae developed by 41 hours-post-fertilization and they are 105-110 μm long. Eight days after fertilization the larvae reached the umbonate stage with a length of 170 μm. Twenty/twenty-five-day-old larvae showed the ciliated foot, typical of the pediveliger stage, with a mean valve length of 200 μm. Thirty days after fertilization the metamorphosis was complete and the post-larvae measured 280 μm (Figure 4). The growth of wedge clam larvae corresponded to a linear model with a determination coefficient (R^2) of 0.74, with daily growth increment of 5.49 μm/day (Figure 5). Figure 6 shows the results obtained at CIMA for the culture of *D. trunculus* and other bivalve species: *R. decussatus* [2], *E. magnus*, *S. marginatus* [9, 10, 16] and *V. corrugata* [5, 20].

The average oocyte size of *D. trunculus* was 75-85 μm, similar to that recorded for other bivalves such as *R. decussatus*, 62-68 μm [21], *V. corrugata*, 65-75 μm [5] and *E. magnus*, 70-90 μm [9], and smaller than *S. marginatus*, which reached until 143-155 μm [8].

The D-shaped larvae developed by 41 hours post-fertilization, slower than the other species: *S. marginatus* (12 hours), *R. decussatus* (26 hours), *V. corrugata* and *E. magnus* (30 hours). However, wedge clam reached the umbone stage after 8 days, as occurs in *R. decussatus* and *V. corrugata*, and later than *S. marginatus* (3 days), and *E. magnus* (7 days) [10].

Donax trunculus reached the pediveliger stage 20-25 days post-fertilization, meaning a slower larval development in comparison with *V. corrugata*, which needs 16-18 days [5], and Solenids, with pediveliger larvae observed between 5 and 14 days [8]. Larval period lasted one month, similar to the value observed for *R. decussatus* in similar conditions in CIMA [21], and longer than *V. corrugata*, *E. magnus* and *S. marginatus*, which showed larval developments around 22-25, 20 and 8 days, respectively [10].

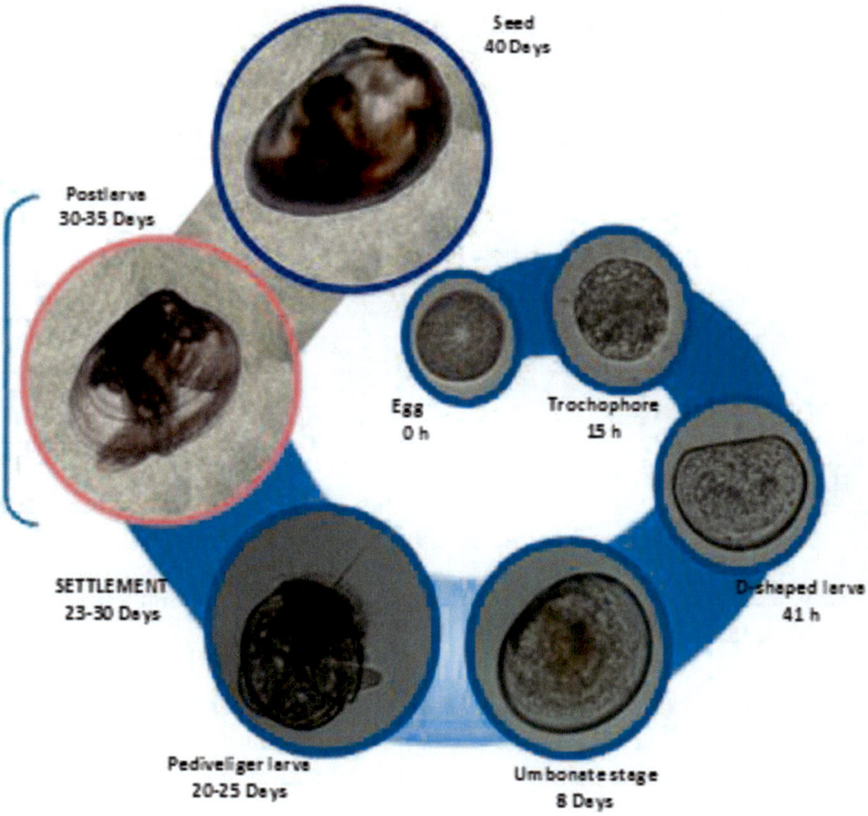

Figure 4. Embryonic and larval development of *D. trunculus* in hatchery.

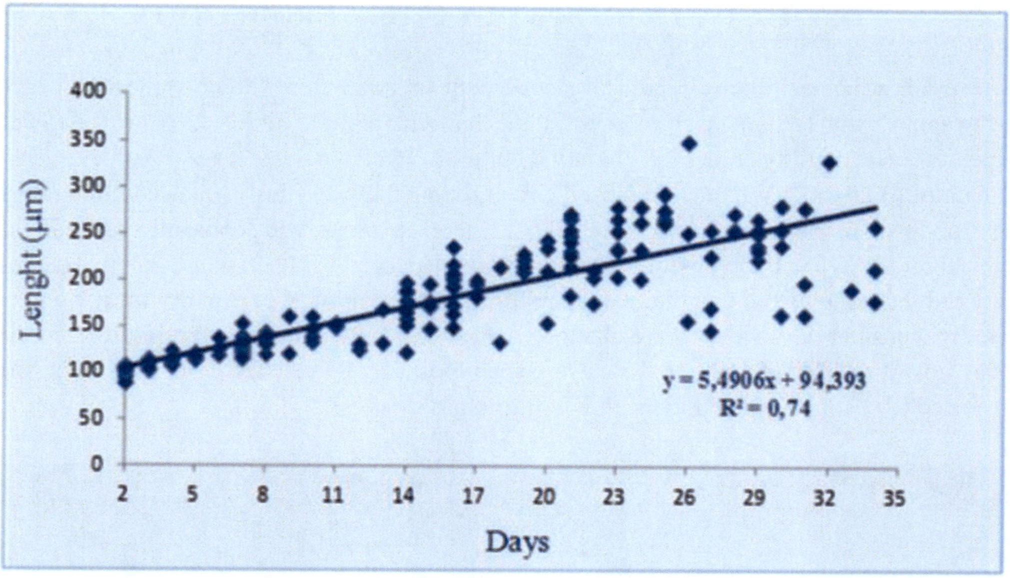

Figure 5. Growth in length of *D. trunculus* larvae.

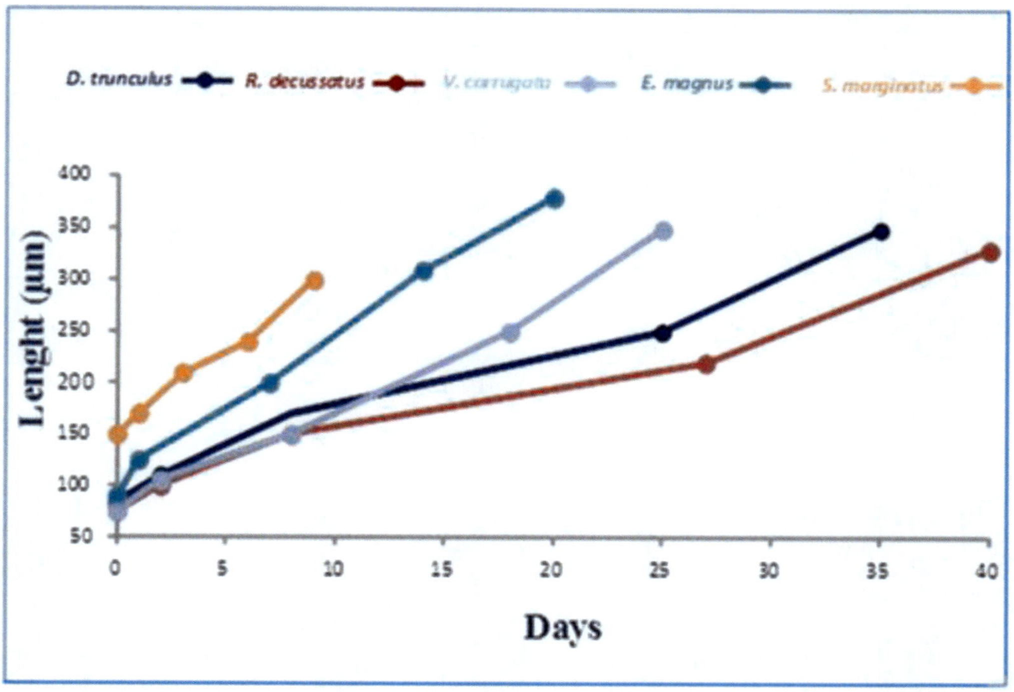

Figure 6. Growth in length (µm) from egg to newly-settled post-larvae for different bivalve species reared at CIMA.

In larval culture *D. trunculus* displayed a growth rate of 5.49 µm/day, fitting to a linear model similar to those observed in other bivalves. However the growth rates of species such as *V. corrugata*, *E. magnus* and *S. marginatus*, are substantially higher, with values of 7.65, 13 and 16 µm/day, respectively [5, 8, 9].

POST-LARVAL AND SPAT CULTURE/PRODUCTION

Post-larvae were initially reared in sieves with inverted flow (down-wellers), in larval rearing tanks, with a thin layer of sand at the bottom (particle size > 170 µm). Seawater temperature was slightly colder than in larval cultures (18±1°C).

Cultures may be maintained in closed circuit, being the water fully renewed three times a week, or in open circuit (Figure 7), in 500-L fiberglass tanks with constant seawater flow (renewal rate 9 L/h) and the phytoplankton mixture dispensed by a peristaltic pump. The post-larvae fed the same mixed diet than larval cultures, at a ration of 2% mean dry meat weight of the individuals per day. Sieves were changed when necessary to increase mesh size according to the post-larval length. After 15 days, when post-larvae reached 750 µm, they were transferred to spat holding cylinders with 200 µm mesh sieves.

Figure 7. Open post-larval culture system.

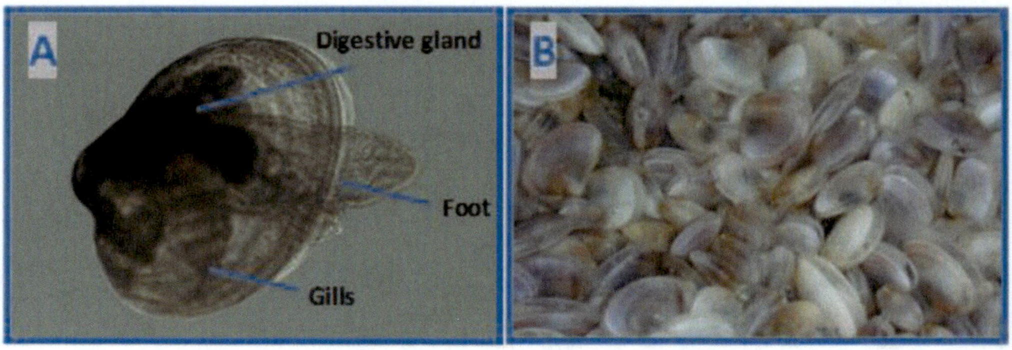

Figure 8. Wedge clam. A) Post-larvae. B) Spat.

The first month post-larvae were translucent, being visible gill filaments, digestive tract and foot (Figure 8A). As post-larvae continued to grow, the shell became harder, increasing the opacity and showing similar colouration to adult wedge clams (Figure 8B).

Under the conditions described above, starting from individuals > 350 μm and for 3 months, growth in length of the post-larvae and spat fitted to a linear model (R^2=0.98) with a growth rate of 0.04 mm/day (Figure 9).

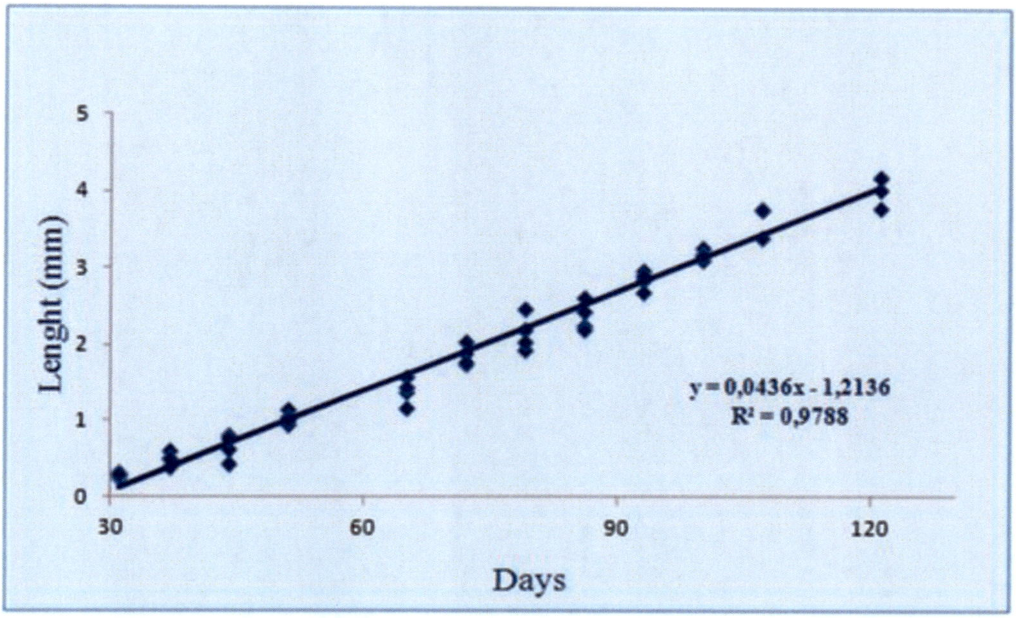

Figure 9. Growth in length of *D. trunculus* in post-larval period.

The average survival was 35%, between 13% and the highest value of 55%. Survival of 90-days spat for *V. corrugata* at CIMA was between 30 and 50% [5], but following an exponential pattern. In Solenids (*E. magnus*, *E. siliqua* y *S. marginatus*), 120-days survival rates were lower, 2-5%, reaching sizes around 20-30 mm [10], though the morphology is different.

INTERMEDIATE CULTURE AND ON-GROWING

Spat were maintained in the hatchery, under the same conditions described above, until they reach a length enough to be retained in mesh screens > 2000-3000 μm. Then they were transferred to natural beds because the cost of maintaining them in a hatchery is too high. Different types of containers or structures may be used in this stage, hanging on floating rafts or directly seeded in the substrate. One suspended structure consisted of four PVC cylinders (20 cm diameter, 22 cm high) perforated with 8 holes in the sides and 1 mm mesh to cover the structure (Figure 10A).

Each one had a 150 μm mesh at the bottom to retain the thin layer of sand yet allow its aeration. Other hanging structures assayed were lantern nets (10 storeys) with solid steel rings (38 cm diameter) and mesh size of 1 mm (Figure 10B).

Other system used 50-L polyethylene bins/buckets with forced water flow (Figure 11).

The base was perforated (2 mm holes) and held a mesh with a size determined according to the spat. The container was closed by a cover, with a PVC periscope tube, a "T" structure and a floater to stabilize the structure. The air injected forced the flow of seawater from the bottom to the upper "T" structure.

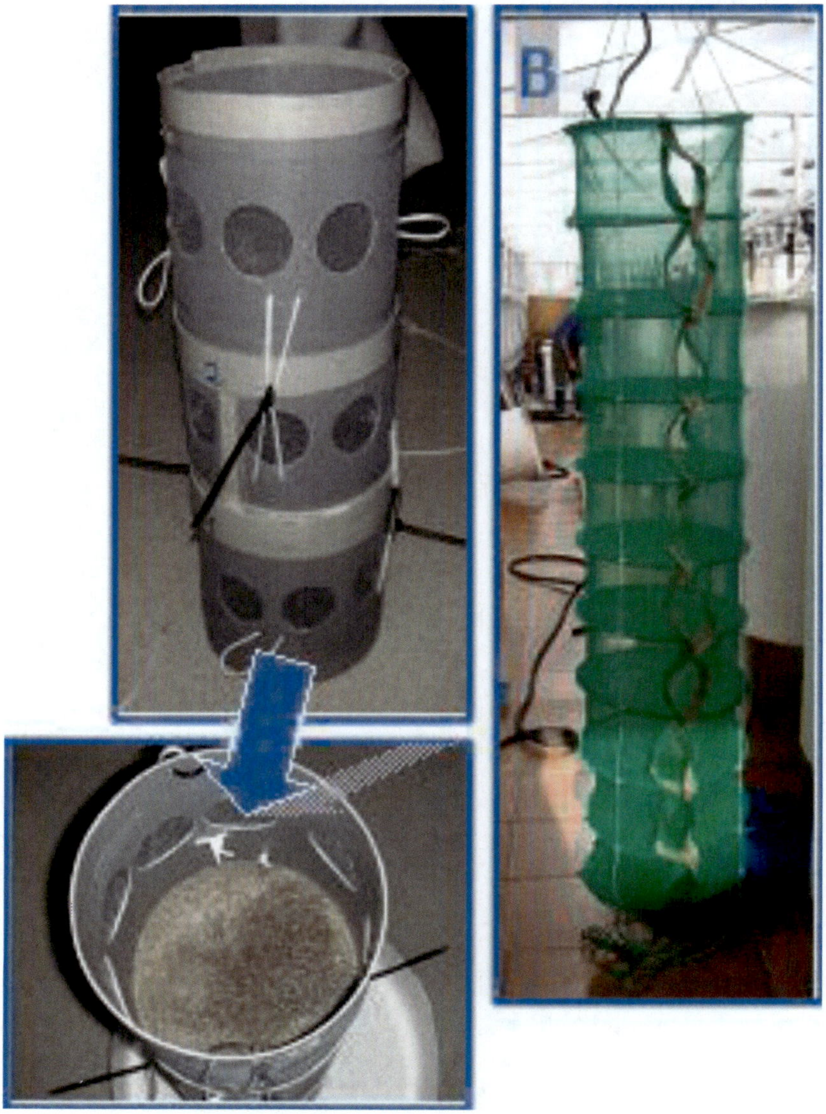

Figure 10. Equipment for pre-growing of wedge clam in natural environment. A) PVC cylinders with sand. B) Lantern nets.

In other pre-growing experience, spat was directly seeded in the sand, in special containers to study growth and survival (Figure 12). Eleven-months-old spat, with an average length of 7.5 mm, were transferred from the hatchery to a natural bed, named "O Lombo das Navallas," in "Ría de O Barqueiro" (43°44′41″N, 7°40′50″W). Individuals were placed in four mesh cages (50 cm x 50 cm x 50 cm) with a stainless steel frame, each of which was fitted with a top cover with a 2 mm mesh net to prevent wedge clam migration. These cages were buried in the sand leaving 20 cm of each one protruding above the surface of the sediments.

Figure 11. Pre-growing in polyethylene containers (left) and *D. trunculus* spat after 1 month (right).

The best results were obtained in polyethylene containers with with forced flow. This structure allowed to culture high number of spat from different initial sizes and guaranteed high survival rates (48% in this study) without malformations. Similar structures had been assayed with good results for other bivalve species, such as *V. corrugata*, *R. decussatus* and *R. philippinarum* [23]. These structures with forced seawater flow by air injection systems showed good results in natural environment and they could imply an important advance in the on-growing culture of bivalves in Galicia.

Density is one of the most important parameters in the culture of bivalve species, influencing the growth and the mortality rates. Our experience with *D. trunculus* showed that cultures performed at highest densities reached the highest survival rates, unlike *R. decussatus* spat, as observed by Martínez et al. [15], which stopped the growth and suffered severe mortalities at high densities.

Other factors such as location, environmental conditions and initial length of seed should be considered, as observed by Cerviño-Otero (2011) with *V. corrugata* spat. Different growth rates from 3.5 mm spat were recorded in the different combinations of culture systems and locations, achieving the best results in floating rafts. Temperature and food are also important parameters [4]. *Donax trunculus* showed a slow growth in winter, thus the pre-growing in spring-summer, with high temperature and phytoplankton availability could favour growth and survival.

Wedge clam spat (7-8 mm) were seeded on natural substrate in "O Lombo das Navallas" in March. Onwards one month high mortalities were recorded (87%), but they reached sizes around 11 mm. The survival rate in the following months was 100%. Severe initial mortalities could be explained by the adaptation to the natural environment or by the initial size of the animals. The monitoring could only be maintained for 5 months, with an average final size of 22 mm. The growth rate was around 3 mm/month.

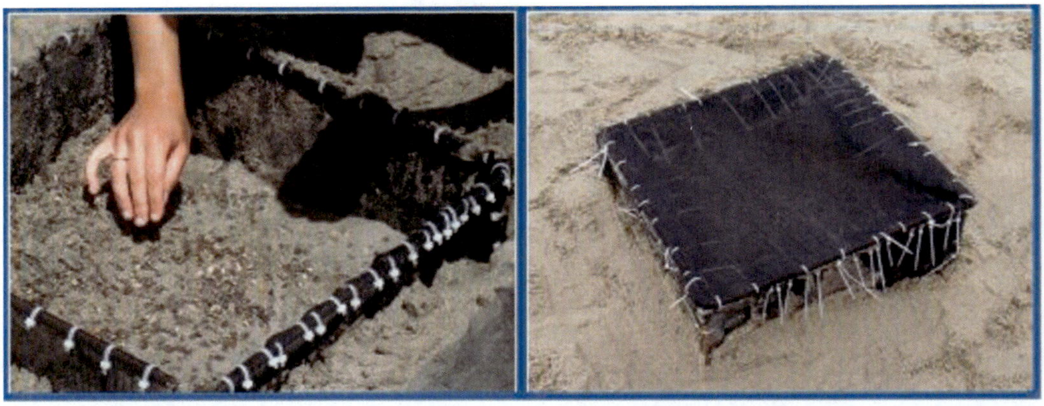

Figure 12. Holding containers for on-growing seed of the wedge-shell clam buried in natural beds.

In this stage, the monitoring until commercial size (35 mm) was complicated due to the difficult to maintain the structures in the same location. The current swept the cages away, being buried in the sand and losing the references. Other systems had been previously assayed in the same natural bed, such as wood boxes or boards enclosing the area where the clams had been seeded. However, the strong hydrodynamic and the consequent movement of substrate resulted in the loss of these structures. Different labelling techniques have been proposed for Donacidae. One is the use of calcein in *Donax hanleyanus*. Calcein is a non-lethal growth marker, emitting a bright green fluorescence band under blue light, with potential for using in long-term studies in natural environment [14]. In other study, Dugan and McLachlan [11] used a metal detector to track change in the position of intertidal *Donax serra* with aluminum tags glued to the shells. Use of appropriate labelling techniques could solve the problem of monitoring *D. trunculus* in outgrowing areas.

In summary, the spat of *D. trunculus* may be obtained in hatchery and seeded in natural beds until the commercial size. Spawns were obtained in hatchery from ripe adults collected between March and July, in a Galician natural bed, and adults were also conditioned successfully at temperatures of 17 and 21°C. The spawning induction was simple, keeping the broodstock dry followed by immersion in seawater at 18°C. The transformation into the D-shaped larvae occurred at an average percentage of 57%. The larval development lasted one month and the growth from post-larvae until 3 mm-spat 90 days. The best results of pre-growing in natural environment were achieved in spring-summer using forced flow systems. Individuals after 5 months seeded in natural substrate reached an average length of 22 mm.

References

[1] Aranda-Burgos, J. A., da Costa, F., Nóvoa, S., Ojea, J., Martínez-Patiño, D. 2014a. Effects of microalgal diet on growth, survival, biochemical and fatty acid composition of *Ruditapes decussatus* larvae. *Aquaculture,* 420-421, 38-48.

[2] Aranda-Burgos, J. A., da Costa, F., Nóvoa, S., Ojea, J., Martínez-Patiño, D. 2014b. Embryonic and larval development of *Ruditapes decussatus* (Bivalvia: Veneridae): a study of the shell differentiation process. *J. Mollus. Stud.,* 80, 8-16.

[3] Bayed, A., Guillou, J. 1985. Contribution à l'etude des populations du genre *Donax*: la population de *D. trunculus* L. (Mollusca, Bivalvia) de Mehdia (Maroc). *Ann. Institut. Océanograph.,* Paris 61, 139-147.

[4] Bayne, B. L., Newell, R. C. 1983. Physiological energetic of marine molluscs. In Saleudin, A. S. M., Wilbur, K. M. (Eds) *The Mollusca* Vol. 4. Academic Press, NY. pp. 407-415.

[5] Cerviño-Otero, A. 2011. Ciclo reproductivo, cultivo en criadero y en el medio natural de la almeja babosa *Venerupis pullastra* (Montagu, 1803). Ph.D. Thesis. Universidad de Santiago de Compostela. 288 pp. (In Spanish).

[6] Chanley, P. 1969. Larval development of the coquina clam, *Donax variabilis* Say, 1822 with a discussion of the structure of the larval hinge in the Tellinacea. *Bull. Mar. Sci.,* 19, 214-224.

[7] da Costa, F., Darriba, S., Martínez-Patiño, D. 2008. Embryonic and larval development of *Ensis arcuatus* (Jeffreys, 1865) (Bivalvia: Pharidae). *J. Mollus. Stud.,* 74, 103-109.

[8] da Costa, F., Martínez-Patiño, D. 2009. Culture potential of the razor clam Solen marginatus (Pennánt, 1777). *Aquaculture,* 288, 57-64.

[9] da Costa, F., Darriba, S., Martínez-Patiño, D., Guerra, A. 2010. Culture possibilities of the razor clam *Ensis arcuatus* (Pharidae: Bivalvia). *Aquac. Res.,* 42, 1549-1557.

[10] da Costa, F., Aranda-Burgos, J. A., Cerviño-Otero, A., Fernández-Pardo, A., Louzán, A., Nóvoa, S., Ojea, J., Martínez-Patiño, D. 2012. Clam hatchery and nursery culture. In da Costa, F. (Ed.). *Clams fisheries and aquaculture*, Nova Pub., NY. pp. 217-254.

[11] Dugan, J. E., McLachlan, A. 1999. An assessment of longshore movement in *Donax serra* Röding (Bivalvia: Donacidae) on an exposed sandy beach. *J. Exp. Mar. Biol. Ecol.,* 234, 111-124.

[12] Gaspar, M. B., Chícharo, L. M., Vasconcelos, P., García, A., Santos, A. R., Monteiro, C. C. 2002. Depth segregation phenomenon in *Donax trunculus* (Bivalvia: Donacidae) populations of the Algarve coast (southern Portugal). *Sci. Mar.,* 66, 111-121.

[13] Guillou, J., Le Moal, L. 1978. Variabilité spatio-temporelle des populations de *Donax* en Baie de Douarnenez. *Haliotis,* 9, 77-88.

[14] Herrmann, M., Lepore, M. L., Laudien, J., Arntz, W. E., Penchaszadeh, P. E. 2009. Growth estimations of the Argentinean wedge clam *Donax hanleyanus*: A comparison between lenght-frequency distribution and size-increment analysis. *J. Exp. Mar. Biol. Ecol.,* 379, 8-15.

[15] Martínez, M., Cerviño, A., Coo, A., García, A., Martínez-Patiño, D., Nóvoa, S., Ojea, J., Sebe, M. P., Casal, S., Fariña, J., Mosquera, P., Rodríguez, R. 1997. Parámetros biológicos y medioambientales del preengorde de almeja fina (*Venerupis decussata*) en sistemas sobreelevados. *Xornadas sobre o desenvolvemento produtivo do marisqueo a pé.* Santiago de Compostela. Spain (In Spanish).

[16] Martínez-Patiño, D. 2002. Estudio de los solénidos, *Solen marginatus* (Pennánt, 1777) y *Ensis siliqua* (Linné, 1758), de los bancos naturales de la Ría de Ortigueira y Ría del Barquero: ciclo gametogénico, composición bioquímica y cultivo larvario. PhD Thesis. Universidad de Santiago de Compostela. 240 pp. (In Spanish).

[17] Martínez-Patiño, D. 2013. Balance de las nuevas especies aportadas a la diversificación de la acuicultura y el marisqueo. *XV Foro dos recursos mariños e da acuicultura das rías galegas,* 15, 95-115 (In Spanish).

[18] Moüeza, M. 1975. Contribution à l'étude de la biologie de *Donax trunculus* L. mollusque lamellibranche dans l'Algérois. Thése Doct. Es. Sciences Nat., Univ. Aix-Marseille. 196 pp.

[19] Moüeza, M. 1976. Contribution á l'étude de la biologie de *Donax trunculus* L. (Mollusque Lamellibranche) mecanisme de la nutrition. *J. Mollus. Stud.,* 42, 17-30.

[20] Nóvoa, S. 2007. Metabolismo lipídico, ácidos grasos en el cultivo larvario de almeja babosa, *Venerupis pullastra* (Montagu, 1803). "Calidad ovocitaria, larvaria y nutricional con una aproximación al uso de la microencapsulación lipídica." Ph.D. Thesis. Universidad de Santiago de Compostela. 244 pp. (In Spanish).

[21] Ojea, J., Pazos, A. J., Martínez, D., Nóvoa, S., García-Martínez, P., Sánchez, J. L., Abad, M. 2008. Effects of temperature regime on broodstock conditioning of *Ruditapes decussatus*. *J. Shellfish Res.,* 27, 1093-1100.

[22] Robinson, A. 1992. Gonadal cycle of *Crassostrea gigas* Kumamoto (Thunberg) in Yaquina Bay, Oregon and optimum conditions for broodstock oysters and larval culture. *Aquaculture,* 106, 89-97.

[23] de Santiago, J. A., Fernández, A., Ruiz, M., Guerra, A. 2008. Preengorde de almeja babosa (*Venerupis pullastra* Montagu, 1803), almeja fina (*Ruditapes decussatus* Linné, 1758) y almeja japonesa (*Ruditapes philippinarum* Adams and reeve, 1850) en tres sistemas de preengorde. *X Foro dos Recursos Mariños e da Acuicultura das Rías Galegas,* 10, 381-387.

In: Oysters and Clams
Editor: Jesús L. Romalde
ISBN: 978-1-63485-074-2

Chapter 3

SHORT-TERM MAINTENANCE OF DEPURATED CLAMS: A PRELIMINARY MODELLING EXPERIMENT

Luci Jesus[1], Jaime Aníbal[1,2,*], Isabel Ratão[1,3] and Eduardo Esteves[1,4,*]

[1]Departamento de Engenharia Alimentar, Instituto Superior de Engenharia, Universidade do Algarve, Faro, Portugal
[2]CIMA-Centro de Investigação Marinha e Ambiental, Universidade do Algarve, Faro, Portugal
[3]MeditBio, Faculdade de Ciências e Tecnologia, Universidade do Algarve, Faro, Portugal
[4]Centro de Ciências do Mar-CCMAR, Faro, Portugal

ABSTRACT

The objective of the depuration process is to eliminate pathogenic microorganisms that contaminate shellfish, e.g., grooved carpet shell or Manila clams, oysters, cockles or mussels. During depuration, for a period that ranges 24 h to 48 h, the animals do not feed although they invest energy in the filtration process (the rational of the depuration process). In short periods, this practice has no significant effect on the animals' condition. However, it is expected that over longer periods of time, in the range of weeks, the impact of depuration is more obvious and important to the physiological condition of the individuals, making them less attractive in terms of nutritional and commercial quality.

Considering the importance of food safety related to the consumption of shellfish and taking into account the pertinent legislation, it was considered interesting from a commercial point of view, to study the possibility of maintaining already depurated clams for longer, up to 3-4 weeks' periods in a parallel system to the depuration system. This would allow circumventing periods of ban on harvesting and still ensuring the condition and safety of bivalves for further marketing and human consumption. Herein, an individual bioenergetics model based in the scope for growth concept was used to predict

[*] Corresponding Authors address. Departamento de Engenharia Alimentar, Instituto Superior de Engenharia, Universidade do Algarve, Campus da Penha, 8005-139 Faro, Portugal. Email: janibal@ualg.pt (JA) and eesteves@ualg.pt (EE).

the condition and growth of clams maintained in a system similar to that used for depuration at different temperatures and concentrations of microalgae. The results allow planning more accurately a follow-up experiment to be carried out at pilot scale.

Keywords: depurated clams, "Scope for growth," *Ruditapes* spp., net production, condition index, bioenergetics model

INTRODUCTION

The objective of the depuration process is to eliminate pathogenic microorganisms that contaminate shellfish, e.g., grooved carpet shell or Manila clams, oysters, cockles or mussels. "Depuration consists of placing shellfish in flowing clean seawater such that the animals resume normal pumping activity and thereby expel contaminants from their gills and intestinal tract over a period of time" [1]. Parameters such as shellfish suitability (in terms of salubrity and condition after harvesting and general handling), physiological conditions (*viz.* dissolved oxygen, loading, shellfish to water ratio, water flow, salinity, temperature, turbidity and disturbances), and infrastructures and operations (design of operating system, basic hygiene draining, batch control, water quality) are critical [2]. During depuration, commonly 24 to 48 h, the animals do not actually feed but they invest energy in the filtration process. In short periods, this practice has no significant effect on the animals' condition. However, it is expected that over longer periods of time, weeks, the impact of the clearance procedure is more obvious and important to the physiological condition of the individual [3], making them less attractive in terms of nutritional and commercial quality.

Taking into consideration the importance of food safety related to the consumption of bivalve mollusks [4], the general framework for controls in the context of international trade provided by the *Codex Alimentarius Commission* [5, 6], and the legislative requirements in Europe [7–9] and in the USA [10] - the two trading blocks that drive many of the controls applied in other exporting countries - it was considered interesting from a commercial point of view (in a collaboration between Universidade do Algarve, Faro, and a depuration and expedition center, PMA Aquacultura Lda., Tavira), to study the possibility of maintaining already depurated clams for longer, up to 3-4 weeks periods in a parallel system to the depuration system. This would allow circumventing periods of ban on harvesting, decided by the competent (national) authorities based upon laboratorial analyses of culture/harvesting water quality and bivalves' composition, that averaged 16.8 consecutive days of complete ban in 2014 [11], and still ensuring the condition and safety of bivalves for further marketing and human consumption.

The growth and net production of shellfish can be alternatively assessed through individual-based bioenergetics models [12] such as those based in the "scope for growth" (SFG) concept [*cf.* 12–16]. Models, in general, are abstractions of reality, often in the form of mathematical equations, that describe or simulate the dynamics among the variables that compose a system. The SFG is a term originally coined by C. E. Warren and G. E. Davis in 1967 which can be summarized as "the surplus of energy available for growth beyond that required for maintenance" [17] and has become prominent in studies of the energetics of filter-feeding bivalves and other invvertebrates [15, 18]. SFG equates to the difference over time between processes of energy acquisition, i.e., assimilation, and energy expenditure, i.e., respiration and growth (Figure 1). Besides routine monitoring, bioenergetics models can be used as low cost modeling tools to preliminarily assess relatively complex systems such has depuration facilities wherein a number of variables act as forcing functions, e.g., temperature, dissolved oxygen, water flow, salinity [2].

In this chapter, an individual bioenergetics model based in the SFG concept was used to predict the growth and condition of clams maintained in a system similar to that used for depuration. Different temperatures and concentrations of rations of microalgae were used as input variables in the modelling experiment. The results will allow a more accurate planning of the needed follow-up, pilot scale experiment that, eventually, will support the maintenance of already depurated clams for longer periods in order to circumvent periods of ban on harvesting and still ensure the condition and safety of bivalves for further marketing and human consumption.

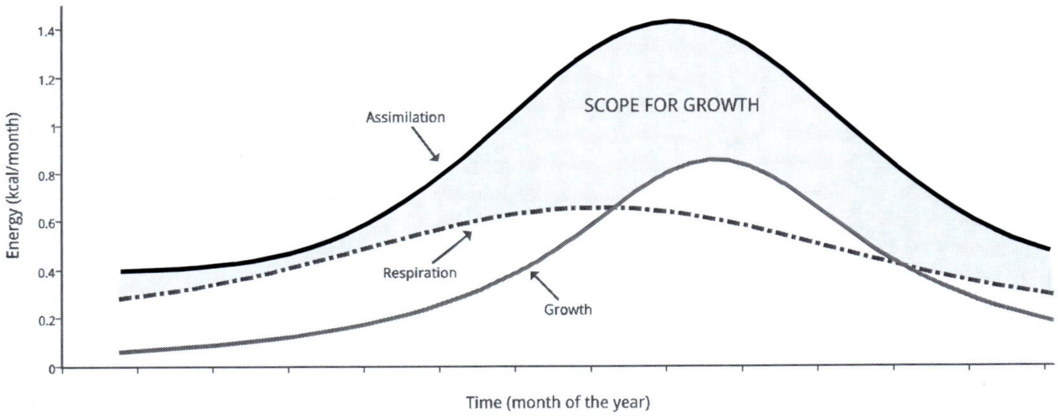

Figure 1. Illustration of the SFG concept a seasonal change in energy (adapted from [12]).

Modelling the Net Production and Condition of Clams

Herein, the model in Flye-Sainte-Marie et al. [19], proposed for culture and fisheries sites of the Manila clam *R. phillipinarum* in Marennes-Oléron Bay (France) and detailed below, was chosen to estimate the net production and condition. The model is based on the SFG concept that is illustrated in Figure 2 and can be (re)expressed via the balanced energy equation [12, 14, 18]:

$$C\text{-}F = A = R + E + P$$

or more succinctly but equivalently as:

$$P = A - (R + E)$$

where (all as energy per unit time) C is the total consumption of food energy, F is the faecal energy loss, A is the absorbed food energy, R is the respiratory energy expenditure, E is the energy lost as excreta and P is the energy available for growth and reproduction (i.e., “scope for growth”). Several studies consistently showed that energy lost via excreta was a small proportion (<5%) of the total energy budget [18, 20]. Thence for routine monitoring, the balanced energy equation can be further simplified to:

$$SFG = A - R$$

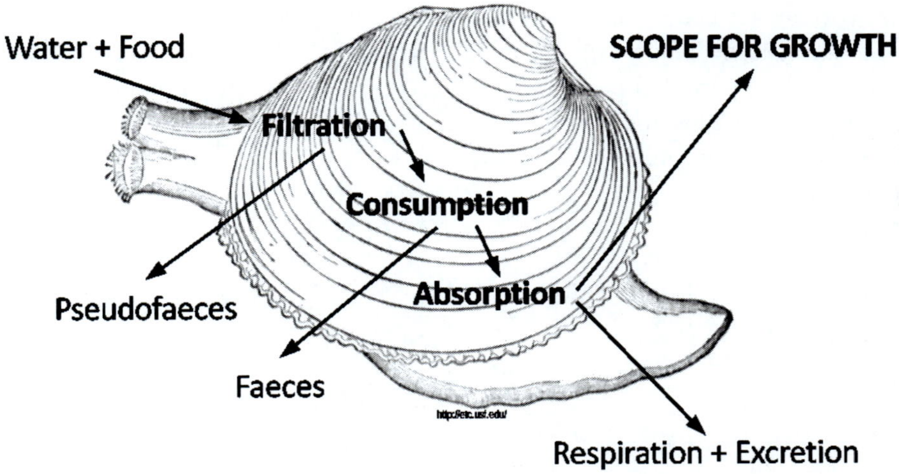

Figure 2. Illustration of the SFG concept as part of energy budget (adapted from [14]).

In this context, two indices were calculated to assess the growth and condition of clams: the net production (i.e., SFG) and the condition index. In order to reflect commercial practices, three size classes of clams were stipulated, 0.1, 0.2 and 0.3 g dry weight (DW) of edible mass, roughly corresponding to small, medium and large-sized whole animals. Estimates of production and condition (i.e., the response variables) were obtained for temperatures ranging from 12 to 20°C and food supply varying between 0.004 to 0.012 g DW L^{-1}, respectively. Temperature and food ration are the (independent) relevant variables i.e., depuration conditions. Furthermore, a maintenance period of 4 weeks (i.e., 28 days), exceeding the longest reported time that the capture of bivalves has been banned, was also considered herein.

Firstly, the net production (P_{net}, g day^{-1}) of clams was estimated based in [19]:

$$P_{net} = I \times A - R$$

where I is the ingestion rate, A is the assimilation rate and R stands for respiration rate. The ingestion rate (I, g day^{-1}) was calculated using:

$$I = Food \times Filt$$

where the amount of food ($Food$, g DW L^{-1}) is given by $Food = Chla \times F_C$ (wherein $Chla$ is the concentration of chlorophyll a, in μg L^{-1}, and F_C is the feeding coefficient, in g DW μg^{-1} Chla, and the filtration rate ($Filt$, L day^{-1}) is obtained from $Filt = [20.049 \times W^{0.257} \times (Filt(T)/Filt(16))] \times 1.44$. The filtration rate ($Filt(T)$, L h^{-1}) is a polynomial function of temperature (T, °C): $Filt(T) = -5.62 \cdot 10^{-3} T^2 + 0.18T - 0.30$ ($Filt(16)$ is the result of this function for a temperature of 16°C).

The Assimilation rate (A, g day^{-1}) was obtained using:

$$A = AE \times Food \times Filt$$

where $AE = AE_0 + (AE_1 \times W)/(K_A + W)$, AE_0, AE_1 and K_A were set to 0.1, 0.6 and 2.74 g respectively, and W is the DW (g) of clams.

The respiration rate (R, g DW day^{-1}) was estimated following:

$$R = [27.88 \times W^{0.85} \times (Resp(T)/(Resp(21))] \times 2.177 \cdot 10^{-3}$$

wherein the temperature-dependent respiration rate ($Resp(T)$, ml O_2 h^{-1}) is a polynomial function: $Resp(T) = -4.75 \cdot 10^{-4} T^2 + 2.02 \cdot 10^{-2} T - 2.43 \cdot 10^{-2}$ ($Resp(21)$ is the result of the function for a temperature of 21°C).

Secondly, the shell weight-based condition index (CI) used herein was calculated as [19]:

$$CI = (C + 1.0562)/0.0124$$

and the condition (C) was calculated from

$$C = (W - W_0)/(W_{max} - W_0)$$

where W_0 is the initial weight of (individual) clams (from $W_0 = a_0 \times L^b$), W is the final weight and W_{max} is the maximum weight obtained from $W_{max} = a_{max} \times L^b$ (where a_0 is $8.52 \cdot 10^{-7}$, a_{max} is $1.70 \cdot 10^{-6}$ and b is 3.728).

Calculations were carried out using Microsoft® Excel and graphs were plotted using plot.ly (Plotly Technol. Inc., 2015).

PHYSIOLOGICAL RESPONSES OF DEPURATED CLAMS DURING SHORT-TERM MAINTENANCE

Estimates of the net production and the condition index for a range of temperatures and food concentrations are presented in Figures 3 and 4 as contour plots.

In terms of P_{net}, smaller clams, 0.10 g DW, are expected to grow (since $P_{net} > 0$) for temperatures between 12°C and 18°C and for algal (food) concentrations above 0.018 g DW L^{-1} (Figure 3). P_{net} can reach 0.016 g DW d^{-1} (at 16°C and 0.028 g DW L^{-1}). Conversely, for lower feeding rations, in the range 0.004-0.010 g DW L^{-1}, clams shall lose weight since $P_{net} < 0$ (-0.0045 to -0.0081). For higher temperatures, e.g., 20°C, higher algal concentrations (>0.022 g DW L^{-1}) are needed to originate $P_{net} > 0$; in fact, $P_{net} = 0.012$ g DW d^{-1} for 0.028 g DW L^{-1}.

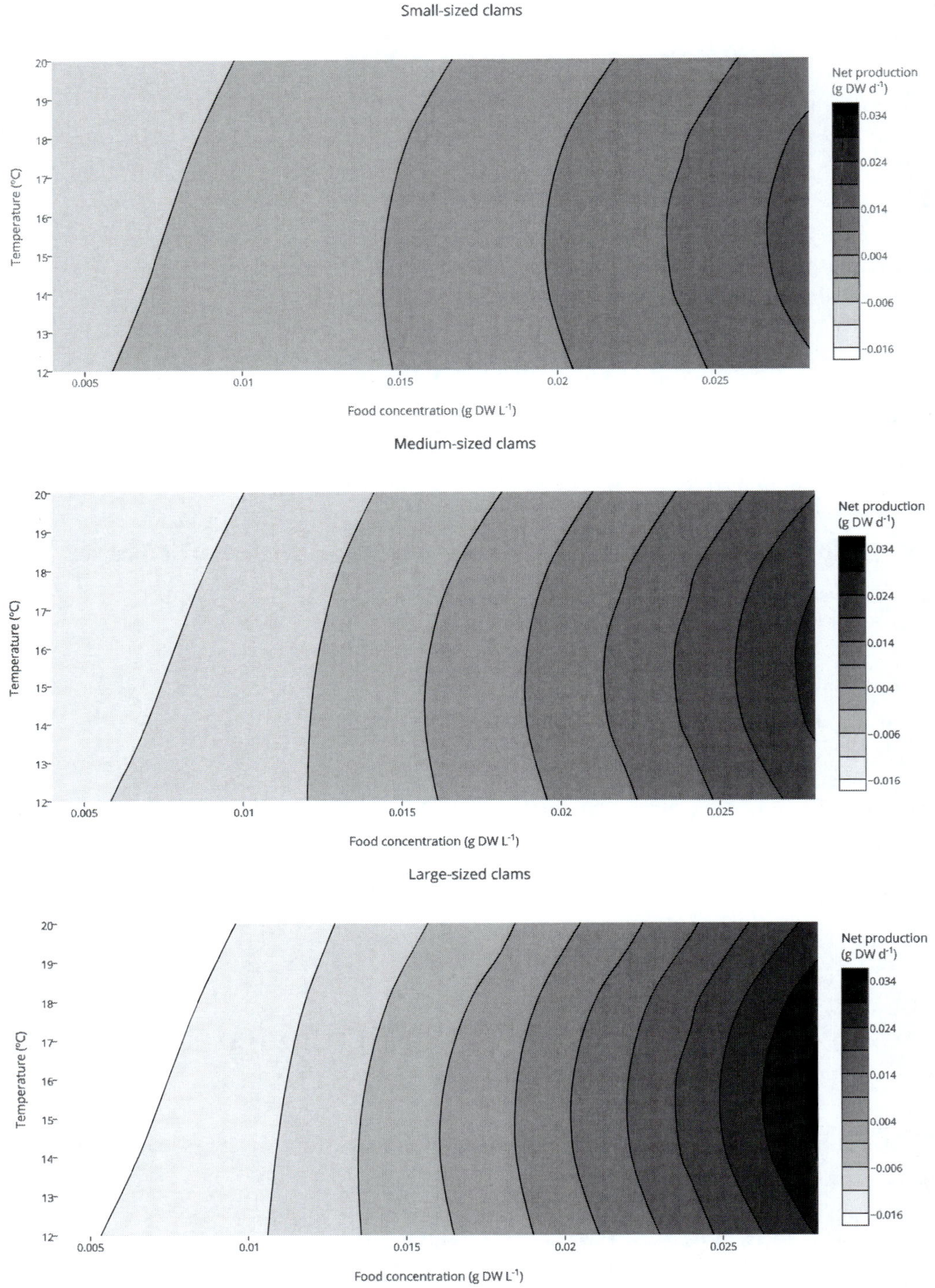

Figure 3. Contour plots of the estimates of the net production for small-, medium- and large-sized clams as a function of temperature and food concentration calculated following the model proposed in Flye-Sainte-Marie et al. [19].

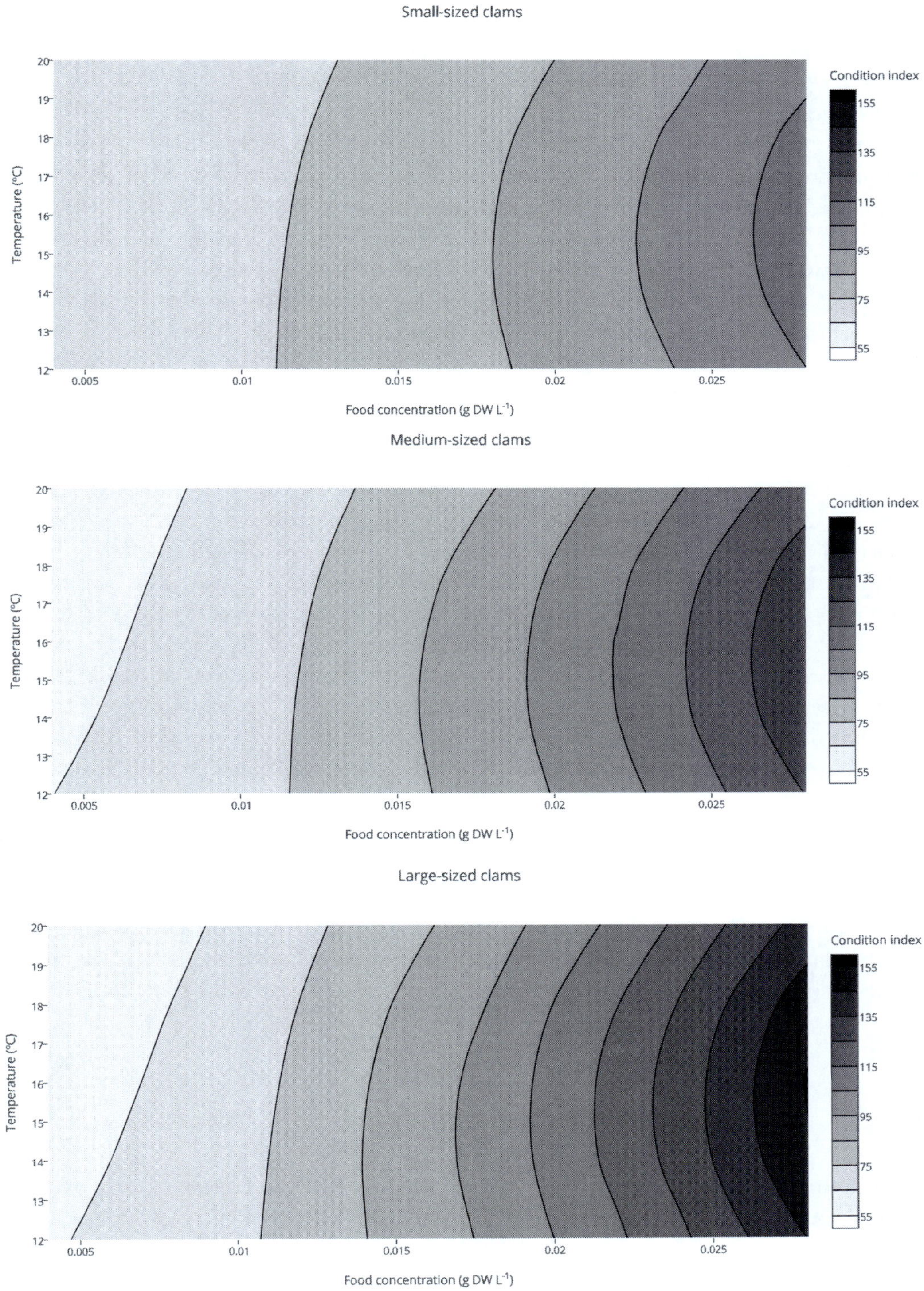

Figure 4. Contour plots of the estimates of the condition index for small-, medium- and large-sized clams as a function of temperature and food concentration calculated following the model proposed in Flye-Sainte-Marie et al. [19].

A similar behavior of the estimates of P_{net} was obtained for medium- and larger-sized clams (0.2 and 0.3 g DW, respectively) (Figure 3) but higher values were calculated. P_{net} can reach 0.021-0.026 g DW d^{-1} for medium-sized clams and 0.029-0.035 g DW d^{-1} for larger clams when specimens are kept in water at 12-18°C and fed > 0.018 g DW L^{-1} of algae.

Expectedly, the condition index results (Figure 4) follow relatively closely those obtained for the net production. Estimates of the index are above 100, i.e., there is a weight gain during the stipulated 28 days' period of maintenance, when algal concentrations exceed 0.028 g DW L^{-1}, 0.022 g DW L^{-1} and 0.022 g DW L^{-1} respectively for smaller-, medium- and larger-sized specimens whatever the water temperatures considered (in the range 12 to 20°C).

The net production and condition index are reliant on the duration of the depuration process. During the first 72 h there is considerable loss of initial fat content, namely of fatty acids [21]. After longer periods of starvation, quite relevant reductions in carbohydrates and proteins have been observed, e.g., in *R. decussatus* and *Venerupis pullastra* [3]. Together with glycogen and, occasionally, proteins, energy reserves are consumed to overcome inanition [3]. Seemingly, the decline in the condition index in settings similar to those maintained during "normal" depuration is different among bivalve species, *viz.* oysters, mussels, cockles and clams [21] suggesting that besides (initial) energy reserves, some species are able to reduce their metabolic rate as a physiological response to starvation [3, 22].

The model applied herein implements the effects on clams' energetic reserves of environmental conditions, specifically water temperature and quality and quantity of feeding (e.g., seston). These conditions are highly variable *in situ* during the year but can be controlled (for temperature) and even conditioned (for microalgal diets) in the depuration facility. The effect of temperature on feeding and growth and the feeding, digestion, and absorption of suspension-feeding bivalves is well documented [*cf.* 12]. On the other hand, carbohydrates, namely glycogen, are considered the main energy source in adult bivalves [3] and are important in gametogenesis [23, 24] and during periods of stress [25] or starvation [3]. Thus, the ability to withstand the depuration process (or even to grow if fed; as estimated herein) is largely dependent upon initial energetic condition of specimens, that has been reported to show strong yearly seasonality in terms of biochemical composition [e.g., 26, 27], and the settings of the depuration process [*cf.* 1, 2], e.g., temperature but also depends upon the nutritional quality of microalgal diets [28] or the concentration and size of suspended particles, more importantly benthic algal cells [29, 30], if feeding is to be considered as a option to maintain already depurated clams during periods of ban.

CONCLUSION

According to the estimates obtained herein, keeping clams at temperatures of 14-16°C and algal concentrations above 0.018 g DW L^{-1} would allow their maintenance for a period of 28 days. Interestingly, the estimated temperature range herein corresponds to that currently in use but the feeding of clams is not common practice and should be further investigated. In fact, food safety, on one hand, and technical and economic issues, on the other hand, are concerns that must be considered. Certified algae, for example phytobloom® (nekton S.A., Portugal), should be used to prepare the feed to avoid the re-contamination of clams already depurated. The duration and period of day when to provide the feed as well as the time needed to purge the tank water from debris after feeding should also be studied and optimized. The results presented in this chapter can be adapted to other species by simply changing model parameters accordingly, if data is available. Notwithstanding, the observed trends and general findings are expected to persist and allow to plan more accurately follow-up experiments at pilot-scale.

REFERENCES

[1] Lee, R., Lovatelli, A., Ababouch, L. 2008. Bivalve depuration: fundamental and practical aspects. *FAO Fish. Tech. Paper 511,* 139 pp.

[2] Lees, D., Younger A., Doré, B. 2010. Depuration and relaying. In: Rees, G., Pond, K., Kay, D., Bartram, J., Santo Domingo (Eds.) *Safe Management of Shellfish and Harvest Waters*. J. World Health Organization (WHO). IWA Publishing, London, UK.

[3] Albentosa, M., Fernández-Reiriz, M. J., Labarta, U., Pérez-Camacho, A. 2007. Response of two species of clams, *Ruditapes decussatus* and *Venerupis pullastra*, to starvation: Physiological and biochemical parameters. *Comp. Biochem. Physiol., Part B: Biochem. Mol. Biol.* 146, 241-249.

[4] Sauvé, G. 2013. *Molluscan Shellfish Safety*. Springer Science & Business Media, Dordrecht.

[5] CAC. 2008. Standard for live and raw bivalve molluscs (CODEX STAN 292-2008). World Health Organization Food and Agriculture Organization of the United Nations, Rome.

[6] CAC. 2012. Code of Practice for fish and fishery products. 2nd edition, World Health Organization Food and Agriculture Organization of the United Nations, Rome.

[7] EU. 2004. Regulation (EC) No. 853/2004 of the European Parliament and of the Council of 29 April 2004 laying down specific hygiene rules for on the hygiene of foodstuffs. *Official Journal of the European Union* L 139/55.

[8] EU. 2005. Commission Regulation (EC) No. 2073/2005 of 15 November 2005 on microbiological criteria for foodstuffs. *Official Journal of the European Union* L 338/1.

[9] EU. 2007. Commission Regulation (EC) No. 1441/2007 of 5 December 2007 amending Regulation (EC) No. 2073/2005 on microbiological criteria for foodstuffs. *Official Journal of the European Union* L 322/12.

[10] FDA. 2007. National Shellfish Sanitation Program Guide for the control of molluscan shellfish (2007 Revision, Printed March 2009). U.S. Department of Health and Human Services, Public Health Service, Food and Drug Administration and Interstate Shellfish Sanitation Conference, 549 pp.

[11] IPMA. 2015. *Histórico da apanha e captura de moluscos bivalves (2014).* http://www.ipma.pt/pt/pescas/bivalves/historico/index.jsp.

[12] Gosling, E. 2015. *Marine Bivalve Molluscs*. John Wiley & Sons, Chichester, UK.

[13] Bayne, B. L. 1998. The physiology of suspension feeding by bivalve mollusks: an introduction to the Plymouth 'TROPHEE' workshop. *J. Exper. Mar. Biol. Ecol.* 219, 1-19.

[14] Dame, R. F. 2011. *Ecology of Marine Bivalves: An Ecosystem Approach*, 2nd Edition, CRC Press, Boca Raton, FL.

[15] Jørgensen, C. B. 1990. *Bivalve Filter Feeding: Hydrodynamics, Bioenergetics, Physiology and Ecology*. Olsen & Olsen, Fredensborg, Denmark.

[16] Smaal, A. C., Widdows, J. 1994. The scope for growth of bivalves as an integrated response parameter in biological monitoring. In: Kramer, K. (Eds.) *Biomonitoring of coastal waters and estuaries*. pp. 247-268. CRC Press, Boca Raton, FL.

[17] Levinton, J. 2015. *Glossary of Marine Biology*. http://life.bio.sunysb.edu/marinebio/glossary.html.

[18] Widdows, J., Staff, F. 2006. Biological effects of contaminants: Measurement of scope for growth in mussels. *ICES Tech. Mar. Environ. Sci.* 40, 30 pp.

[19] Flye-Sainte-Marie, J., Jean, F., Paillard, C., Ford, S., Powell, E., Hofmann, E., Klinck, J. 2007. Ecophysiological dynamic model of individual growth of *Ruditapes philippinarum*. *Aquaculture* 266, 130-143.

[20] Widdows, J. 1993. Marine and estuarine invertebrate toxicity tests. In: Calow, P. (Ed.) *Handbook of Ecotoxicology. Volume 1.* pp. 145-166. Blackwell, Oxford.

[21] Ruano, F., Ramos, P., Quaresma, M., Bandarra, M. N., Fonseca, I. P. 2012. Evolution of fatty acid profile and Condition Index in mollusc bivalves submitted to different depuration periods. *Rev. Port. Ciên. Vet.* 107 (581-582), 75-84.

[22] Randall, D., Burggren, W., French, K. 2001. *Eckert: Animal Physiology*, 5th Edition. W. H. Freeman, New York.

[23] Joaquim, S., Matias, D., Matias, A. M., Moura, P., Roque, C., Chícharo, L., Gaspar, M. B. 2014. Biochemical and energy dynamics throughout the reproductive cycle of the striped venus *Chamelea gallina* (Mollusca, Bivalvia). *Invert. Reprod. Develop.* 58, 284–293.

[24] Karray, S., Smaoui-Damak, W., Rebai, T., Hamza-Chaffai, A. 2015. The reproductive cycle, condition index, and glycogen reserves of the cockles *Cerastoderma Glaucum* from the Gulf of Gabès (Tunisia). *Environ. Sci. Pollut. Res.* doi:10.1007/s11356-015-4337-6.

[25] Sheehan, D., McDonagh, B. 2008. Oxidative stress and bivalves: a proteomic approach. *Invert. Surv. J.* 5, 110-123.

[26] Aníbal, J., Esteves, E., Rocha, C. 2011. Seasonal variations in gross biochemical composition, percent edibility, and condition index of the clam *Ruditapes decussatus* Cultivated in the Ria Formosa (South Portugal). *J. Shellfish Res.* 30, 17-23.

[27] Rodriguez-Moscoso, E., Pazo, J. P., Garcia, A., Cortés, F. F. 1992. Reproductive cycle of Manila clam, *Ruditapes philippinarum* (Adams and Reeve 1850) in Ria of Vigo (NW Spain). *Sci. Mar.* 56, 61-67.

[28] Albentosa, M., Camacho, A.P., Beiras, R. 1996. The effect of food concentration on the scope for growth and growth performance of *Ruditapes decussatus* (L.) seed reared in an open-flow system. *Aquac. Nutr.* 2, 213-220.

[29] Sobral, P., Widdows, J. 2000. Effects of increasing current velocity, turbidity and particle-size selection on the feeding activity and scope for growth of *Ruditapes decussatus* from Ria Formosa, southern Portugal. *J. Exper. Mar. Biol. Ecol.* 245, 111-125.

[30] Velasco, L. A., Navarro, J. M. 2002. Feeding physiology of infaunal (*Mulinia edulis*) and epifaunal (*Mytilus chilensis*) bivalves under a wide range of concentrations and qualities of seston. *Mar. Ecol. Prog. Ser.* 240, 143-155.

In: Oysters and Clams
Editor: Jesús L. Romalde

ISBN: 978-1-63485-074-2

Chapter 4

MICROBIOTA ASSOCIATED TO CLAMS AND OYSTERS: A KEY FACTOR FOR CULTURE SUCCESS

***Sabela Balboa*[*], *Aide Lasa, Diego Gerpe, Ana L. Diéguez and Jesús L. Romalde*[*]**
Departamento de Microbiología y Parasitología, CIBUS-Facultad de Biología, Universidade de Santiago de Compostela, Santiago de Compostela, Spain

ABSTRACT

The study of the microbiota associated to clams and oysters is essential not only to know the sanitary status of mollusc populations but also to design new strategies for the prevention and control of the disease outbreaks.

Bacterial pathogens are the main cause of the repetitive mortality episodes that seriously decimate the mollusc production worldwide both in hatcheries and natural beds. A number of *Vibrio* species have been described as causative of mortality episodes since the early 1960s, being responsible over recent years of severe mortality outbreaks. However, the introduction of new species to intensive shellfish culture including gastropods and cephalopods lead to the appearance of new pathogens belonging to the genera *Pseudomonas* or *Serratia* among others that can constitute a problem for the better-established cultures of clam and oyster.

At the other end of the spectrum, bacteria with antimicrobial activities are becoming essential to control pathogenic bacteria and therefore improve larvae survival. In the last years, different species belonging to *Phaeobacter*, *Pseudoaltermonas* or *Roseobacter* genera have been identified and tested in the laboratory and in the field to be used as probiotic (extended definition for aquaculture) with promising results.

The use of next generation sequencing techniques had provided a window into the complex world of microbial communities, helping the researchers to understand their interactions and potentially allowing the improvement of shellfish production.

* Corresponding authors address: Departamento de Microbiología y Parasitología. CIBUS-Facultad de Biología. Universidade de Santiago de Compostela. Campus Vida s/n. 15782, Santiago de Compostela. Spain. Email: sabela.balboa@usc.es; jesus.romalde@usc.es.

This review aims to give an overview of the state of art on the composition of the microbiota associated to clam and oyster cultures, and its putative role on the success of their culture.

Keywords: microbiota, *Vibrio*, bivalves, metagenomics, diversity, pathogens, probiotic

INTRODUCTION

Coastal and estuarine environments are growing areas of bivalve molluscs whose culture became an important industry in many countries, due to the increasing importance of these animal as protein for human consumption. Due to their filter-feeding habit, bivalves accumulate a rich and diverse bacterial microbiota [1, 2, 3]. The study of such microbiota is essential to know the sanitary status, as well as to determine the pathobiological bases of the periodic disease outbreaks affecting bivalve populations. In fact, this is a critical point in the culture of molluscs, the repeated episodes of mortality due to bacterial infections that usually decimates de the production.

The studies analysing the diversity, distribution and density of marine bacteria associated with those bivalve molluscs are scarce and mostly focussed on the characterization of bacteria with pathogenic potential for clams and oysters, due to the tedious work involved in the analysis of numerical taxonomy for the study of the bacterial communities.

The first studies regarding the bacteria associated to molluscs date back to the 1960´s and ever since most of them are only based on the culturable fraction of the microbiota. Two kinds of microbiota can be described associated with bivalve cultures: the resident microbiota, stable in the host regardless the influence of the environment, and the transition microbiota, which vary according to the environmental conditions. Some authors only recognises the transition microbiota [4, 5]. A variety of functions have been assigned to these bacterial communities, including food, source of vitamins and growth factors, a key role in defence preventing the colonization by bacterial pathogens or elimination of toxic substances [3, 6, 7].

This review intends to give an overview of the studies performed on the bacterial communities associated to the culture of two of the molluscs with highest commercial interest, clam and oysters.

CULTURABLE MICROBIOTA ASSOCIATED TO CLAM AND OYSTER ADULTS

The first studies analysing the bacterial communities associated to bivalve cultures date back to the sixties with Pacific oyster *Crassotrea gigas* [8], studying for the first time the natural microbiota associated to the animals. Since then, several authors have focussed their work in the analisys of the microbiota associated to these aquatic animals and the environment where they live [1, 3, 9].

Classic studies examined bacteria populations present in the molluscs from a public health hazard point of view, analyzing the potential human pathogens that may be concentrated in the bivalves due to their filtering behaviour [10, 11, 12]. More recently several studies focussed on the isolation and identification of pathogens for commercially important mollusc species as will be discussed in this chapter.

The first study analysing the microbiota associated to bivalve species was carried out by Colwell and Liston [8] using Pacific oyster (*Crassostrea gigas*) located in three different areas. They found a high proportion of Gram-negative bacteria (more than 80%) with predominance of the genera *Pseudomonas* and *Vibrio* but also of *Flavobacterium* and *Achromobacter*. In general, the detected bacteria were those considered psychrophilic marine bacteria that have been physiologically adapted to survive within the bivalve.

Similar results were obtained when *C. gigas* were grown in different conditions, i.e., floating trays in different geographical locations and seawater tanks were studied [9], being the most abundant genera the same found by Colwell and Liston [8]. Besides the indigenous bacterial communities, they studied the role of the predominant bacteria associated to mortality outbreaks, identifiying *Pseudoalteromonas haloplanktis* (formerly *Pseudomonas enalia*) as responsible for the infection.

Lovelace et al. [13] analysed the microbiota associated to *C. gigas* and *Crassostrea virginica* culturated in two different geographical locations, demonstrating that the bacterial load (in terms of total heterotrophic bacteria) was similar between locations, although with some qualitative differences. They found, moreover, a seasonal trend in the the abundance of both *Vibrio* and *Pseudomonas*, being slightly higher in the warm months. On the other hand, these authors observed a correlation between the microbiota found in water and in the animals, although bacteria counts were one to two logarithms less in the water than in the animals. Other studies reported similar results, showing less bacteria in the sediment or in the seawater than in the flesh of the animals.

Sugita et al. [14] went further in the study of the bacterial communities of molluscs by studying the composition of the microbiota from different species including *Ruditapes philippinarum*, *Mactra veneriformis*, *Mytilus corusus*, *C. gigas*, *Phacosoma japonicum* and *Scapharca broughtonii*, all from the same geographical location (Tokyo bay) and at different growth stages of their vital cycle as well as in different organs. In the wild oysters, a predominance of *Vibrio*, *Pseudomonas* and *Achromobacter* was observed, as well as the presence of *Escherichia*, *Neisseria* and some Gram-positive genera as *Corynebacterium*, *Streptococcus* and *Micrococcus*. The bacterial counts in the digestive tract were higher than in the flesh and higher, in turn, to that found in water and sediment, although the composition was very similar.

Other authors performed comparisons among different species of molluscs and those with the surronding water, founding simiral results. Kueh and Chan [1] studied the microbiota of oysters (*C. gigas*), clams (*Anadara* (*Scapharca*) *cornea*) and the mussel *Perna viridis* and the surronding water showed a dominance of the the aerobic bacteria respect of the anaerobic in both water and animals. Regarding the heterotrophic bacterial load, the counts from oyster ranged between 10^4 to 10^7 CFU/g or ml, being the highest numbers found in the gut, while the load of the remaining organs was comparable with the surronding water. At genera level, in oysters over 30% were identified as presumptive *Pseudomonas*, followed by *Vibrio* and to a lesser extent, *Aeromonas* and *Alcaligenes,* that interestingly were never found in water where the predominant organisms were coliforms. They suggested, in agreement with other authors that the microbiota in association with the bivalves is less intimate than in higher animals [5] and therefore, bacteria found in the molluscs is a reflection of the bacteria found in water although they recognises a commensal microorganisms to some extent. This comensal bacteria would be adapted to the specific physicochemical conditions of the animals. In that sense, they described the microbiota of the water as a heterogeneus array of microorganisms whereas only the more halophilic groups would be associated to molluscs.

The similarities between the microbiota found in water and bivalves have been extensively documented, not only at a descriptive level but also at functional one. Thus, Prieur et al. [3] described the significant proportion of isolates with lipolytic and proteolytic activity both in water and bivalves in a study performed in Manila clam (*R. philippinarun*) and mussel (*Mytilus edulis*) in a experimental hatchery. They applied numerical taxonomy techniques to the study of bacterial populations present in bivalves, comparing the different species and each species with its surronding water. As was described before, the proportion of gram-negative fermentative bacteria in the animal flesh was higer than in the surronding water [3, 15], and mostly composed by *Photobacterium* and *Vibrio*. When compared both sets of isolates (animal plus surronding water), the species were almost identical, indicating the tight relationship between animal and water.

The taxonomic studies by numerical analysis of phenotypic traits became an essential tool for the analysis of the health status of the molluscs. In this kind of analysis, the results of the phenotypic test performed by classical test and/or miniaturized systems such as API are coded as binary responses (1/0) and the data matrix is constructed to be further analysed by statistical tests which allows more accurate comparisons. Using this tool, several analyses have been performed in the 90s before the advent of the molecular techniques.

Spain is nowadays one of the main producers of shellfish aquaculture with a continuous growing since the nineties decade. The growing industry got the attention of numerous research groups, many of which focussed on the sanitary state of the cultures from a microbiological point of view.

The first works held in Spain studied the microbiota associated to Mediterranean oyster and the surrounding water from 1989 to 1990 [9, 16]. They grouped all the isolates obtained after the one-year survey according to their fermentative or oxidative metabolism and analysed them separately. However they did not compare the bacterial populations present in the oyster with those in the water. In a first approach, the facultatively anaerobic Gram-negative bacteria were studied, describing a total of 28 phena comprising 370 isolates. Half of the isolates corresponded to autochtonous marine *Vibrionaceae* bacteria, being the predominant groups, as described by other authors, *V. splendidus* and *V. harveryi* but remaining a lot of the isolates as unidentified species. Moreover, they found a high heterogeneity of the *V. splendidus* clade [9]. In a second study [16], they analysed the anaerobic bacteria, isolating a total of 245 strains that were grouped in 46 phena. Those that were identified included typical marine eubacteria such as *Alteromonas*, *Shewanella*, *Delaya*, *Marinomonas*-like, *Pseudomonas*, *Flavobacterium*, *Agrobacterium*-like and *Oceanospirillum*. It is interesting to point out the absence of genera as *Cytophaga* and *Hyphomonas* as well as some typical oligotrophs as *Caulobacter*, although the growth conditions used were not the optimal for those groups [16].

Montilla et al. [17] performed a comparision among different mollusc species, including clams (*R. decussatus* and *R. philippinarum*), *Ostrea edulis* and *C. gigas* in a survey performed monthly during one year in the Delta del Ebro (Spain). They found that the two species of clams accumulated the higher amount of *Vibrio* species while oysters showed lower load. The predominant was *V. fluvialis* while *V. alginolyticus* was the minoritary vibrio in the samples.

The dominance of *V. fluvialis* is discordant with other studies cited in this review, and might be explained by the lower salinity values described as optimal for this species and the fact that the samples were taking from a nursery in comparison with other works in related geographical areas but sampled in the open Mediterranean sea. Same explanation can be applied to the lack of *V. vulnificus* in the sample, usually found in oyster samples [18, 19, 20, 21].

A more accurate study of the oyster microbiota was performed by Pujalte et al. [22] that studied the difference in bacterial populations in different seasons in a one-year survey using *O. edulis* cultivated by the long line method, that is in the open Mediterranean sea. They found that the bacterial counts were higher in warm months than in cold ones, as expected, founding a reduction as high as 2 logarithms depending on the season (from 3.7 x 10^6 CFU/g in July to 3.2 x 10^4 CFU/g in January) on both marine heterotrophic bacteria and *Vibrio*. Counts obtained in the surronding water were at least two orders of magnitude less than in the animal flesh. The halophilic fermentative bacteria were predominant during most part of the year in *O. edulis*. In the warm season, from March to October, they found that the heterotrophic bacterial fraction was dominated by halophilic, facultatively anaerobic bacteria (mostly *Vibrionaceae*) while during the cold season the microbiota corresponded mainly to Gram-negative bacteria belonging to undescribed α-*Proteo*bacteria phyla. Same taxa and in similar proportions were also found in water. Interestingly, an alternation in the *Vibrio* population was found in oysters, related with changes in temperature. Thus, *V. splendidus* was the predominant at temperatures lower than 20ºC while *V. harveyi* was the main group during warm months. In that sense, other authors found that the predominant *Vibrio* species in oysters cultivated in the Delta del Ebro (Tarragona, Spain) was *V. fluvialis* [17]. As in other studies, the proportion of *Vibrio* with hydrolytic capabilities was very high. A possible explanation for this finding is that such capabilities, helping to breakdown the filtered organic macromolecules, might constitute an advantage for both organisms *Vibrio* and oyster.

To the date, the most complete study on the culturable microbiota associated with cultured clams in natural beds has been carried out by our research group, using phenotypic traits and the sequencing of 16S rRNA gene and hosekeeping genes [23]. The study included two different clam species (*R. philippinarum* and *R. decussatus*) in Galicia (NW of Spain), analysing several geographical locations and/or seasonal distribution.

The bacteria density showed a seasonal distribution with higher counts during the warmer season but with similar values for all the locations analysed both for total heterotrophic bacteria and *Vibrio*, results that are in agreement with previous reports in other parts of Spain, namely the Delta del Ebro and the Mediterranean sea [9, 17, 22, 24, 25]. The number of total bacteria and *Vibrio* were similar for the different organs analysed as well as the extrapaleal fluid, being always temperature-dependent. However, the bacterial load was higher in the gonad of the clams in certain locations [23], which could be of special importance due to the possible vertical transmission of potential pathogens. Bacterial diversity was very similar among geographic locations and similar in turn, to that found by Arias et al. [25] that, although they found a seasonal trend, were not able to find any specific bacterial diversity with any geographic location.

The dominant taxon was by far the genus *Vibrio*, accounting for approximately 60% of the isolates in every clam species and/or different location analysed. Those isolates were grouped in 32 different phena by numerical taxonomy on the basis of the Unweighted Pair Group Method with Arithmetic mean (UPGMA) with a clear dominance of *V. splendidus* group, constituting around 50% of the vibrios and was subdivided in two different phena, corresponding to the formerly known as biovar I and II. Other dominant species were *V. lentus*, *V. alginlyticus*, *V. diazotrophicus*, *V. aestuarianus*, *V. pelagius* and *V. superstes* [26]. The minoritary phena corresponded to species as *V. mediterranei* or *V. pectenicida* and more than 40 strains remained unidentified. Again, these results were similar to those obtained in other studies [9, 22, 24, 27]. Selected strains were subjected to Amplified Fragment Length Polymorphism (AFLP) including the type species of known vibrios but only 23% of the identificacion at species level was coincident with the phenotypic results, being the *V. splendidus* group the most variable [26].

Further studies based on the sequence of 16S rRNA gene and Multilocus Sequence Analysis (MLSA) enable the description of 9 new species of *Vibrionaceae*, namely *Aliivibrio finisterrensis*, *V. breoganii*, *V. gallaecicus*, *V. celticus*, *V. atlanticus*, *V. artabrorum*, *V. toranzoniae* and *V. cortegadensis* [28-34].

Bacteria with aerobic metabolism, i.e., oxidative bacteria, constituted around 35% of the associated microbiota that could be grouped in 25 different clusters. The predominant genera were *Pseudoaltermonas* and *Shewanella* (52.8 and 16% respectively) and at less extent *Tenacibaculum*, *Psycrobacter*, *Alteromonas*, *Polaribacter*, *Luteimonas*, *Marinomonas*, *Lacinutrix* or *Cobetia* [23]. This diversity was similar to that found in oysters [22, 35, 36]. As in the case of *Vibrio*, some seasonal distribution could be observed. Some Gram-positive isolates were found too mainly belonging to *Nesterenkonia* or *Arthrobacter* [23].

The marine environment is extremely diverse, with huge variations in pressure and temperature. Nevertheless, life, especially microbial life, thrives throughout the marine biosphere and microbes have adapted to all the divergent environments present. Using traditional culturing techniques, only between the 0.01 and 1% of the total microorganisms present can be described [37-39] and still, bacteria from phyla *Proteobacteria*, *Firmicutes*, *Bacteroidetes* and *Actinobacteria* has been described.

Few authors have compared the microbiota associated with the molluscs with that present in the surrounding water, obtaining different results. Prieur et al. [3] studied the microbiota present in different molluscs cultures, including *C. gigas* in comparision with the surrronding water and concluded that the animal posses higher amounts of *Vibrio* and *Photobacterium* than the surrounding water, results that are in agreement with Martin [15] when they compared the mussel microbiota with this found in the water. Similar microbiota associated to oyster flesh and water, i.e., *Photobacterium* and *Vibrio*. On the contrary, Pujalte et al. [22] found a very similar microbiota in the oyster and the surrounding water from Mediterranean sea all over the year, although vibrios were found in higher concentration in the oysters, and their adhesion to oyster clam tissues are probably enhanced by the capabilities of producing hydrolytic exoenzymes. Those are the expected results considering that the bivalves are filtering animals.

Culturable Microbiota Associated to Clam and Oyster Hatcheries

Hatcheries are nowadays the main source of seed for the aquacultures of many bivalve molluscs with high economic value, more specifically for clams and oysters. Seed obtainment by induced spawning has been studied from a technical point of view in order to establish the optimum conditions for its performance. The influence of environmental factors including salinity, temperature, culture density has been the aim of numerous works. Microbiological aspects are of outstanding importance due in part to potential poor management practices that can cause stress to the animals, making them more susceptible to disease outbreaks. On the other hand, bacteria are easily introduced through inlet water (even when is treated), food sources or even through vertical transmission from broodstock. Moreover the destabilization of the microbiota through disinfection creates the optimum condition for opportunistic pathogens to colonize the system.

However, and despite of the numerous descriptions of outbreaks in hatcheries, the scarcity of systematic and deep studies, and therefore the lack of knowledge in the bacterial populations associated to these cultures have led to the search of partial solutions usually focussed on the elimination of the microbiota from the culture seawater. In general, the few studies regarding the microbiota of this vital stage are centered in the *Vibrio* isolates since they are the cause of the mortality episodes. Moreover, usually this studies are held in the hatchery comparing the microbiota of the different compartments and the water rather than the animals themselves.

Lodeiros et al. [40] set the *Vibrio* concentration as low risk in 10^3 CFU/ml. Similarly, Widman et al. [41] set this value for bay scallop (*Argopecten irradians irradians*) in 10^4 vibrios/ml. It has been proposed that that in the with great scallop (*Pecten maximus*) most of the bacteria found in tanks come from algal cultures used to feed the animals, being composed by a 73% *α-Proteobacteria*, more specifically *Rhodobacter* and *Roseobacter* groups [42].

The marine environment showed higher variability than the water tank. The *γ-Proteobacteria* group was the predominant although less abundant than in the water tank (37%) with a 7% of *Vibrio*, close to the CBF group (*Cytophaga-Flavobacteria-Bacteroides*) that constituted the 33% of the microbiota, with lower representation from the Gram-positive and *α-Proteobacteria* (14 and 16% respectively). This is a typical distribution of the bacteria in the faunal assemblages in marine environment, according with other studies based on culturable bacteria where they found that *γ-Proteobacteria* tend to dominate [42] although few studies have found CFB group and *α-Proteobacteria* as the major constituents [44, 45].

Schulze et al. (2006) studied the balance between pathogenic and potentially probiotic bacteria in fish farms and hatcheries from different bivalve species, namely Manila clams (*Ruditapes philippinarum*), Pacific oysters (*C. gigas*), geoduck clams (*Panopea abrupta*), abalone (*Haliotis kamtschatkana*) and Pacific scallps (*Patinopecten yessoensis*), analysing larvae and/or broodstock as well as macroalgae, sea water and sea sediment. They isolated 598 strains, being able to group them in 172 phylotypes on the basis of Denaturing Gradient Gel Electrophoresis (DGGE) genotyping and 16S rRNA gene sequencing. They found that the majority of the bacteria found in the hatchery environment were *γ-Proteobacteria* (76% of the phylotypes) of which approximately 45% were *Vibrio*. They also found *α-Proteobacteria* and Gram-positive bacteria and only one isolate from each *ε-Proteobacteria* and CBF group. This results are similar to those found in other studies of fish farms. In seawater, macroalgae and sea sediment from the surrounding marine environment CFB were found.

Prado et al. [47] analysed two different batches of razor clam (*Solen marginatus*) with different survival rates. The batch with better survival rate showed lower bacterial load and only around 20% of the isolates were identified as *Vibrio* while the batch sufferering mortalities showed higher bacterial load with more than 75% of *Vibrio*, all of them belonging to the Splendidus clade. In the sea water from both culture tanks the heterotrophic bacterial counts were between $1x10^4$ to $1.3x10^5$ CFU/ml, being all the isolates oxidative in the case of the healthier tank while in the one showing mortalities around 10^2 CFU/ml were *Vibrio*.

BACTERIAL PATHOGENS

Most studies regarding the bacteria composition of the molluscs, have been focused on the characterization of bacteria with pathogenic potential for oysters and clams. As it was mentioned before, one of the main problems in aquaculture of molluscs is the repetitive episodes of mortality at both in initial culture stages in the hatchery and in natural beds. In most of the cases it has been demonstrated that the problems are caused by bacterial pathologies, being the main aetiological agents members of the genus *Vibrio* [48-50], although some oxidative bacteria has been also associated to mortality outbreaks in larvae, juveniles and adults [51-67].

From 1990s the diversity of *Vibrio* species associated with bivalve species in different geographical areas has been the subject of numerous studies. The main common conclusions obtained from these studies were that environmental parameters, such as variations in the water temperature and salinity, can influence the diversity of *Vibrio* spp. in the environment, as well as the physiological state of the bivalve and its susceptibility to bacterial infections. But in all cases, the predominant species associated with bivalves from different geographical location and/or climates were *V. splendidus*, *V. alginolyticus*, *V. harveyi* or any combination of these species [22, 24-27, 48, 68-70].

More recently, *V. cyclitrophicus*, *V. splendidus* and *V. alginolyticus* have been described as the predominant in carpet shell clams and Manila clams, constituting the 66.6% of isolates. Other species such as *V. fluvialis*, *V. vulnificus* and *V. mimicus* have also been associated with molluscs [68, 71]. It is important to point out that these studies identified the *Vibrio* strains using only phenotypic methods and, therefore, the real diversity present in bivalves could be underestimate. In fact the high variability present in certain groups such as *V. splendidus* and *V. harveyi* groups make impossible to discriminate species based only in phenotypic methods [72-74].

On the other hand, the influence of environmental parameters as temperature and salinity are key in the diversity and alternance of *Vibrio* populations [22, 25, 75-77].

As was mention before, hatcheries are nowadays the main source of seed for aquaculture for many bivalves such as oysters or clams. Although the influence of environmental factors including salinity, temperature or culture density has been extensively studied, there is a lack of knowledge on the microbiological aspects due to the shortage of the studies, most of the time focussed on a specific outbreak, and even in those studies the causative agent remains unidentified [66, 78-86]. In the hatchery, although all the larval stages are vulnerable, the temporary fixation of the larvae on the bottom of the tank is a particularly delicate moment since the larvae is in exposed to high concentration of potential pathogenic bacteria associated with the tank surface, moribund larvae or organic detritus.

The main cause of mortalities due to bacterial diseases in all larval stages is the genus *Vibrio*. The first report of a mortality outbreak in a hatchery caused by a bacteria belonging to the genus *Vibrio* was described by Guillard [87] in USA during an outbreak of the clam larvae *Mercenaria mercenaria* that killed 70% of the population and was denominated "Bacillary necrosis" but is now known under the more general name vibriosis. Larvae showed a disrupted velum and internal tissues. The bacteria causing this pathology were identified later as *V. tubiashii* [88] and *V. alginolyticus* [89]. Since then, the disease has been described in a plethora of bivalves, including *C. virginica*, *Ostrea edulis*, *M. mercenaria* as well as in *A. irradians* and *Teredo navalis*, and is always associated with swarms of bacteria appearing on the margins of the larvae that became progressively denser [89].

Several *Vibrio* species have been described as the causative agent of this larvae disease, sometimes mistaken due to the use of classical biochemical tests and the inherent variability to this genus, as is the case of the original description, that those isolates belonging to *V. tubiashii* species were originally identified as *V. anguillarum*. However, in the last decades some other species have been associated with this pathology, as is the case of *V. neptunius*, *V. ostreicida*, affecting oyster, besides the *V. pectenicida* in scallop larvae. Moreover, numerous species have been described as opportunistic pathogens causing vibriosis in larvae as *V. pectenicida*, *V. neptunius*, *V. coralliilyticus*, *V. ostreicida*, *V. neptunius* [66], *V. harveyi* and *V. aestuarianus* [90-94].

Other genus that has been associated to mortalities in bivalves is *Pseudomonas*, recognised as pathogen to different molluscs including clams (*M. mercenaria*), oysters (*O. edulis* and *C. virginica*) and pectenids, although in most cases were isolated together with different *Vibrio* spp [87, 95, 96].

Members of the genera *Altermonas* have also been associated to mortalities of oyster larvae (*C. gigas*) in a hatchery and bacteria belonging to genus *Moraxella* have occasionally been associated to outbreaks in *Tridacna gigas* larvae. Finally, isolates from the genus *Aeromonas* and *Plesiomonas* were isolated from moribund larvae of *T. gigas* and its pathogenitciy was subsequently demonstrated in a experimental challenge [96, 97].

Juveniles and Adults

Juvenile and adult bivalves are less susceptible to vibriosis, and in general to other pathogens, than larvae since they have developed immune system and therefore, the resistance to bacterial infection significantly increases in the adult stage [59].

The two main diseases affecting adult oyster and clam are Summer Mortality (SM) and Brown Ring Disease (BRD) respectively.

The SM is a classic oyster disease that affects mainly populations of juveniles of Pacific oyster (*C. gigas*) and it is usually associated with stress conditions, mainly high water temperatures. The disease was associated to high levels of a *Vibrio* sp. [98], later identified as *V. splendidus* [54, 67, 99]. Further, the analysis of the haemolymph of moribund oysters using phenotypic and molecular methods showed that the prevalent species causing this disease was *V. aestuarianus* subsp. *francensis* [56, 61, 100], *V. splendidus* [54, 101]. More recently, *V. harveyi* was proposed as a possible agent to this disease since was detected in most samples of an outbreak occurred in 2008 and was able to produce mortality in experimental challenges [102].

As reviewd in the last chapter of this book, the most accepted theory today is that the SM can not be attributed to a specific pathogen (bacterial or viral) but rather is a combination with the physiological state of the host, environmental factors and the presence of different opportunistic infectious *Vibrio* species [60, 65, 103]..

In clams, the most usual adult disease and the only one with demonstrated bacterial origin, is the BRD and it has been considered one of the main limiting factors in the culture of Manila clam in Europe but it was also detected in Korea and Japan [65, 104-106]. The causative agent of this disease is *V. tapetis* that can infect different hosts although with different susceptibility. Thus, *V. tapetis* has been isolated from species of clams, including *R. philippinarum*, *R. decussatus*, *Venerupis aurea*, *M. mercenaria*, oyster *C. virginica* and cockle (*Cerastorderma edule*) but the infectivity was only probed to *R. philippinarum* and to lesser extent to *R. decussatus* [65].

V. tapetis colonizes the periostracal lamina and alters the calcification process on the inner surface of the valves leading to the appearance of a characteristic brown deposit, consisting on conchiolin, between the edge of the shell and the pallial line.

As in the case of SM, environmental factors play a key role in the establishment and development of the disease, more specifically, temperature and salinity. In that sense, the BRD is more prevalent in spring and winter as the optimal temperature for the growth of *V. tapetis* is 15ºC and in experimental challenges with Manila clams demonstrated that the disease was more severe at 20 ppt than at 40 ppt [65]. On the other hand, it has been demonstrated that there are families of molluscs more resistant to the disease than others. As it was mentioned above, Manila clams are the most susceptible species and among them, there are families more resistant. In an experiment performed by Drummond et al. [107] using Irish and Galician clams, demonstrated that after the inoculation with the type strain, the Galician clams experimented increased mortalities and the BRD prevalence was significantly higher. Allam et al. [108] showed that American Manila clams were more resistant to the disease than French Manila clams, showing the American ones less clinical signs and higher recovery capabilities.

The diagnosis of the disease is currently based on the examination of the characteristic brown ring deposit and the reparation stages visible on the shell [65]. For an early diagnosis of the disease, specific PCR detection protocols targeting the 16S rRNA sequence have been designed [109].

Recently two subspecies for the pathogen have been described on the basis of their phenotypic, serological and genetic intraspecific variability related with the geographical origin, *V. tapetis* subsp. *tapetis* and *V. tapetis* subsp. *britannicus*, whose isolates are restricted to the British Islands [110]. Moreover, three main subgroups were described using MLSA and 2D-PAGE techniques related with the host origin [111].

Besides these well-described diseases, in the last decades, the pathogenic potential of recently described vibrios has been demonstrated. As it was stated before, *V. splendidus* clade is a complex and highly variable group that has been traditionally associated with mortality outbreaks. In the last 15 years, new species have been described within this group, being nowadays composed by 16 species, four of which were demonstrated to be potential pathogens for clams and/or oysters in experimental challenges or in natural episodes of mortality.

That is the case of *V. crassostreae*, first isolated from the haemolymph of diseased *C. gigas* [112] whose pathogenicity was further demonstrated by Gay et al. [54], in the same species. Similarly, *V. celticus* showed pathogenic potential for adult Manila and carpet shell clams although it was originally isolated from healthy carpet shell clams [30].

Similarly, *V. fortis* isolated from a range of hosts, including diseased larvae oysters with confirmed pathogenic potential using animal models as rainbow trout and *Artemia* nauplii [113]. The same pathogenicity models were used to demonstrate the virulence abilities of *V. brasiliensies*, isolated from healthy lion´s paw scallop (*Nodipecten nodosus*) larvae, belonging to the *V. orientalis*-clade [114]. Other non-Splendidus vibrios isolated from molluscs have been proposed as pathogen for other aquatic organisms, as *V. lentus*, isolated from oysters [115], pathogenic for octopus (*Octopus vulgaris*) and turbot (*Scophthalmus maximus*) that excrete a lethal extracellular protease similar to one produced by Splendidus-like strains that has been proposed to be of potential risk for reared aquatic organisms [117]. Finally, *V. neptunius*, originally isolated from healthy scallop larvae, rotifers and turbot larvae [114] as further isolated associated to a mortality episode of *O. edulis* larvae and its pathogenicity confirmed in experimental tests causing mortalities as high as 100% [66].

It is of special importance the so-call Harveyi clade or core group, that includes several species responsible for episodes of vibriosis in different molluscs, namely, *C. gigas* [93] and *R. decussatus* [58].

Other bacterial genera were described as pathogen for molluscs as is the case of members of the gram-positive genera *Nocardia*. *Nocardia crassostreae* is the causative agent of Nocardiosis, affecting oysters. It was originally described as "Focal Necrosis," "Multiple Abscesses" or "Fatal Inflammatory Bacteriemia" [117-120] and later renamed as "Pacific Oyster Nocardiosis." The disease affects mainly to Pacific oyster and has been detected in Japan [121, 122], North America, Canada [123-129] and Europe [130] but was also to flat oyster (*O. edulis*), which seems to be more susceptible to the disease [131]. The disease is usually developed when the temperature of the water is high (usually over 20ºC) and it has even linked to the summer mortalities [127, 128, 131]. *Nocardia* is a genus comprising over 80 species, often soil saprophytic communities that are widespread in the marine sediments and can act as opportunistic pathogens [127, 132, 133].

Roseovarius crassostreae was first described as an undescribed species from oysters affected from the so-called Juvenile Oyster Disease (JOD) [134]. When it was first described, it was placed in the family *Roseovarius* (α-*Proteobacteria*) that at the time was considered as symbiotic microorganism with a variety of marine plants and invertebrates, with no pathogenic representatives [135]. Its role as pathogen was established by experimental challenge [59, 136] and later on confirmed with a natural outbreak by Maloy et al. [136], and the disease was renamed as *Roseovarius* oyster Disease. The only outbreak caused by this pathogen was recorded in hatcheries of *C. virginica* in United States [137-139]. Usually occurs at high temperatures and moderate salinities, causing death in about 1-2 weeks with clinical signs similar to those of BRD in clams [140].

Regarding the intracellular prokaryotes present in bivalves only few studies were done. More specifically *Rickettisia*-like organisms (RLO), have been described a variety of aquatic organisms, including molluscs. In fact, ever since their first description in 1977 [141], RLO infections have been described in approximately 25 mollusc species [142], including different species of clams and oysters [23, 143-147]. RLO are obligated intracellular gram negative, rod-shaped microorganisms directly transmited between host usually with water-borne transmission [148].

They have been associated to high mortalities of farmed molluscs, usually within the epithelial cells of the mantle, digestive gland, gills and conective tissues [146, 149]. More specifically, mortality caused by RLO organisms has been described in *Crassostreae ariakensis* in Spain [150], in *Tegillarca granosa* in China during Spring and Autumm, reaching mortalities up to 100% [151]. Another mortality case was reported in a farmed juvenile pearl oyster (*Pinctada maxima*) with mortality rates occurred in 4 to 6 months [142]. Other cases described also associated to mortalities in clam (*Polititapes rhomboides*) in Spain [145], tropical pearl oyster (*P. maxima*) and Australian oyster (*C. ariakensis*) in China [146, 152]. In all these cases, the only pathogen observed was a RLO organism and therefore, proposed as the causative agent. Although to the date, any systematic study has been done in terms of prevalence, in the comparative study performed in Galicia, higher prevalence of RLO has been found in *R. decussatus* than *R. philippinarum* [23].

PROBIOTICS

The control of pathogens in mollusc cultures is a complicated task. Since bivalves are filter feeders, the constant flow of seawater through the animal body difficults the establishment of a stable bacterial community in the digestive tract, which makes easy the spread and proliferation of pathogens. For that reason, the environment, basically the inlet seawater, determines the microbiota of larval cultures. And this environment is extremely diverse, containing a plethora of obligate and opportunistic pathogens but also beneficial and neutral bacteria.

The classical methods are focused on the elimination of most or even the entire bacterial population by different methods, such as filtration, pasteurization, ozone and UV radiation of water and tanks [153]. The use of chemoterapeutic agents is another approach that has been widely used in mollusc hatcheries but with inconsistent results besides the consequent risk of potential appearance of resistance and their transference to human pathogens [154]. Vaccination constitutes an excellent alternative, used successfully in fish farms, but is not applicable to molluscs since these organisms do not posses a real immunological system.

The final objective of the methods cited above is to eliminate the majority or the entire microbiota from the water within the hatcheries, but this approach seems not to be reasonable since the lack of a balanced microbiota can favour the colonization of the system by non-desirable microorganisms due to the lack of natural competitors. Besides, at least one part of the bacteria communities may play a beneficial role on the larval development [49]. It is precisely this fact that lead to the development of the use of probiotics in the aquaculture. A probiotic suitable for being used in aquaculture can be defined as live microbial additive with a beneficial effect on the host, modifying the microbiota associated with the host or the environment, ensuring an optimal use of the feed or improving its nutritional value, improving the host response against the disease, or getting a better quality of its environment [153].

One of the desirable characteristics of a probiotic strain is being isolated from the environment where is going to be used, in that case a hatchery [97, 153]. Therefore, the study of the whole microbiota, and not only the pathogens is a key point in the successful of the cultures.

The majority of the studies on probiotics in aquaculture were performed on fish or crustaceans and only a few are focussed on molluscs. Few studies were focussed on pectenids, such those carried out by Lodeiros et al. [155] using an antibiotic-producing marine bacteria on a scallop (*Pecten ziczac*) hatchery or the application of *P. haloplanktis* in *Argopecten purpuratus* larvae, being the latter very active against *Vibrio* [97].

Gibson et al. [156] and Longeon et al. [157] described the efficacy of *Aeromonas media* and *Pseudoalteromonas* spp to be applied in larval cultures of *C. gigas* among others. In the last years, *Phaeobacter* is becoming more popular as probiotic due to their great spectrum of inhibition against pathogenic bacteria for aquaculture [158]. Prado et al. [159] used an isolate from *P. gallaeciensis* to test its activity in water, phytoplankton and larval oyster (*O. edulis*) and clam (*R. philippinarum*) cultures showing a high inhibitory power against *Vibrio*.

Metagenomics

As can be stated by all the works mentioned above, the microbiota of the molluscs is mainly composed by bacteria belonging to the γ-*Proteo*bacteria division, mostly *Vibrio*, *Pseudomonas,* and to a less extent α-*Proteobacteria.*

Nevertheless, it is important to point out that all these studies involved the isolation and culture of the bacteria that, besides being a tedious work, underestimate the bacterial variability. Romero and Espejo [160] calculated that the colonies grown on agar plates constituted less than 0.01% of the total variability present in the sample.

Only a few studies have used non-culturable technologies to study the microbiota of oyster and scallop. Of them, we can distinguish two types of methodologies, a first approach using techniques as DDGE, *in situ* hybridizations, Restriction Fragment Length Polymorphisms (RFLP), Ribosomal Intergenic Spacer Analysis (RISA) or the sequencing of fragments of 16S rRNA gene and a second one, more recent and accurate, involving pyrosequencing.

Several studies have used techniques such as DDGE but using the amplicons of 16S rDNA from bacteria previously isolated and therefore, not describing the whole microbiota. In that sense, Romero et al. [161] analysed the bacterial populations associated to the oyster *Tiostrea chilensis* after depuration using RFLP and the sequencing of the 16S-23S rRNA intergenic region and found that *Arcobacter* was the most prevalent bacteria in oysters.

Hernández-Zarate and Olmos-Soto [162] analysed different tissues of the Pacific oyster (*C. gigas*) collected from a Mexican farm and the bacteria present in the samples were analysed by *in situ* hybridization. This study showed a much larger bacterial diversity than was thought and broke some paradigms as the fact that *Vibrio* was the major genus present in cultured molluscs or the demonstration that the gills showed greater bacterial growth than digestive gland or gonads. These results are in contrast with those of Sugita et al. [14] that found more bacterial counts (using culturable methods) in the flesh of the animal, and with those of Romalde et al. [23] that found higher amount of bacteria in the gonad of reared clams in certain locations. In terms of diversity, they found that the most abundant bacteria in gonads and digestive glands were *γ-Proteobacteria* and Gram-positives with a low G+C content, while in the gills besides these two division, Gram-positives with high G+C and groups of CFB were also found, although the first two groups represented the 95% of the actively growing bacteria. Species-specific PCR confirmed the presence of these taxa and in addition, they detected members of α and β-*Proteobacteria*, more specifically *Pseudomonas* and *Bacillus*.

In a study using RFLP and sequencing of the 16S RNA gene using the digestive gland of Sydney rock oysters (*Saccostrea glomerata*), Green and Barnes [163] described the bacterial variability present in healthy oysters in comparison with individuals infected with the protozoan (*Marteilia sydneyi*). A total of 23 different species from at least eight bacterial divissions were identified from the un-infected *S. glomerata*. Only half of the 16S rDNA sequences represented by 12 Operational Taxonomic Units (OTUs) were affiliated with previously described species with more than 97% similarity. The remaining 11 sequences corresponded novel phylotypes. The un-infected *S. glomerata* showed bacterial divisions belonging to *Firmicutes* (44%), *Proteobacteria* (36%), *Cyanobacteria* (7%) and *Spirochaete* (5%). The remaining bacteria groups such as *Actinobacteria*, *Chloroflexi*, *Chlorophyta* and *Fusobacteria* made up a total of 8%. The digestive gland of Sidney rock oyster harboured a larger bacterial diversity showing 23 pylotypes from eight bacterial divisions (α and γ-*Proteobacteria*, *Firmicutes*, *Spirochaetes*, *Chlorophyta*, *Cyanobacteria*, *Actinobacteria*, *Fusobacteria* and *Chloroflexi*), being the *α-Protobacteria* and *Firmicutes* the dominant phyla, that constituted novel taxa related to the intracellular microorganisms *Mycoplasma* and *Rickettsiales*. These results contrast with the classic studies where *γ-Proteobacteria* were the dominant taxa as was exposed before, which makes sense considering that the most abundant bacteria in oyster are unculturable [161]. A comparison between healthy animals and oysters infected with the protozoan *M. sydneyi* was also performed. In the diseased animals they found a clear dominance of the *α-Proteobacteria* phylotipe corresponding to RLO. This, together with the fact that diseased animals did not show the presence of *Cyanobacteria* (associated with the feeding process), suggests that once the oyster stop feeding the colonization by intracellular bacteria is enhanced, allowing the oysters starve to death.

Boardman et al. [166] used culture-independent methods in comparison with classic ones for the study of the microbiota associated to healthy and JOD-affected *C. virginica*. They found, as could be expected, that *R. crassostreae* was the dominant, if not the unique, bacteria present in the sampled oysters although it is important to point out that the sampling for the comparison consisted in the uptake of a sample from the inner valve of the oyster with a swab, and it is precisely this part of the animal where the bacteria is attached during the development of the infection. One interesting data from this work was the fact than only half of the sequences obtained were assigned to already defined OTU, confirming therefore the underestimation of the bacterial diversity in the culturable-based methods.

Zurel et al. [167] studied the microbiota associated to the gill of the oysters *Chama pacifica* and *Chama savignyi* in Israel by ARISA, determining that the most abundant division was, as expected, *Proteobacteria* and more specifically γ-*Proteobacteria*. The dominant phylotype in the gills was *Ocenospirales* that was also found in environmental samples. This phylotype was associated by previous studies to different aquatic animals in the same area, being most probably part of the allochtonous bacteria of the oysters.

To the date, the most complete work regarding oyster microbiota are those performed by Trabal and coworkers [168, 169]. In two different works they analysed the resident microbiota associated to three different species of oysters, (*Crassostrea corteziensis*, *C. gigas* and *Crassostrea sikamea*) cultivated in the same hatchery and planted in two different geographical locations with different environmental conditions, one with low pollution and minor human activity and the other one with contamination from agricultural activities and industrial sewage effluents.

In a first approach Trabal et al. [168], and using RFLP, Temperature Gradient Gel Electrophoeresis (TGGE) and the 16S rRNA gene sequencing from the bands obtained from TGGE analysed the bacteria associated to *C. corteziensis* and *C. gigas* at three different growth stages (postlarvae, juvenile and adult). The bacterial communities associated with the oysters at each grow-out site were similar if not identical, through the growing and were influenced by the environment. At postlarvae stage the microbiota was very similar in both species, being therefore, associated to the food that, at this point, was controlled in the laboratory.

There were differences among the microbiota associated to both oysters even when grown at the same site and *C. gigas* was the most diverse. After the oysters were planted in different locations, the microbiota changed, although *Burckholderia cepacia* remained associated with the oysters regardless the species and the geographical location. Moreover, each species showed a particular species or group of species from postlarvae to adult stage regardless the geographical location. Thus, *C. gigas* showed a permanent population of *Bacillus* and *C. corteziensis* belongin to *Cristispira* and *Spirochaeta* (*Sprirochaetaceae*) when planted in the clean area.

In terms of sequences they found that the main bands (OTUs) obtained from TGGE gels belonged to the phyla *Proteobacteria* and *Firmicutes*, more specifically to the genera *Brukhorderia* and *Borrelia* and to the Gram-positve bacteria *Bacillus* and *Propionibacterium*. *Ruegeria*, *Paracoccus*, *Methylobacterium* and *Vibrio* were also present, being *B. cepacia* the most abundant species.

The second study was using pyrosequencing techniques based on the V3-V5 regions of the 16S rRNA gene studying the resident microbiota of *C. corteziensis*, *C. gigas* and *C. sikamea* [169]. They found 13 phyla from the oyster microbial community that were dominated by *Proteobacteria*, *Bacteroidetes*, *Actinobacteria* and *Firmicutes*. The most abundant phylum was *Proteobacteria* through all the life stages and at different cultivation sites.

In the postlarvae stage, the α and *γ-Proteobacteria* were the most abundant while the adulsts showed variation sin the abundance of each *Proteobacteria* classes that varied by species and sit. In adults, the most abundant class was *γ-Proteobacteria* followed by α *β – Proteobacteria*. *Bacteroidetes* was the second most abundant phyla in the oyster microbiota and the most common in the postlarvae but was reduced in the adults. The low G+C Gram-positive bacteria *Firmicutes* was present in all life stages although less abundant than those mentioned above. *Fusobacteria*, *Tenericutes*, *Acidobacteria*, *Chlorobi*, *Deinococcus*, *Thermus*, *Spirochaetes*, *Thermotoga* and *Verrucomicrobia* were detected at low amount and varied according with the geographical location.

In total, they identified 243 bacterial genera, a great bacterial diversity never reported before for molluscs. The adult and postlarvae stages showed diferences in the proteobacteria community structures The predominant genera in postlarvae stage was *Neptuniibacter* followed by *Marinicella*, *Rhodovulum* and *Oceanicola*. In the adults at both locations were *Burkholderia* and *Escherichia/Shigella* species although *Burkholderia* was also abundant in postlarvae stage.

The highest abundance of *Bacteroidetes* was found at postlarvae stage, being the most abundant the class *Flavobacteria*, result in agreement with the work carried out by Fernández-Piquer et al. [170] which compared the microbiota associated to *C. gigas* under different storage conditions. In both postlarvae and adults, representatives of *Actinobacteria* were found. In contrast with all the previous studies based on culturable methods, *Vibrio* and *Pseudomonas* were not abundant neither in adults nor in postlarvae.

During the postlarvae stage the microbiota had a uniform bacterial composition, regardless of the host and more influenced by environmental causes as the food. The microbiota of adult *C. gigas* showed more intraespecific variability and was not influenced by the environment in contrast with *C. corteziensis* and *C. sikamea.* They found differences in the composition of the bacterial community associated with oysters at different stages but not between different species of oysters grown in the same location, showing therefore a common core microbiota depending on the cultivation site. Interestingly, they found the presence of one bacteria common to all stages in every location, *B. cepacia.* This bacteria has been recently described to be present in the Pacific oyster [170] showing probiotic activity against *V. alginolyticus* and *V. harveyi*, and may be the reason of the strong attachment.

A more recent work studying the microbiota associated to *C. virginica* using the region V4 of 16S rRNA gene in the gulf of Mexico showed a total of 28 bacteria phyla, with *Cyanobacteria* and *Photobacterium* as the most abundant one in the oyster tissues while in the surrounding water was *Cyanobacteria* and *Pelagibacter* [171].

More recently, Lockmer et al. [172] studied the microbiota of different tissues of Pacific oysters and the surrounding water in different locations using the V1-V2 regions of the 16S rRNA gene by pyrosequencing. They found that the overall richness was higher in seawater but the relative OTUs richness was higher in oysters, mainly belonging to *Vibrionaceae*. The most abundant communities in seawater *(α- Proteobacteria* and *Flavobacteriaceae*) were also those found in the oysters, especially in haemolymph, although with lower abundances.

Comparing the tissues, haemolymph possessed a distinctive bacterial community, mainly composed by *ε-Proteobacteria* (*Arcobacter*), *Flavobacteria*, *Fusobacteria* (*Psychriliobacter*) and *γ-Proteobacteria* (*Oceanospirillaceae*, *Vibrionaceae*) and fewer *Spirochaetes* (*Brachyspirae*) and β *–Proteobacteria*. Among the rest of the tissues, the biggest difference was the high abundance of *Mollicutes* (*Mycoplasma*) in the gut.

The diversity found within the oyster, the positive correlation between the number of culturable vibrios and the species richness inside the tissue, together with the similarity of the haemolymph and seawater communities, allowed the authors to propose that haemolymph is not only a reflection of the seawater microbiota but also the entry way for the *Vibrio* that can easily spread through solid tissues and create a systemic infection [172].

Conclusion

The emergence and re-emergence of marine pathogens, as well as the colonization of new niches has led to the increase of outbreaks of different diseases over the last decades. In that sense, it has been previously described that the global warming is playing an important role in the changes and/or unbalances in the microbiota present in the oceans and, therefore, in the organisms living there.

Disturbances of fine-tuned interactions among host, microbiota and the environment may have lethal consequences for the host [173], since resident microbiota is involved in the immune defense mechanisms by preventing colonization and the establishment of pathogens. As stability is important for community functioning, environmental disturbances (including rising temperature) may have important adverse consequences.

Of special importance are the infections caused by opportunistic pathogens as those from genus *Vibrio*. Some of the effects of increasing sea surface temperature on the biology and ecology of vibrios have been recently reported [174], including changes in bacterial growth rate, culturability, expression of pathogenicity traits, and interactions with aquatic organisms and abiotic surfaces.

The biocomplexity of the interactions between bacteria and environment in a climate change scenario is today poorly understood. Further multidisciplinary studies are needed in order to advance in the knowledge of the occurrence and spread of pathogenic bacteria on temporal and spatial scales, on the interactions of bacteria with other organisms in relation with rising temperature, on the effects of climate change on host susceptibility, etc.

Such studies would allow the development of helpful models to predict, prevent and control new and/or re-emerging diseases and therefore, to protect the molluscan aquaculture sector worldwide.

REFERENCES

[1] Kueh, C. S. W., Chan, K. Y. (1985). Bacteria in bivalve shellfish with special reference to the oyster. *J. Appl. Bacteriol.* 59, 41-47.

[2] Murchelano, R. A., Brown, C. (1970). Heterotrophic bacteria in Long Island Sound. *Mar. Biol.* 7, 1-6.

[3] Prieur, D., Mével, G., Nicolas, J. L., Plusquellec, A., Vigneulle, M. (1990). Interactions between bivalve molluscs and bacteria in the marine environment. *Oceanogr. Mar. Biol. Annu. Rev,* 28, 277-352.

[4] Jorquera, M. A., Silva, F. R., Riquelme, C. E. (2001). Bacteria in the culture of the scallop *Argopecten purpuratus* (Lamarck, 1819). *Aquacult. Int.* 9, 285–303.

[5] Savage, D. C. (1977) Microbial ecology of the gastrointestinal tract. *Ann. Rev. Microbiol.* 31, 107-133.

[6] Mchenery, J. G., Birbeck, T. H. (1986). Inhibition of filtration in *Mytilus edulis* L. by marine vibrios. *J. Fish Dis.* 9, 257-261.

[7] Seguineau, C., Laschi- Loquerie, A., Moal, J., and Samai, J. F. (1996). Vitamin requirements in great scallop larvae. *Aquacult. Int.* 4, 315-324.

[8] Colwell, R. R., and Liston, J. (1960). Microbiology of shellfish. Bacteriological study of the natural flora of Pacific oysters (*Crassostraea gigas*). *Appl. Microbiol.* 8, 104-109.

[9] Ortigosa, M., Garay, E., Pujalte, M.J. (1994) Numerical taxonomy of *Vibrionaceae* isolated from oysters and seawater along an annual cycle. *Syst. Appl. Microbiol.* 17, 216–225.

[10] Chan SS, Ho KW (1993) Bacterial contamination of the Pacific oysters (*Crassostrea gigas*) cultured in Deep Bay, Hong Kong. *Environ. Technol.* 14, 861–867.

[11] Dupont J, Jehl-Pietri C, Mnard D (1992) Comparative study of bacterial and viral faecal contamination in shellfish: demonstration of seasonal variations. *Biomed. Lett.* 47:329–335.

[12] Fraiser MB, Koburger JA (1983) Incidence of *Salmonella* in clams, oysters, and mullet. *J. Food. Prot.* 47, 343–345.

[13] Lovelace, T. E., Tubiash, H., Colwell, R. R. (1968) Quantitative and qualitative commensal bacterial flora of *Crassostrea virginica* in Chesapeake Bay. *Proc. Nat. Shellfish. Ass.* 58, 82–87.

[14] Sugita, H., Tanaami, H., Kobashi, T., Deguchi, Y. (1981) Bacterial flora of coastal bivalves. *Bull. Jpn. Soc. Sci. Fish* 47, 655–661.

[15] Martin, Y. (1976) Importance des bactéries chez les mollusques bivalves. *Haliotis* 7, 97–103.

[16] Ortigosa, M., Garay, E., Pujalte, M. J. (1995) Numerical taxonomyofaerobic, Gram-negative bacteria associated to oysters and surrounding seawater in Mediterranean coast. *Syst. Appl. Microbiol.* 17, 588–600.

[17] Montilla, R., Palomer, J., Fuster, M. C., Viñas, V. (1995) Taxonomy and protein fingerprinting of halophilic *Vibrio* isolates from bivalves of the Ebre delta. *Can. J. Microbiol.* 41, 64–69.

[18] Hlady, W. G., Mullen, R. C., Hopkins, R. S. (1993) *Vibrio vulnificus* from raw oysters leading cause of reported deaths from foodborne illness in Florida. *J. Florida Med. Ass.* 80, 536-538.

[19] Hor, L. I., Goo, C. T., Wan, L. (1995) Isolation and characterization of *Vibrio vulnificus* inhabiting the marine environment of the southwestern area of Taiwan. *J. Biomedical Sci.* 2, 384–389.

[20] O'Neill, K. R., Jones, S. H., Grimes, D. J. (1990) Incidence of *Vibrio vulnificus* in northern New England water and shellfish. *FEMS Microbiol. Lett.* 72, 163–168.

[21] Wright, A. C., Hill, R. T., Johnson, J. A., Roghman, M-C., Colwell, R. R., Morris, J. G. (1996) Distribution of *Vibrio vulnificus* in the Chesapeake Bay. *Appl. Environ. Microbiol.* 62, 717–724.

[22] Pujalte, M.J., Ortigosa, M., Macián, M.C., Garay, E. (1999). Aerobic and facultative anaerobic heterotrophic bacteria associated to Mediterranean oysters and seawater. *Int. Microbiol.* 2, 259-266.

[23] Romalde, J. L., Diéguez, A. L., Doce, A., Lasa, A., Balboa, S., López, C., Beaz-Hidalgo, R. (2013). Advances in the knowledge of the microbiota associated with clams from natural beds. In F. da Costa (Eds.) *Clam Fisheries and Aquaculture*. New York: Nova Science Publishers. pp. 163-190.

[24] Montilla, R., Palomar, J., Santmarti, M., Fuste, C., Viñas, M. (1994). Isolation and characterization of halophilic *Vibrio* from bivalves bred in nurseries at the Ebre Delta. *J. Invertebr. Pathol.* 63, 178-181.

[25] Arias, C. R., Macián, M. C., Aznar, R., Garay, E., Pujalte, M.J. (1999). Low incidence of *Vibrio vulnificus* among *Vibrio* isolates from sea water and shellfish of the western Mediterranean coast. *J. Appl. Microbiol.* 86, 125-134.

[26] Beaz-Hidalgo, R., Cleenwerck, I., Balboa, S., De Watcher, M., Thompson, F. L., Swings, J., De Vos, P., Romalde, J. L. (2008). Diversity of vibrios with reared clams in Galicia (NW Spain). *Syst. Appl. Microbiol.* 31, 215-222.

[27] Castro, D., Pujalte, M. J., Lopez-Cortes, L. Garay, E., and Borrego, J.J. (2002). Vibrios isolated from the cultured manila clam (*Ruditapes philippinarum*): numerical taxonomy and antibacterial activities. *J. Appl. Microbiol.* 93, 438-447.

[28] Beaz-Hidalgo R., Cleenwerck I., Balboa S., Prado S., de Vos P., and Romalde J. L. (2009). *Vibrio breoganii* sp. nov., a non motile alginolytic marine bacterium within the *Vibrio halioticoli* clade. *Int. J. Syst. Evol. Microbiol.* 59, 1589-1594.

[29] Beaz-Hidalgo R., Doce A., Pascual J., Toranzo A. E., Romalde J. L. (2009). *Vibrio gallaecicus* sp. nov., isolated from cultured clams in north-western Spain. *Syst. Appl. Microbiol.* 32, 111-117.

[30] Beaz-Hidalgo, R., Diéguez, A. L., Cleenwerck, I., Balboa, S., Doce, A., de Vos, P., Romalde, J. L. (2010). *Vibrio celticus* sp. nov., a new *Vibrio* species belonging to the Splendidus clade with pathogenic potential for clams. *Syst. Appl. Microbiol.* 33, 311-315.

[31] Beaz-Hidalgo R., Doce A., Balboa S., Barja J. L., Romalde J. L. (2010). *Aliivibrio finisterrensis* sp. nov., isolated from Manila clams, *Ruditapes philippinarum* and emended description of the genus *Aliivibrio*. *Int. J. Syst. Evol. Microbiol.* 60, 223-228.

[32] Diéguez, A. L., Beaz-Hidalgo, R., Cleenwerck, I., Balboa, S., De Vos, P., Romalde, J. L. (2011). *Vibrio atlanticus* sp. nov. and *Vibrio artabrorum* sp. nov. isolated from clam (*Ruditapes decussatus* and *R. philippinarum*). *Int. J. Syst. Evol. Microbiol.* 61, 2406-2411.

[33] Lasa, A., Diéguez, A. L., Romalde, J. L. (2013). *Vibrio toranzoniae* sp. nov., a new member of the Splendidus clade in the genus *Vibrio*. *Syst. Appl. Microbiol.* 36, 96-100.

[34] Lasa, A., Diéguez, A. L., Romalde, J. L. (2014). Description of the new species *Vibrio cortegadensis* sp. nov., isolated from clams. *Antonie van Leeuwenhoek* 105, 335-341.

[35] Beleneva, I. A., Zhukova, N. V. Maslennikova, E. F. (2003). Comparative study of microbial communities from cultured and natural populations of the mussel *Mytilus trossulus* in Peter the Great Bay. *Microbiology.* 72, 472-477.

[36] Cavallo, R. A., Acquaviva, M. I., Stabili, L. (2009). Culturable heterotrophic bacteria in seawater and *Mytilus galloprovincialis* from a Mediterranean area (Northern Ionian Sea-Italy). *Environ. Monit. Assess.* 149, 465-475.

[37] Amann, R. I., Ludwig, W., Schleifer, K. H. (1995) Phylogenetic identification and in situ detection of individual microbial cells without cultivation. *Microbiol. Rev.* 59, 143–169.

[38] Green, B. D., Keller, M. (2006). Capturing the uncultivated majority. *Curr. Opin. Biotechnol.* 17, 236-240.

[39] Leadbetter, J. R. (2003) Cultivation of recalcitrant microbes: cells are alive, well and revealing their secrets in the 21st century laboratory. *Curr. Opin. Microbiol* 6, 274-281.

[40] Lodeiros, C., Bolinches, J., Dopazo, C. P., Toranzo, A. E. (1987) Bacillary necrosis in hatcheries of *Ostrea edulis* in Spain. *Aquaculture.* 65, 15–29.

[41] Widman, J. C. (2001) Manual for hatchery culture of the bay scallop, *Argopecten irradians irradians*. Milford, Conn.: National Oceanic and Atmospheric Administration, National Marine Fisheries Service, Northeast Fisheries Science Center: Connecticut Sea Grant College Program.

[42] Sandaa, R. A., Magnesen, T., Torkildsen, L., Bergh, Ø. (2003) Characterisation of the bacterial community associated with early stages of great scallop (*Pecten maximus*),

using denaturing gradient gel electrophoresis (DGGE). *Syst. Appl. Microbol.* 26, 302-311.

[43] Eilers, H., Pernthaler, J., Glöckner, F. O., Amann, R. (2000) Culturability and in situ abundance of pelagic bacteria from the North Sea. *Appl. Environ. Microbiol.* 66, 3044-3051.

[44] Field, K. G., Gordon, D., Wright, T., Rappe, M., Urback, E., Vergin, K., Giovannoni, S. J. (1997) Diversity and depth-specific distribution of SAR11 cluster rRNA genes from marine planktonic bacteria. *Appl. Environ. Microbiol.* 63, 63-70.

[45] Kasai, Y., Kishira, H., Syutsubo, K., Harayama, S. (2001) Molecular detection of marine bacterial populations on beaches contaminated by the Nakhodka tanker oil-spill accident. *Environ. Microbiol.* 3, 246-255.

[46] Schulze, A. D., Alabi, A. O., Tattersall–Sheldrake, A. R., Miller, K. M. (2006) Bacterial diversity in marine hatchery: balance between pathogenic and potentially probiotic bacterial strains. *Aquaculture.* 256, 50–73.

[47] Prado, S., Dubert, J., da Costa, F., Martínez-Patiño, D., Barja, J. L. (2014) Vibrios in hatchery cultures of the razor clam, Solen marginatus (Pulteney). *J. Fish Dis.* 37, 209-217.

[48] Beaz-Hidalgo, R., Balboa, S., Romalde, J. L., Figueras, M. J. (2010). Diversity and pathogenicity of *Vibrio* species in cultured bivalve molluscs. *Environ. Microbiol. Reports* 2, 34-43.

[49] Romalde, J. L., Barja, J. L. (2010). Bacteria in molluscs: good and bad guys. Current Research, Technology and Education Topics. In A. Mendez-Vilas (Eds). Applied Microbiology and Microbial Biotechnology. Formatex Research Center, pp 136-147.

[50] Romalde, J. L., Diéguez, A. L., Lasa, A., Balboa, S. (2013). New *Vibrio* species associated to molluscan microbiota: a review. *Front. Microbiol.* 4:413. eCollection.

[51] Allam, B., Paillard, C. Ford, S. E. (2002). Pathogenicity of *Vibrio tapetis*, the etiological agent of brown ring disease in clams. *Dis. Aquat. Org.* 48, 221-231.

[52] Anguiano-Beltrán, C., Lizárraga-Partida, M. L., Searcy-Bernal, R. (2004). Effect of *Vibrio alginolyticus* on larval survival of the blue mussel *Mytilus galloprovincialis*. *Dis. Aquat. Org.* 59, 119-123.

[53] Estes, R. M., Friedman, C. S., Elston, R. A., and Herwig, R. P. (2004). Pathogenicity testing of shellfish hatchery bacterial isolates on Pacific oyster *Crassostrea gigas* larvae. *Dis. Aquat. Org.* 58, 223-230.

[54] Gay, M., Berthe, F. C. J., Le Roux, F. (2004). Screening of *Vibrio* isolates to develop an experimental infection model in the Pacific oyster *Crassostrea gigas*. *Dis. Aquat. Org.* 59, 49-56.

[55] Gay, M., Renault, T., Pons, A. M., Le Roux, F. (2004). Two *Vibrio splendidus* related strains collaborate to kill *Crassostrea gigas*: taxonomy and host alterations. *Dis. Aquat. Org.* 62, 65-74.

[56] Garnier, M., Labreuche, Y., García, C., Robert, M., and Nicolas, J. L. (2007). Evidence for the involvement of pathogenic bacteria in summer mortalities of the Pacific oyster *Crassostrea gigas*. *Microbiol. Ecol.* 53, 187-196.

[57] Garnier, M., Laubreche, Y., Nicolas, J. L. (2008). Molecular and phenotypic characterization of *Vibrio aestuarianus* subsp. *francensis* subsp. nov., a pathogen of the oyster *Crassostrea gigas*. *Syst. Appl. Microbiol.* 31, 358-365.

[58] Gómez-León, J., Villamil, L., Lemos, M. L., Novoa, B. (2005). Isolation of *Vibrio alginolyticus* and *Vibrio splendidus* from aquacultured carpet shell clam (*Ruditapes decussatus*) larvae associated with mass mortalities. *Appl. Environ. Microbiol.* 71, 98-104.

[59] Gómez- León, J., Villamil, L., Salger, S. A., Sallum, R. H., Remacha-Triviño, A., Leavitt, D. F., Gómez-Chiarri, M. (2008). Survival of eastern oysters *Crassostrea virginica* from three lines following experimental challenge with bacterial pathogens. *Dis. Aquat. Org.* 79, 95-105.

[60] Labreuche, Y., Lambert, C., Soudant, P., Boulo, V., Huvet, A., Nicolas, J. L. (2006). Cellular and molecular hemocyte responses of the Pacific oyster, *Crassostrea gigas*, following bacterial infection with *Vibrio aestuarianus* strain 01/32. *Microbes Infect.* 8, 2715-2724.

[61] Labreuche, Y., Soudant, P., Goncalves, M, Lambert, C., Nicolas, J. L. (2006). Effects of extracellular products from the pathogenic *Vibrio aestuarianus* strain 01/32 on lethality and cellular immune responses of the oyster *Crassostrea gigas*. *Dev. Comp. Immunol.* 30, 367-379.

[62] Lee, K. K., Ping-Chung, L., Huang, C. Y. (2003). *Vibrio parahaemolyticus* infections for both humans and edible mollusc abalone. *Microbes Infect.* 5, 481-485.

[63] Liu, P. C., Chen, C. Y., Huang, C. Y., Lee, K. K. (2000). Virulence of *Vibrio parahaemolyticus* isolated from cultured small abalone, *Haliotis diversicolor supertexta*, with withering syndrome. *Lett. Appl. Microbiol.* 31, 433-437.

[64] Liu, P. C., Chen, Y. C. Lee, K. K. (2001). Pathogenicity of *Vibrio alginolyticus* isolated from diseased small abalone *Haliotis diversicolor supertexta*. *Microbios.* 104, 71-77.

[65] Paillard, C. (2004). A short-review of brown ring disease, a vibriosis affecting clams, *Ruditapes philipinarum* and *Ruditapes decussatus*. *Aquat. Living Resour.* 17, 467-475.

[66] Prado, S., Romalde, J. L., Montes, J., Barja J. L. (2005). Pathogenic bacteria isolated from disease outbreaks in shellfish hatcheries. First description of *Vibrio neptunius* as an oyster pathogen. *Dis. Aquat. Org.* 67, 209-215.

[67] Waechter, M., Le Roux, F., Nicolas, J. L., Marissal, E., Berthe, F. (2002). Characterization of pathogenic bacteria of the cupped oyster *Crassostrea gigas*. *C. R. Biol.* 325, 231-238.

[68] Caballo, R. A., Stabili, L. (2002). Presence of vibrios in seawater and *Mytillus galloprovincialis* (Lam.) from the Mar Piccolo of Taranto (Ionian Sea). *Water Res.* 36, 3719-3729.

[69] Guisande, J. A., Montes, M., Farto, R., Armada, S. P., Pérez, M. J., Nieto T. P. (2004). A set of tests for the phenotypic identification of culturable bacteria associated with Galician bivalve mollusc production. *J. Shellfish Res.* 23, 599-610.

[70] Lafisca, A., Pereira C. S., Giaccone, V., Rodriguez, D. D. P. (2008). Enzymatic characterization of *Vibrio alginolyticus* strains isolated from bivalves harvested at Venice lagoon (Italy) and Guanabara Bay (Brazil). *Rev. Inst. Med. Trop. S. Paulo.* 50, 199-202.

[71] Maugeri, T. L., Caccamo, D. Gugliandolo, C. (2000). Potentially pathogenic vibrios in brackish waters and mussels. *J. Appl. Microbiol.* 89, 261-266.

[72] Le Roux, F., Austin, B. (2006). *Vibrio splendidus*. In: F. L. Thompson, B. Austin, J. Swings (Eds), *The Biology of Vibrios*. pp. 285-296, ASM Press edition, Washington.

[73] Pascual, J., Macián, M. C., Arahal, D. R., Garay, E., Pujalte, M. J. (2010). Multilocus Sequence Analysis of the central clade of the genus *Vibrio* using the 16S rRNA, *recA, pyrH, rpoD, gyrB, rctB* and *toxR* genes. *Int. J. Syst. Evol. Microbiol.* 60, 154-165.

[74] Thompson F. L., Gevers D., Thompson C. C, Dawyndt P., Naser S., Hoste B., Munn, C. B., Swings, J. (2005). Phylogeny and molecular identification of vibrios on the basis of multilocus sequence analysis. *Appl. Environ. Microbiol.* 71, 5107-5115.

[75] Jones, M. K., Oliver, J. D. (2009). *Vibrio vulnificus*: Disease and Pathogenesis. *Infect. Immun.* 77, 1723-1733.

[76] Kaspar, C. W., Tamplin, M.L. (1993). Effects of temperature and salinity on the survival of *Vibrio vulnificus* in seawater and shellfish. *Appl. Environ. Microbiol.* 59, 2425-2429.

[77] Motes, M. L., De Paola, A., Cook, D. W., Veazey, J. E., Hunsucker, J. C., Garthright, W. E., Blodgett, R. J., Chirtel, S. J. (1998). Influence of water temperature and salinity on *Vibrio vulnificus* in Northen Gulf and Atlantic Coast Oysters (*Crassostrea virginica*). *Appl. Env. Microbiol.* 64, 1459-1465.

[78] Brown, C. (1981) A study of tw shellfish-pathogenic *Vibrio* strains isolated from a Long Island hatchery during a recent outbreak of disease. *J. Shellfish Res.* 1, 83-87.

[79] Brown C, Losee E. (1978) Observations on natural and induced epizootics of vibriosis in Crassostrea virginica larvae. *J. Invertebr. Pathol.* 31, 41-47.

[80] Elston, R. (1984) Prevention and management of infectious disease in intensive mollusc husbandry. *J. Worl Maricul. Soc.* 15, 284–300.

[81] Elston, R., Leibovitz, L. (1980). Pathogenesis of experimental vibriosis in larval American oysters, *Crassostrea virginica. Can. J. Fish Aquat. Sci.* 37, 964-978.

[82] Di Salvo, L. H., Blecka, J., Zebal, R. (1978) Vibrio anguillarum and larval mortality in a California coastal shellfish hatchery. *Appl. Environ. Microbiol.* 35, 219-221.

[83] Prieur, D., Carval, Y. J. P. (1979) Bacteriological and physico-chemical analysis in a bivalve hatchery: Techniques and preliminary results. *Aquaculture.* 17, 359-374.

[84] Sugumar G, Nakai T, Hirata Y, Matsubara D, Muroga K. (1998) *Vibrio splendidus* biovar II as the causative agent of bacillary necrosis of Japanese oyster *Crassostrea gigas* larvae. *Dis. Aquat. Organ.* 33:111–118.

[85] Tubiash, H. S. (1975) Bacterial pathogens associated with cultured bivalve mollusk larvae. In: Smith WL, Chanley CMH, eds. Culture of Marine Invertebrate Animals. New York, NY: Plenum Press. pp. 61-71.

[86] Tubiash, H. S., Chanley, P. E., Leifson, E. (1965). Bacillary necrosis, a disease of larval and juvenile bivalve molluscs. I. Etiology and epizootiology. *J. Bacteriol.* 90, 1036-1044.

[87] Guillard, R. R. L. (1959). Further evidence of the destruction of bivalve larvae by bacteria. *Biol. Bull.* 117, 258-266, 1959.

[88] Hada, H. S., West, P. A., Lee, J. V., Stemmler, J., Colwell, R. R. (1984) *Vibrio tubiashii* sp. nov., a pathogen of bivalve mollusks. *Int. J. Syst. Bacteriol.* 34, 1–4.

[89] Tubiash, H. S., Colwell, R. R., Sakazaki, R. (1970) Marine vibrios associated with bacillary necrosis, a disease of larval and juvenile bivalve mollusks. *J. Bacteriol.* 103, 271–272.

[90] Genard B, Miner P, Nicolas JL, Moraga D, Boudry P, Pernet F, Tremblay R. (2013) Integrative study of physiological changes associated with bacterial infection in Pacific Oyster larvae. *PLoS One* 8(5):e64534.

[91] Kesarcodi–Watson A, Miner P, Nicolas JL, Robert R. (2012) Protective effect of four potential probiotics against pathogen–challenge of the larvae of three bivalves: Pacific oyster (*Crassostrea gigas*), flat oyster (*Ostrea edulis*) and scallop (*Pecten maximus*). *Aquaculture* 344–349:29–34.

[92] Prado S, Dubert J, Romalde JL, Toranzo AE, Barja JL. (2014). *Vibrio ostreicida* sp. nov., a new pathogen of bivalve larvae. *Int. J. Syst. Evol. Microbiol.* 64, 1641–1646.

[93] Saulnier D, de Decker S, Haffner P, Cobret L, Robert M, Garcia C. (2010) A large–scale epidemiological study to identify bacteria pathogenic to Pacific Oyster *Crassostrea gigas* and correlation between virulence and metalloprotease–like activity. *Microb. Ecol.* 59, 787–798.

[94] Thompson, F. L., Li, Y., Gomez-Gil, B., Thompson, C. C., Hoste, B., Vandemeulebroecke, K., Rupp, G. S., Pereira, A., De Bem, M. M., Sorgeloos, P., Swings J. (2003a). *Vibrio neptunius* sp. nov., *Vibrio brasiliensis* sp. nov. and *Vibrio xuii* sp. nov., isolated from the marine aquaculture environment (bivalves, fish, rotifers and shrimps) *Int. J. Syst. Evol. Microbiol.* 53, 245-252.

[95] Helm, M. M., Smith, F. M. (1971) Observations on a bacterial disease in laboratory-cultured larvae of the European flat oyster, *Ostrea edulis* L. *Int. Counc. Explor. Sea CM 1971/K: 10*, 9.

[96] Sutton, D. C., Garrick, R. (1993) Bacterial disease of cultured giant clam Tridacna gigas larvae. *Dis. Aquat. Org.* 16, 47-53.

[97] Riquelme, C., Hayashida, G., Araya, R., Uchida, A., Satomi, M., Ishida, Y. (1996) Isolation of a native bacterial strain from the scallop *Argopecten purpuratus* with inhibitory effects against pathogenic vibrios. *J. Shell- fish Res.* 15, 369–374.

[98] Lipp, P. R., Brown, B., Liston, J., Chew, K. (1976). Recent findings on the summer diseases of Pacific oysters. *Proc. Nat. Shellfish Ass.* 65, 9-10.

[99] Lacoste, A., F. Jalabert, S. Malham, A. Cueff, F. Gelebart, C. Cordevant, M. Lange, S. A. Poulet. (2001) A *Vibrio splendidus* strain is associated with summer mortality of juvenile oysters *Crassostrea gigas* in the Bay of Morlaix (North Brittany, France)." *Dis. Aquat. Organ.* 46, 139-145.

[100] Azandégbé A, Garnier M, Andrieux–Loyer F, Kérouel R, Philippon X, Nicolas JL. (2010) Occurrence and seasonality of *Vibrio aestuarianus* in sediment and *Crassostrea gigas* haemolymph at two oyster farms in France. *Dis. Aquat. Organ.* 91, 213–221.

[101] Le Roux F, Gay M, Lambert C, Waechter M, Poubalanne S, Chollet B, Nicolas JL, Berthe F. (2002) Comparative analysis of *Vibrio splendidus*–related strains isolated during *Crassostrea gigas* mortality events. *Aquat. Living Resour.* 15, 251–258.

[102] Allain, G., Arzul, I., Chollet, B., Cobret, L., de Decker, S., Faury, N., Ferrand, S., François, C., Garcia, C., Haffner, P., Joly, J. P., Michel, J., Miossec, L., Morga, B., Nicolas, J. L., Omnes, E., Pépin, J. F., Robert, M., Saulnier, D., Schikorski, D., Segarra, A., Tourbiez, D., Renault, T. 2009. Summer mortality outbreaks of French Pacific oysters, *Crassostrea gigas* in 2008: research and detection of pathogens. Proc. 14th EAFP International Conference. Diseases of Fish and Shellfish. Prague, Czech Republic. pp. 127.

[103] Pruzzo C, Gallo G, Canesi L. (2005) Persistence of vibrios in marine bivalves: The role of interactions with haemolymph components. *Environ. Microbiol.* 7, 761–772.

[104] Drummond, L., Mulcahy, M. F., Culloty, S. C. (2010) A survey of the health status of the Manila clam *Ruditapes philippinarum* in Ireland with specific reference to brown ring disease. *Aquacult. Int.* 18, 787-800.

[105] Matsuyama, T., Sakai, T., Kiryu, I., Yuasa, K., Yasunobu, H., Kawamura, Y., Sano, M. (2010) First isolation of *Vibrio tapetis*, the etiological agent of brown ring disease (BRD), in Manila Clam *Ruditapes philippinarum* in Japan. *Fish Pathol.* 45, 77-79.

[106] Park, K.-I., Paillard, C., Le Chevalier, P., Choi, K-S. (2006) Report on the occurrence of brown ring disease (BRD) in Manila clam, *Ruditapes philippinarum*, on the west coast of Korea. *Aquaculture* 255, 610-613.

[107] Drummond, L. C., Balboa, S., Beaz, R., Mulcahy, M. F., Barja, J. L., Culloty, S. C., Romalde, J. L. (2007) The susceptibility of Irish-grown and Galician-grown Manila clams, *Ruditapes philippinarum*, to *Vibrio tapetis* and brown ring disease. *J. Invertebr. Pathol.* 95, 1-8.

[108] Allam, B., Ashton-Alcox, K. A., Ford, S. E. (2001) Haemocyte parameters associated with resistance to brown ring disease in *Ruditapes* spp. clams. *Develop. Comp. Immunol.* 25, 365-375.

[109] Balboa, S., Doce, A., Diéguez, A. L., Romalde, J. L. (2011). Evaluation of different species-specific PCR protocols for the detection of Vibrio tapetis. *J. Invert. Pathol.* 108, 85-91.

[110] Balboa, S., Romalde, J. L. (2013). Multilocus sequence analysis of *Vibrio tapetis*, the causative agent of Brown Ring Disease: Description of *Vibrio tapetis* subsp. *britannicus* subsp. nov. *Syst. Appl. Microbiol.* 36, 183-187.

[111] Balboa, S., Bermudez-Crespo, J., Gianzo, C., Romalde, J.L. (2011) Proteomics and multilocus sequence analysis confirm intraspecific variability of *Vibrio tapetis*. *FEMS Microbiol. Lett.* 324, 1-87.

[112] Faury N, Saulnier D, Thompson FL, Gay M, Swings J, Le Roux F. (2004) *Vibrio crassostreae* sp. nov., isolated from the haemolymph of oysters (*Crassostrea gigas*). *Int J Syst Evol Microbiol* 54, 2137–2140.

[113] Thompson, F. L., Thompson, C. C., Hoste, B., Vandemeulebroecke, K., Gullian, M., Swings J. (2003). *Vibrio fortis* sp. nov. and *Vibrio hepatarius* sp. nov. isolated from aquatic animals and the marine environment. *Int. J. Syst. Evol. Microbiol.* 53, 1495-1501.

[114] Thompson, F. L., Li, Y., Gomez-Gil, B., Thompson, C. C., Hoste, B., Vandemeulebroecke, K., Rupp, G. S., Pereira, A., De Bem, M. M., Sorgeloos, P., Swings, J. (2003) *Vibrio neptunius* sp. nov., *Vibrio brasiliensis* sp. nov. and *Vibrio xuii* sp. nov., isolated from the marine aquaculture environment (bivalves, fish, rotifers and shrimps). *Int. J. Syst. Evol. Microbiol.* 53, 245–252.

[115] Macián MC, Ludwig W, Aznar R, Grimont PAD, Schleifer KH, Garay E, Pujalte MJ. (2001) *Vibrio lentus* sp. nov., isolated from Mediterranean oysters. *Int. J. Syst. Evol. Microbiol.* 51:1449–1456.

[116] Farto, R., Armada, S. P., Montes, M., Perez, M. J., Nieto, T. P. (2006). Presence of a lethal protease in the extracellular products of *Vibrio splendidus-Vibrio lentus* related strains. *J. Fish Dis.* 29, 701-707.

[117] Imai, T., Numachi, K., Oizumi, J., Sato, S. (1965) Studies on the mass mortality of the oyster in Matsushima Bay. II. Search for the cause of mass mortality and the possibility

to prevent it by transplantation experiment. *Bull. Tohoku Reg. Fish. Res. Lab.* 25, 27–38.

[118] Imai, T., Mori, K., Sugawara, Y., Tamate, H., Oizumi, J., Itikawa, O. (1968) Studies on the mass mortality of oysters in Matsushima bay VII. Pathogenic investigation. Tohoku *J. Agric. Res.* 19, 250–265.

[119] Koganezawa, A., (1974) Present status of studies on the mass mortality of cultured oysters in Japan and its prevention. In: Proceedings of the Third U.S.–Japan Meeting on Aquaculture, Tokyo, Japan, pp. 29–34.

[120] Elston, R. A., Beattie, J. H., Friedman, C., Hedrick, R., Kent, M.L. (1987) Pathology and significance of fatal inflammatory bacteraemia in the Pacific oyster, *Crassostrea gigas*. *J. Fish Dis.* 10, 121–132.

[121] Fujita, T., Matsubara, T., Hirokawa, H., Araki, F. (1953) On the inflammatorious changes of the *Ostrea gigas* in Hiroshima Bay. *Bull. Jpn. Soc. Fish. Sci.* 19, 766– 770.

[122] Fujita, T., Matsubara, T., Hirokawa, H., Araki, F. (1955) On the inflammatorious changes of the *Ostrea gigas* in Hiroshima Bay-II. *Bull. Jpn. Soc. Fish. Sci.* 20, 1063–1065.

[123] Sinderman, C.J., Rosenfield, A. (1967) Principal diseases of commercially important marine bivalve *Mollusca* and *Crustacea*. Fish. Bull. U.S. *Fish Wildlife Serv.* 66, 335–385.

[124] Elston, R. (1986) Occurrence of branchial rickettsiales-like infections in two bivalves molluscs, *Tapes japonica* and *Patinopecten yessoensis*, with comments on their significance. *J. Fish Dis.* 9, 69–71.

[125] Katkansky, S. C., Warner, R. W. (1974) Pacific Oyster Disease and Mortality Studies in California. Marine Resources Technical Report, vol. 25. California Dept. of Fish and Game, Long Beach.

[126] Glude, J. B. "A summary report of Pacific coast oyster mortality investigations 1965–1972." *Proceedings of the Third US–Japan Meeting on Aquaculture at Tokyo, Japan*. Vol. 28. 1974.

[127] Friedman, C. S., Beattie, J. H., Elston, R. A., Hedrick, R. P. (1991) Investigation of the relationship between the presence of a Gram-positive bacterial infection and summer mortality of the Pacific oyster, *Crassostrea gigas* Thunberg. *Aquaculture* 94, 1–15.

[128] Friedman, C. S., Beaman, B. L., Chun, J., Goodfellow, M., Gee, A., Hedrick, R. P. (1998) *Nocardia crassostreae* sp. nov., the causal agent of nocardiosis in Pacific oysters. *Int. J. Syst. Bacteriol.* 48, 237–246.

[129] Lindsay, C. E. (1969) Summary of Progress of Pacific Oyster Mass Mortality Investigations. 21st Annual Report of the Pacific Marine Fisheries Commission for the Year 1968, vol. 1967–68, pp. 36–38.

[130] Engelsma, M. Y., Roozenburg, I., Joly, J. P. (2008) Isolation of *Nocarida crassostreae* from Pacific oyster *Crassostrea gigas* in Europe. *Dis. Aquat. Organ.* 80, 229–234.

[131] Bower, S. M., Goh, B., Meyer, G. R., Carnegie, R. B., Gee, A. (2005) Epizootiology and detection of nocardiosis in oysters. In: Walker, P. et al. (Eds.), Diseases in Asian Aquaculture V. Asian Fisheries Society, Fish Health Section, Manila, pp. 249– 262.

[132] Lee, L. H., Azman, A. S., Zainal, N., Eng, S. K., Fang, C. M., Hong, K., Chan, K. G. (2014) *Novosphingobium malaysiense* sp. nov. isolated from mangrove sediment. Int. *J. Syst. Evol. Microbiol.* 64, 1194–1201.

[133] Xu, J., Wang, Y., Xie, S. J., Xiao, J., Ruan, J. S., (2009) *Streptomyces xiamenensis* sp. nov., isolated from mangrove sediment. *Int. J. Syst. Evol. Microbiol.* 59, 472–476.

[134] Boettcher, K. J., Barber, B. J., Singer, J. T. (1999) Use of antibacterial agents to elucidate the etiology of juvenile oyster disease (JOD) in *Crassostrea virginica* and numerical dominance of an alpha-proteobacterium in JOD-affected animals. *Appl. Environ. Microbiol.* 65, 2534–2539.

[135] Buchan, A., González, J. M., Moran, M. A. (2005) Overview of the marine *Roseobacter* lineage. *Appl. Environ. Microbiol.* 71, 5665–5677.

[136] Maloy, A. P., Ford, S. E., Karney, R. C., Boettcher, K. J. (2007) Roseovarius crassostreae, the etiological agent of Juvenile Oyster Disease (now to be known as Roseovarius Oyster Disease) in Crassostrea virginica. *Aquaculture* 269, 71–83.

[137] Bricelj, V. M., Ford, S. E., Borrero, F.J., Perkins, F. O., Rivara, G., Hillman, R. E., Elston, R. A., Chang, J. (1992) Unexplained mortalities of hatchery-reared, juvenile oysters, *Crassostrea virginica* (Gmelin). *J. Shellfish Res.* 11, 331–347.

[138] Davis, C. V., Barber, B. J. (1994) Size-dependent mortality in hatchery-reared populations of oysters, *Crassostrea virginica*, Gmelin 1791, affected by juvenile oyster disease. *J. Shellfish Res.* 13, 137–142.

[139] Ford, S. E., Tripp, M. R. (1996). Diseases and defense mechanisms. The eastern oyster *Crassostrea virginica*, 581-660.

[140] Paillard, C., Maes, P., Oubella, R. (1994) Brown ring disease in clams. *Annu. Rev. Fish Dis.* 4, 219–240.

[141] Harshbarger, J. C., Chang, S. C., Otto, S. V. (1977) *Chlamydiae* (with phages), mycoplamas, and rickettsiae in Chesapeake Bay Bivalves. *Science* 196, 666–668.

[142] Wu, X., Li, D., Pan, J., Jaing, J. (2003) Studies on rickettsia like organism (RLO) disease of tropical marine pearl oyster – epidemiological investigation of RLO disease in juvenile populations of maricultured *Pincatda maxima*. *Acta Oceanol. Sin.* 22, 421–435.

[143] Norton, J. H., Shepherd, M. A., Long, H. M., Prior, H. C. (1993). Parasites of the giant clams (*Tridacnidae*). Biology and mariculture of giant clams. pp. 18-23, Australian Centre for International Agricultural Research. Canberra, a.c.t. (Australia).

[144] Norton, J. H., Shepherd, M. A., Abdon-Nagutt, M. R., Lindsay, S. (1993). Mortalities in the giant clam *Hippopus hippopus* associated with rickettsiales-like organisms. *J. Invertebr. Pathol.* 62, 207-209.

[145] Villalba, A., Carballal, M.J., López, C., Cabada, A., Corral, L. Azevedo, C. (1999) Branchial rickettsia-like infection associated with clam *Venerupis rhomboides* mortality. *Dis. Aquat. Org.* 36, 53-60.

[146] Wu, X., Pan, J. (1999) Studies on rickettsia-like organism disease of the tropical marine pearl oyster I: the fine structure and morphogenesis of *Pinctada maxima pathogen rickettsia-like organism. J. Invertebr. Pathol. 73, 162–172.*

[147] Wu, X., Pan, J. (2000) An intracellular prokaryotic microorganism associated with lesions in the oyster, *Crassostrea ariakensis*. Gould. *J. Fish Dis.* 23, 409–414.

[148] Fournier, P. E., Raoult, D. (2009) Current knowledge on phylogeny and taxonomy of *Rickettsia* spp. *Ann. N. Y. Acad. Sci.* 1166, 1–11.

[149] Renault, T., Cochennec, N. (1994) *Rickettsia*-like organisms in the cytoplasm of gill epithelial cells of the Pacific oyster *Crassostrea gigas*. *J. Invertebr. Pathol.* 64, 160–162.

[150] Azevedo, C., Villalba, A. (1991) Extracellular giant rickettsiae associated with bacteria in the gill of *Crassostrea gigas* (*Mollusca, Bivalvia*). *J. Invertebr. Pathol.* 58, 75–81.

[151] Zhu, Z., Xu, T., He, Z., Wu, X., Wu, L., Meng, Q., Huang, J. (2012) *Rickettsia*-like organism infection associated with mass mortalities of blood clam, Tegillarca granosa, in the Yueqing Bay in China. *Acta Oceanol. Sin.* 31, 106–115.

[152] Sun, J., Wu, X. (2004) Histology, ultrastructure, and morphogenesis of a rickettsia-like organism causing disease in the oyster, *Crassostrea ariakensis* Gould. *J. Invert. Pathol.* 86, 77-86.

[153] Verschuere, L., Rombaut, G., Sorgeloos, P., Verstraete, W. (2000) Probiotic bacteria as biological control agents in aquaculture. *Microbiol. Mol. Biol. Rev*. 64, 655–671.

[154] Dubert, J., Osorio, C. R., Prado, S., Barja, J. L. (2015). Persistence of Antibiotic Resistant *Vibrio* spp. in Shellfish Hatchery Environment. *Microb. Ecol.* doi:10.1007/s00248-015-0705-5.

[155] Lodeiros, C., Freites, L., Fernández, E., Vélez, A., Bastardo, J. (1989) Efecto antibiótico de tres bacterias marinas en la supervivencia de larvas de la viera Pecten ziczac infectadas con el germen *Vibrio anguillarum. Boletín instituto Oceanográfico de Venezuela. Universidad Oriente*, *28*, 165-169.

[156] Gibson, L. F., Woodworth, J., George, A. M. (1998) Probiotic activity of *Aeromonas media* on the Pacific oyster *Crassostrea gigas*, when challenged with *Vibrio tubiashii*. *Aquaculture* 169, 111–120.

[157] Longeon, A., Peduzzi, J., Barthelemy, M., Corre, S., Nicolas, J. L., Guyot, M. (2004) Purification and partial identification of novel antimicrobial protein from marine bacterium *Pseudoalteromonas* species strain X153. *Mar. Biotechnol.* 6, 633–641.

[158] Ruiz-Ponte, C., Samain, J. F., Sa´ nchez, J. L., Nicolas, J. L. (1999) The benefit of a *Roseobacter* species on the survival of the scallop larvae. *Mar. Biotechnol.* 1, 52–59.

[159] Prado, S., Romalde, J. L., Barja, J. L. (2010) Review of probiotics for use in bivalve hatcheries. *Vet. Microbiol.* 145, 187-197.

[160] Romero, J., Espejo, R. (2001) The prevalence of noncultivable bacteria in oysters (*Tiostrea chilensis*, Philippi, 1845). *J. Shellfish Res.* 20, 1235-1240.

[161] Romero, J., Garcia-Varela, M., Laclette, J. P., Espejo, R. T. (2002) Bacterial 16S rRNA gene analysis revealed that bacteria related to *Arcobacter* spp. constitute an abundant and common component of the oyster microbiota (*Tiostrea chilensis*). *Microb. Ecol.* 44, 365–371.

[162] Hernández-Zárate, G., Olmos-Soto, J. (2006) Identification of bacterial diversity in the oyster *Crassostrea gigas* by fluorescent in situ hybridization and polymerase chain reaction. *Appl. Microbiol.* 100, 664–667.

[163] Green, T. J., Barnes, A. C. (2010) Bacterial diversity of the digestive gland of Sydney rock oysters, *Saccostrea glomerata* infected with the paramyxean parasite, *Marteilia sydney*. *Appl. Microbiol.* 109, 613–622.

[164] Boardman, C. L., Maloy, A., Boettcher, K. (2008). Localization of the bacterial agent juvenile oyster disease (*Roseovarius crassostreae*) within affected eastern oysters (*Crassostrea virginica*). *J. Invertebr. Pathol.* 97, 150-158.

[165] Romero, J., Garcia-Varela, M., Laclette, J. P., Espejo, R. T. (2002). Bacterial 16S rRNA gene analysis revealed that bacteria related to Arcobacter spp. constitute an abundant and common component of the oyster microbiota (*Tiostrea chilensis*). *Microb. Ecol.* 44, 365-371.

[166] Boardman, C. L., Maloy, A. and Boettcher, K. (2008). Localization of the bacterial agent juvenile oyster disease (*Roseovarius crassostreae*) within affected eastern oysters (*Crassostrea virginica*). *J. Invertebr. Pathol.* 97, 150-158.

[167] Zurel, D., Benayahu, Y., Or, A., Kovacs, A., Gophna, U. (2011). Composition and dynamics of the gill microbiota of an invasive Indo-Pacific oyster in the eastern Mediterranean Sea. *Environ. Microbiol.* 13, 1467-1476.

[168] Trabal, N., Mazón-Suástegui, J. M., Vázquez-Juárez, R., Asencio-Valle, F., Morales-Bojórquez, E., Romero, J. (2012). Molecular analysis of bacterial microbiota associated with oysters (*Crassostrea gigas* and *Crassostrea corteziensis*) in different growth phases at two cultivation sites. *Microb. Ecol.* 64, 555-569.

[169] Trabal,N., Mazón-Suástegui, J. M., Vázquez-Juárez, R., Ascencio-Valle, F., Romero, J. (2014). Changes in the composition and diversity of the bacterial microbiota associated with oysters (*Crassostrea corteziensis*, *Crassostrea gigas* and *Crassostrea sikamea*) during commercial production. *FEMS Microb. Ecol.* 88, 69-83.

[170] Fernandez- Piquer, J., Bowman, J. P., Ross, T., Tamplin, M. L. (2012). Molecular analysis of the bacterial communities in the live Pacific oyster (*Crassostrea gigas*) and the influence of postharvest temperature on its structure. *J. Appl. Microbiol.* 112, 1134-1143.

[171] Chauhan, A., Wafula, D., Lewis, D. E., Pathak, A. (2014). Metagenomic assessment of the Eastern oyster-associated microbiota. *Genome Announc.*2(5), e01083-14.

[172] Lokmer, A., Kuenzel, S., Baines, J., Wegner, K. M. (2015). The Role of Tissue specific Microbiota in Initial Establishment Success of Pacific oysters. *Environ. Microbiol.* DOI 10.1111/1462-2920.13163.

[173] Sison-Mangus, M. P., Mushegian, A. A., Ebert, D. (2015) Water fleas require microbiota for survival, growth and reporduction. *ISME J* 9, 59-67.

[174] Vezzulli, L., Pezzati, E., Brettar, I., Höfle, M., Pruzzo, C. (2015) Effects of global warming on *Vibrio* ecology. Microbiol Spectrum 3, VE-0004-2014.

In: Oysters and Clams
Editor: Jesús L. Romalde
ISBN: 978-1-63485-074-2

Chapter 5

MICROBIOLOGICAL APPROACHES TO UNDERSTAND THE THREATS FOR THE WEDGE-SHELL CLAM (*DONAX TRUNCULUS*) IN NATURAL BEDS

Juan L. Barja[1,*], Javier Dubert[1], Jaime Montes[2], José G. Olveira[1], Susana Prado[1] and Dorotea Martínez-Patiño[2]
[1]Departamento de Microbiología y Parasitología e Instituto de Acuicultura, Universidade de Santiago de Compostela, Santiago de Compostela, Spain
[2]Centro de Investigacions Mariñas (CIMA), Consellería do Mar, Xunta de Galicia, Spain

ABSTRACT

The production of wedge-shell clam, *Donax trunculus*, in the coast of Galicia (NW Spain) has suffered a significant reduction during the last ten years without an apparent cause. With the aim to understand the origin of the decline of natural beds, a wide research program was set-up. During a period of two years, analyses were performed on water, sediment and animals sampled on the different areas where historic records of the *D. trunculus* presence were already known. A variety of physicochemical, toxicological and biological analyses were carried out during joint samplings. In this chapter only the microbiological part of this wide study is reported.

Besides the recognition of a number of parasites being present with different prevalence rates on the population of clams, the early conclusions of the study on the possible biological causes for the reduction of the *D. trunculus* populations allow us, at present, to discard an infectious origin as a primary cause of the dramatic reduction of the wedge clam populations. Although some questions remain still unresolved, a significant parasitosis, as well as the presence of herpesvirus and *V. tapetis* were discarded.

Keywords: wedge clam, *Donax trunculus,* bacteria, parasites, herpesvirus

* E-mail: juanluis.barja@usc.es.

Introduction

The wedge clam *Donax trunculus* is an Atlantic-Mediterranean warm-temperate species of bivalve. This suspension-feeding clam is a dominant macrofaunal species on shallow bottoms. Its geographical distribution extends from the French Atlantic to the Senegal and the Mediterranean Sea. It inhabits high energy environments with strong wave actions and with numerous suspended particles [20]. The biology and population dynamics were studied by different authors [10, 15, 16].

However, publications about the presence of parasites on *Donax trunculus* are scarce. Some reports referred to the damage in *D. trunculus* from the trematode *Meiogymnophallus strigatus* (Digenea: Gymnophallidae) [3] and *Rickettsia*-like organisms [8]. Ramón et al. [20] reported the parasitic castration of *D. trunculus* caused by the trematode *Bacciger bacciger* (Digenea: Faustulidae), although the low prevalence of infection and low levels of parasitic castration suggested that the development of the population was not significantly affected by this infection. However, de Montaudoin et al. [9] pointed again to *B. bacciger* infection as possible cause of the difficulty of a *D. trunculus* population in sustaining itself. Recently, Carella et al. [6] described the inflammatory response to infection by metacercariae of *Postmonorchis* sp. inq. (Digenea: Monorchiidae). Regarding to microbiological studies, there is a total lack of references on the presence of bacteria and virus which may adversely affect the health of its populations. This clam represents a commercially important resource, due to its high economical value, reaching prices up to 30 €/kg. In Galicia (NW Spain, Figure 1) there were historic natural beds with important production based on extraction: Vilarrube (Vil), Lombo das Navallas (LdN) and Praia América (PAm). Unfortunately, these populations have suffered a continuous decline from an overall production of 40 T/year in 90’s of the last century, until the current 15 T/year, all extracted from Vilarrube.

Therefore, a monitoring program has been organized and developed during two years to investigate the causes of the decline of natural beds in Galicia and to propose measures to overcome this trend. With these aims, a multidisciplinary team was assembled, covering the different topics to be considered: the biology of the species, water quality, currents in the shoreline, characteristics of sediments, ecotoxicology of the environment, effect of river effluents and run-offs, genetic fitness of the populations, as well as the possible presence of bacterial, viral and parasitic pathogens.

In this chapter, we present specifically the first study carried out to detect the presence of microbial and parasitic agents as a possible cause for the sharp decline of the wedge-shell clam in the natural environment. The other aspects of the study would be considered elsewhere. The work was focused to assess the health status of the *D. trunculus* populations in Galicia, by culture, PCR and histopathological procedures.

Methods

Donax trunculus specimens were collected over two consecutive years, 2013 and 2014, mainly in two geographical locations chosen on the basis of their historic production capacity: Vilarrube (Vil, 43° 38.6’ N, 8° 4.6’ W), nowadays the only with exploitation of this resource in Galicia, and Praia América (PAm, 42° 07´ N 8° 49´ W).

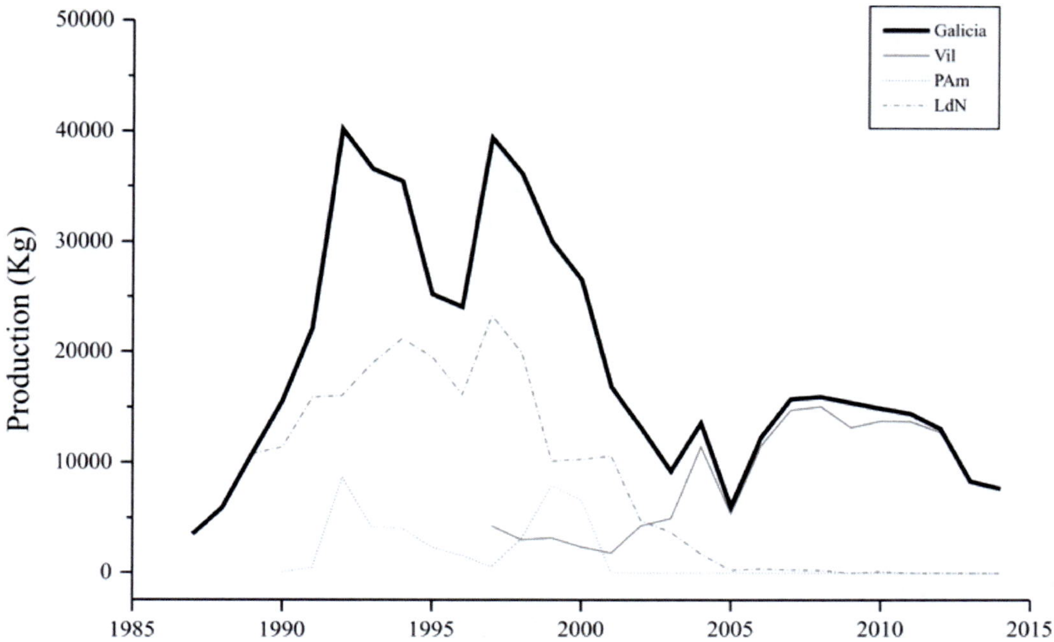

Figure 1. Production of wedge clam in Galicia (thick line) and contribution of each historic natural bed.

For parasitological studies, a total of 716 individuals were processed for routine histopathology. The soft tissues of at least 30 clams (depending on the availability of the resource) from each location were fixed in Davidson's fixative for 24-48 hours. Transverse sections, approximately 5 mm thick, were excised, including the gill, gonad, digestive gland, mantle, nephridia and foot. The samples were embedded in paraffin wax, sliced in 5 μm sections and stained with haematoxylin and eosin (H&E). Parasite prevalence and infection intensity were evaluated for the different samples.

For microbiological studies, the clams were washed with sterile seawater (SSW), opened under aseptic conditions, weighted and homogenized in SSW (dilution 1/10). Appropriated decimal dilutions were prepared and spread on culture plates. The bacteriological media used were Marine Agar (MA, Difco), for marine heterotrophic bacteria, and Thiosulphate Citrate Bile Sucrose (TCBS, Oxoid), which is selective for vibrios. The plates were incubated at 20°C for 24-48 hours in the case of TCBS and 5-7 days for MA. Then, the number of colonies was recorded and the different types selected. The isolates were preserved for further studies in Marine Broth with glycerol (15% v/v) by freezing at -80°C. Bacterial counts were expressed as colony forming units per gram (cfu/g), corresponding to Marine Heterotrophic Bacteria (MHB, MA medium) and Presumptive *Vibrio* counts (PV, TCBS medium).

A set of 10 biochemical and physiological tests were performed with these bacterial isolates. On the basis of the results and the groups formed, a number of bacterial strains were selected for the sequencing of the 16S rDNA gene. All the methods used are described elsewhere [19].

Additionally, the homogenates were used for further analyses of known bivalve pathogens. One aliquot was used to perform the assay for the presence of *Vibrio tapetis*, the causal agent of the Brown Ring Disease, by the PCR procedure of Balboa et al. [2].

A second aliquot of each sample was used to determine the presence of Ostreid herpesvirus (OsHv-1) by PCR and real time-PCR following the recommended procedures by the OIE [18]. Two pairs of primers were used: DPFor: 5'-ATT-GAT-GAT-GTG-GAT-AAT-CTG-TG-3'/ DPRev: 5'-GGT-AAA-TAC-CAT-TGG-TCT-TGT-TCC-3' that amplifies a 179 bp fragment of the ORF 100, responsible for the codification of the viral polymerase, and also the pair C9: 5'-GAG-GGA-AAT-TTG-CGA-GAG-AA-3'/ C10: 5'-ATC-ACC-GGC-AGA-CGT-AGG-3' that amplifies a 179 bp fragment on the ORF4. Each pair has different melting temperatures but both are able to detect the OsHV-1 and the more pathogenic variant OsHV-1 μvar. As positive control, a homogenate of oyster infected with the OsHV-1 was used.

RESULTS

Histopathology/Parasitology

The analysis of the slides for histopathology showed the presence of different protista, including ciliates (Figure 2) and gregarines (*Nematopsis*-like) (Figure 3), and *Rickettsia*-like organisms (RLOs) (Figure 4), in all the samplings. In both locations and for all the parasites found, the highest prevalence was reached in spring. The similar results obtained in both populations suggested a seasonal pattern, more than the influence of the geographical location.

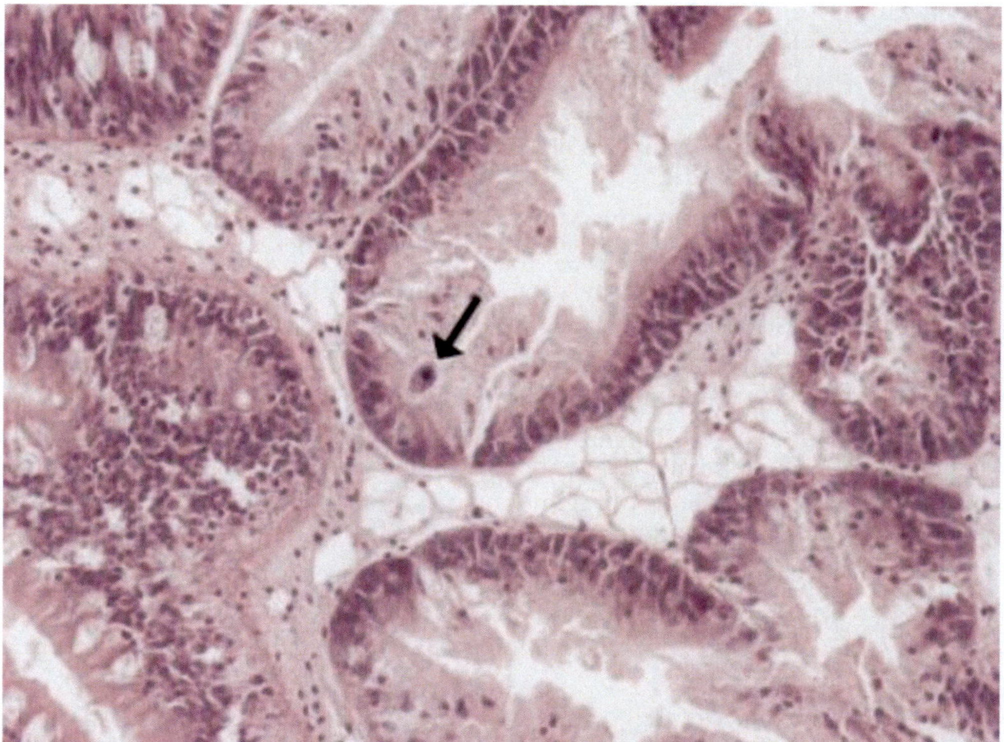

Figure 2. Histological section of an unidentified ciliate (arrow) in digestive tubules of wedge clam (200x).

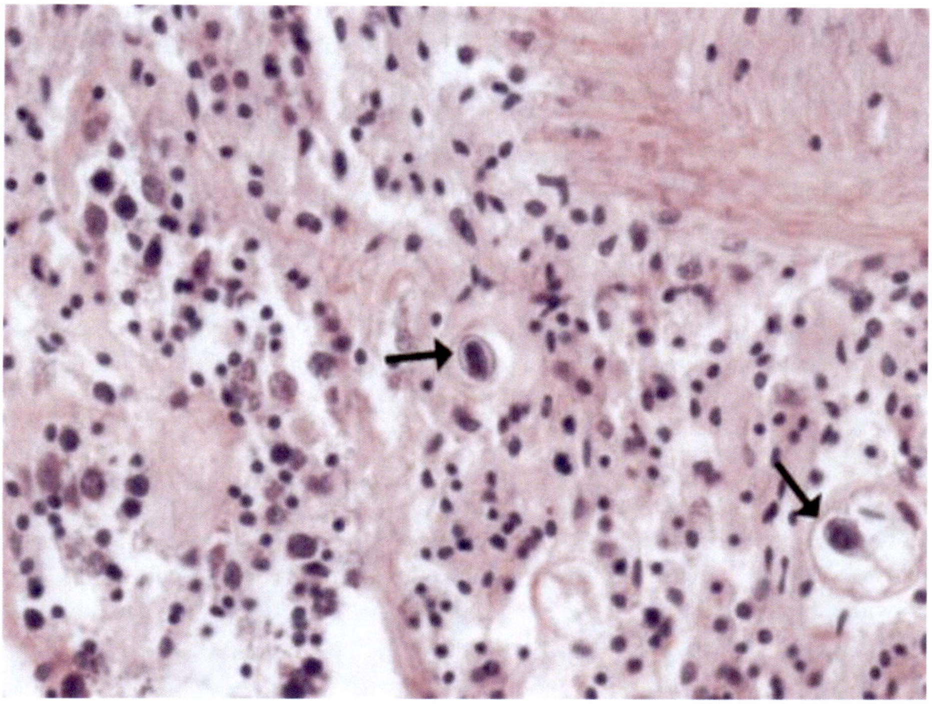

Figure 3. Histological section of gregarines (*Nematopsis*-like) (arrow) in the connective tissue of the digestive gland of wedge clam (400x).

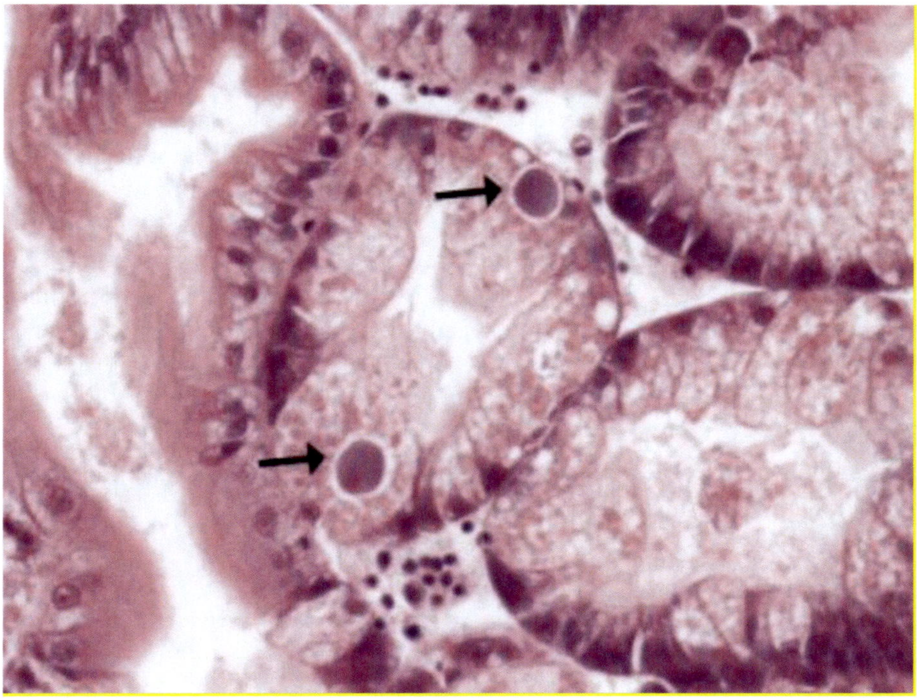

Figure 4. Histological section of Rickettsia-like organisms (arrow) in digestive tubules of wedge clam (200x).

Regarding to these parasitological results, it is important to point-out that, despite of high prevalence in some moments of the year, the intensity of parasitation was always low.

The ranges of prevalence for ciliates were 14-30% (Vil) and 12-43% (PAm) (Figure 5A).

Gregarines appeared in 17-48% of the individuals from Vilarrube and 8-36% in Praia América (Figure 5B). Genus *Nematopsis* has been described in different species of bivalve molluscs, usually associated with a focal hemocyte infiltration without measurable health effects (Bower et al. 1994). However, there are some discrepancies about the significance of *Nematopsis* infections in bivalves, and pathological consequences of the infection have been suggested in *Cerastoderma edule* [1] and *Perna viridis* [23]. Oocysts of gregarines belonging to *Nematopsis* genus has been described in a population of *Donax vittatus* in Galicia, with a prevalence of 95.23%, mainly located in gills and mantle, as expected since the infective stage get into the host through these organs [22].

In our study, as expected according to the low intensity of infection, no obvious host damage was observed.

The prevalence of RLOs was 17-52% (Vil) and 9-46% (PAm) (Figure 5C). RLOs have been only one-time related to *D. trunculus* mortalities, in France, with prevalence between 20 and 46% [8]. Although the results obtained in both locations were in the same range, as occurred with the other parasites above mentioned, no pathological effects were observed.

Microbiology

Wedge clam batches from Vilarrube were processed in autumn, winter and spring, to study the evolution of the bacterial load in this population (Figure 6). The average counts of presumptive vibrios remained similar, ranging from 4.7 to 9.0 x 10^3 cfu/g, regardless the environmental conditions. The main difference among samples was the slight decrease of MHB observed in winter (1.4 x 10^4 cfu/g) respect to the values corresponding to autumn (2.1 x 10^5 cfu/g) and spring (1.8 x 10^5 cfu/g), with the subsequent increase of average ratio PV:MHB, 1:3 in winter (1:37 in autumn, 1:20 in spring). The values of MHB followed a similar pattern to *Rickettsia*-like organisms and gregarines, suggesting again a seasonal fluctuation.

The sampling effort in autumn allowed to obtain three classes according to the size of the clams collected: big (average 32 mm), medium (23 mm) and small-sized (15 mm) clams. A clear pattern could not be established, only a slight decrease of PV load with the increase in size (Figure 7). Anyway, the results of MHB implied that the highest ratio PV:MHB corresponded to medium-sized clams (1:5), followed by small clams (1:18), while big clams showed the lowest ratio (1:238).

Selected colonies from TCBS and MA plates were isolated and further characterized to obtain a preliminar identification.

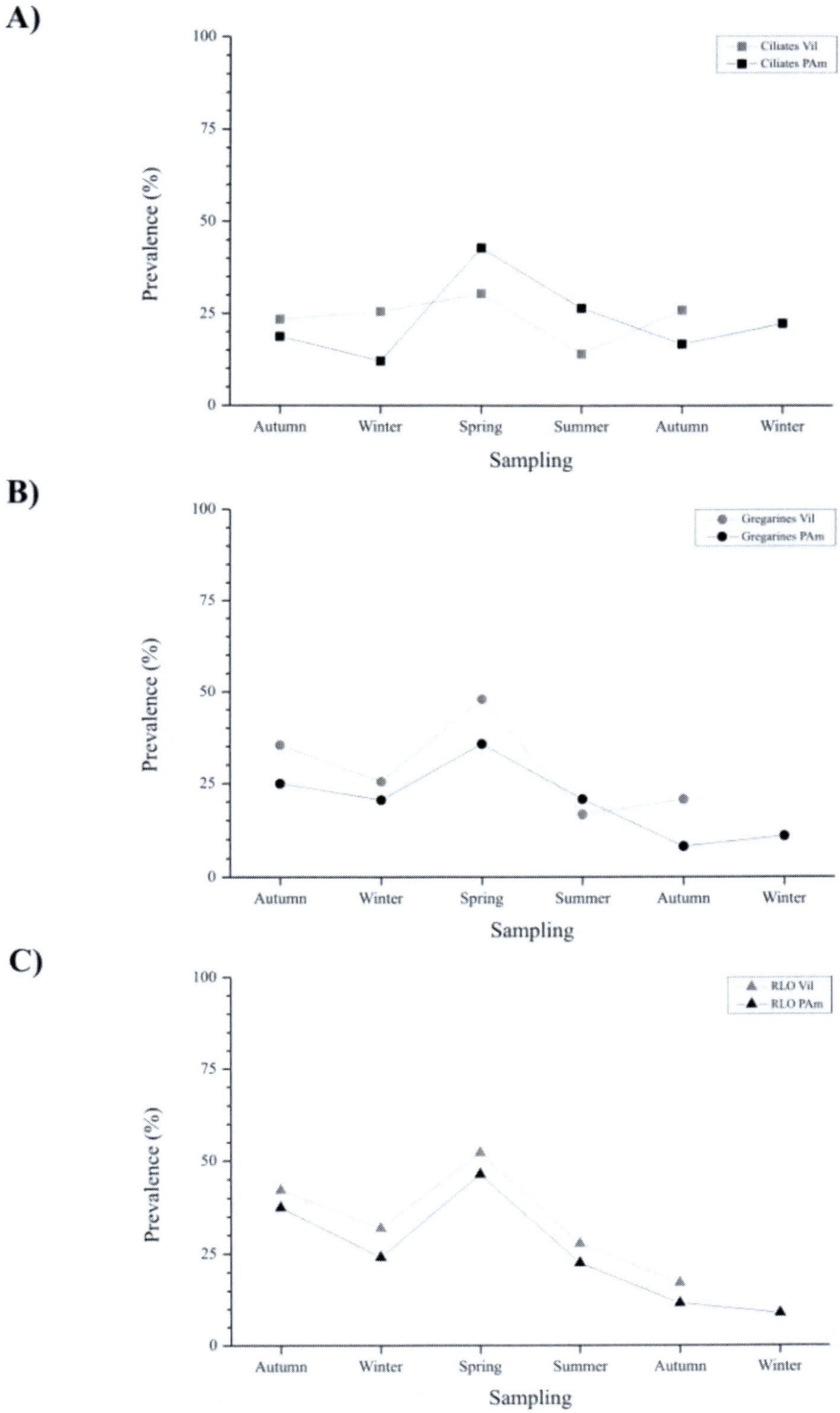

Figure 5. Prevalence of (A) ciliates, (B) gregarines (*Nematopsis*-like) and (C) Rickettsia-like organisms in the different samplings from Vilarrube (Vil) and Praia América (PAm).

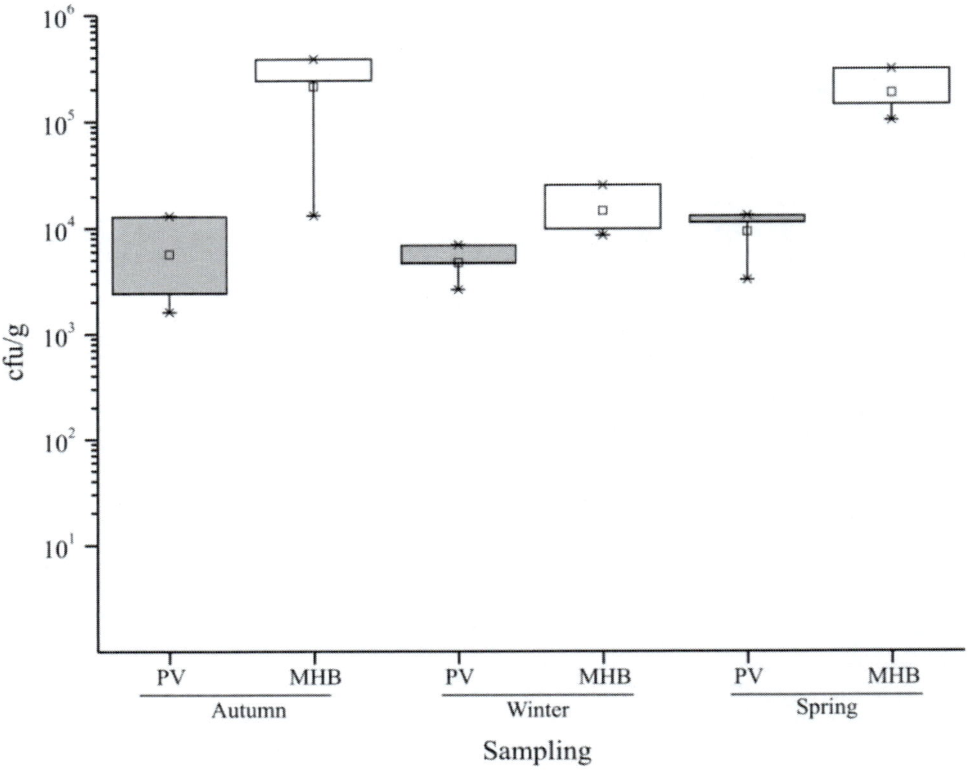

Figure 6. Counts of presumptive vibrios (PV, filled boxes) and marine heterotrophic bacteria (MHB) in the different samplings of *D. trunculus* from Vilarrube.

In autumn samples, the PV were identified as members of the clade Splendidus within genus *Vibrio* (gamma-proteobacteria, 'Vibrionales'). Similar isolates were found in winter, as codominant types of colony with *Shewanella* (gamma-proteobacteria, Alteromonadales, Shewanellaceae). In spring, both genera were identified again in TCBS samples, though the specific composition of vibrios was different. Regarding to MA isolates, genus *Ruegeria* (alpha-proteobacteria, Rhodobacterales) was dominant in autumn and winter, in this last sampling with *Microbulbifer* (gamma-proteobacteria, Alteromonadales, Alteromonadaceae).

As observed in PV, the variability in spring was higher, being identified alpha-proteobacteria (genus *Octadecabacter*, Rhodobacterales), Bacteriodetes (genus *Psychroserpens*, Flavobacteriales) and even members of the clade Splendidus within genus *Vibrio*.

Interestingly, *Vibrio splendidus*-like bacteria were isolated in all the samplings, as the dominant or codominant type of colony on TCBS plates from samples collected in autumn and winter. And other members of the clade Splendidus of genus *Vibrio*, related to the species *V. atlanticus*-*V. tasmaniensis*, were also isolated in all the samplings. The Splendidus clade has been associated with epidemic outbreaks of marine bivalves, affecting all the stages, from hatchery cultures to on-growing in natural environment [11–14, 19, 21]. Recently, an epidemiological survey showed high diversity and massive exchange of virulence factors within the clade, being impossible to discriminate virulent from non-virulent strains [17]. Considering those facts, these bacterial populations related to the Splendidus clade should be monitored. Although the levels recorded seemed to be harmless, the possibility of a seasonal proliferation should be taken into account. Moreover, wedge clams from natural beds are used as broodstock in hatchery, where the pathogenic potential of this clade is high. Therefore, the risk of vertical transmission from breeders to larval cultures, as previously demonstrated in other bivalves [19], should be also considered.

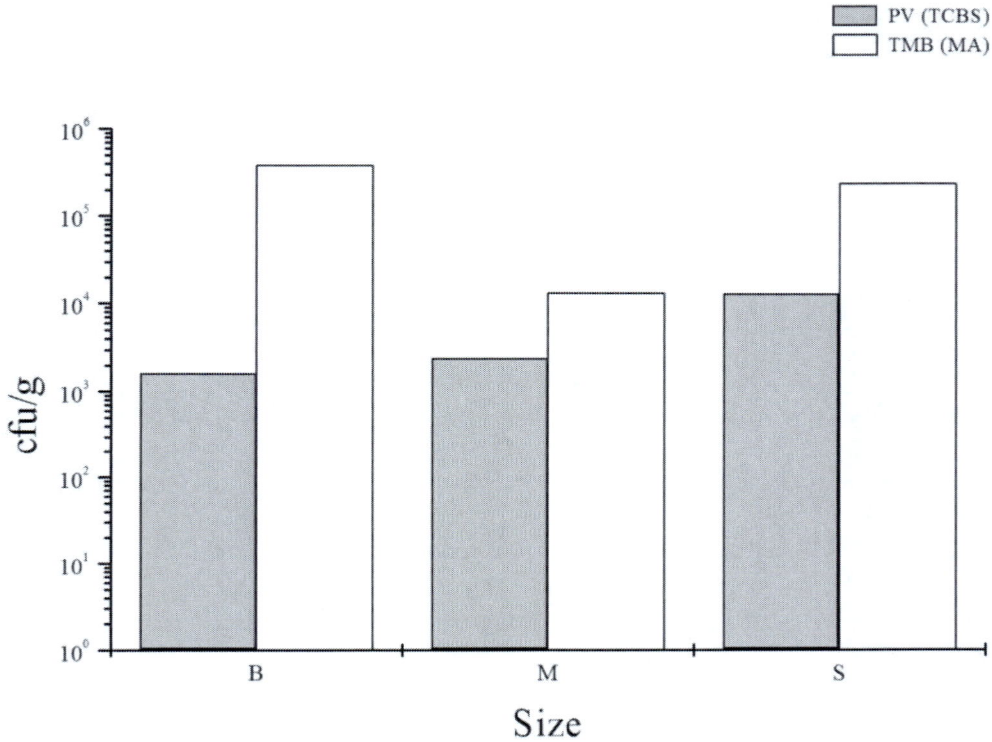

Figure 7. Counts of presumptive vibrios (PV, filled bars) and marine heterotrophic bacteria (MHB) in the different classes of sizes of *D. trunculus* from Vilarrube.

As expected, *Vibrio tapetis* was not detected by PCR in any sample, considering that no signs of disease were observed in the shells. Regarding to known virological diseases in bivalves, since herpesvirus has been associated to mortalities in different species of molluscs in the wild [4], one of the key points of the survey was focused on these agents. All the samples processed for the presence of OsHV-1 and OsHV-1 μvar showed negative results, by any procedure (PCR and real time-PCR) and regardless of the geographic origin (Vil and PAm). The results confirmed the low incidence of this pathogen in the wedge clam, considering that its presence in *Donax* has been only reported by Cochennec-Laureau et al. [7], associated to one episode of mortalities but with low viral load.

CONCLUSION

The results from this first integrated research on the possible infectious origin of the reduction of *Donax trunculus* populations may be summarized in a lack of a clear pathogen responsible for the problem. Further studies are needed to complete these preliminary results, especially to clarify the role of bacterial populations and particularly RLOs.

Acknowledgments

The study was supported by grant 2013/2015 from the Consellería do Mar (Xunta de Galicia) and in part by grant BAPOHATCH from the Ministerio de Economía y Competitividad (Gobierno de España).

The authors thank to J.D. Cerdeira for the data on wedge clam production, and to L. Rodríguez and F. Febrero for the sampling.

References

[1] Azevedo, C., Cachola, R., 1992. Fine structure of the apicomplexa oocyst of *Nematopsis* sp. of two marine bivalve molluscs. *Dis. Aquat. Org.*, 14, 69-73.

[2] Balboa, S., Doce, A., Diéguez, A. L., Romalde, J. L., 2011. Evaluation of different species-specific PCR protocols for the detection of *Vibrio tapetis*. *J. Invertebr. Pathol.*, 108, 85-91.

[3] Bartoli, P., 1983. The larval stages of *Meiogymnophallus strigatus* (MV Levour 1908) n. comb. (Trematoda: Gymnophallidae). *Ann. Parasitol. Hu. Comp.*, 58, 227-242.

[4] Batista, F. M., Arzul, I., Pepin, J.-F., Ruano, F., Friedman, C. S., Boudry, P., Renault, T., 2007. *J. Virol. Meth.*, 139, 1-11.

[5] Bower, S. M., McGladdery, S. E., Price, I. M., 1994. Synopsis of infectious diseases and parasites of commercially exploited shellfish. *Annu. Rev. Fish Dis.*, 4, 1-199.

[6] Carella, F., Culurgioni, J., Aceto, S., Fichi, G., Pretto, T., Luise, D., Gustinelli, A., De Vico, G., 2013. *Postmonorchis* sp.inq. (Digenea:Monorchiidae) metacercariae infecting natural beds of wedge clam *Donax trunculus* in Italy. *Dis. Aquat. Org.*, 106, 163-172.

[7] Cochennec-Laureau, N.,Baud, J.-P., Pepin, J.-F., Benabdelmouna, A., Soletchnik, P., Bacque, C., Garcia, C., Arzul, I., Boudry, P., Huvet, A., Pernet, F., Bachere, E., Bédier, E., Petton, B., Gaussem, F., Stanisiere, J.-Y., Degremont, L., 2011. Les surmortalités des naissains d'huîtres creuses *Crassostrea gigas*: acquis des recherches en 2010. Rapport: RST/LER/MPL/11.07. IFREMER, France.

[8] Comps, M., Raimbault, R., 1978. Infection rickettsienne de la glande digestive de *Donax trunculus* Linné. *Sci. Pêche, Bull. Inst. Pêches Marit.*, 281, 11-12.

[9] de Montaudoin, X., Bazairi, H., Mlik, K. A., González, P., 2014. *Bacciger bacciger* (Trematoda: Fellodistomidae) infection effects on wedge clam *Donax trunculus* condition. *Dis. Aquat. Org.*, 111, 259-267.

[10] Fernández, J., Otero, J. and De Coo, A.1984 Contribución al estudio de la coquina (*Donax trunculus*) en Galicia. *Actas do IV Simposio Ibérico do Estudo do Bentos Mariño,* 2, 133-142.

[11] Garnier, M., Labreuche, Y., García, C., Robert, M., Nicolas, J.-L., 2007. Evidence for the involvement of pathogenic bacteria in summer mortalities of the Pacific oyster *Crassostrea gigas*. *Microb. Ecol.,* 53, 187-196.

[12] Gay, M., Berthe, F. C. J., Le Roux, F., 2004. Screening of vibrio isolates to develop an experimental infection model in the Pacific oyster *Crassostrea gigas*. *Dis. Aquat. Org.*, 59, 49-56.

[13] Gómez-León, J., Villamil, L., Lemos, M. L., Novoa, B., Figueras, A., 2005. Isolation of *Vibrio alginolyticus* and *Vibrio splendidus* from aquacultured carpet shell clam (*Ruditapes decussatus*) larvae associated with mass mortalities. *Appl. Environ. Microbiol.*, 71, 98-104.

[14] Le Roux, F., Gay, M., Lambert, C., nicolas, J. L., Gouy, M., Berthe, F., 2004. Phylogenetic study and identification of *Vibrio splendidus*-related strains based on *gyrB* gene sequences. *Dis. Aquat. Org.*, 58, 143-150.

[15] Manca Zeichen, M., Agnesi, S., Mariani, A., Maccaroni, A. and Ardizzone, G. D. 2002. Biology and Population Dynamics of *Donax trunculus* L. (Bivalvia: Donacidae) in the South Adriatic Coast. *Estuar. Coast Shelf S*, 54, 971-982.

[16] Mazé, R. A., Laborda, A. J., 1988. Aspectos de la dinámica de población de *Donax trunculus* L. (Bivalvia: Donacidae) en la ría de El Barquero (Lugo, NW Spain). *Investigación Pesquera*, 52, 299-312.

[17] Nasfi, H., Travers, M. A., de Lorgeril, J., Habib, C., Sannie, T., Sorieul, L., Gerard, J., Avarre, J. C., Haffner, P., Tourbiez, D., Renault, T., Furones, D., Roque, A., Pruzzo, C., Cheslett, D., Gdoura, R., Vallaeys, T., 2015. A European epidemiological survey of *Vibrio splendidus* clade shows unexplored diversity and massive exchange of virulence factors. *World J. Microbiol. Biotechnol.*, 31, 461-475.

[18] OIE, 2013. Manual of Diagnostic Tests for Aquatic Animals. World Organisation for Animal Helath (OIE), Paris.

[19] Prado, S., Dubert, J., da Costa, F., Martínez-Patiño, D., Barja, J. L., 2014. Vibrios in hatchery cultures of the razor clam, *Solen marginatus* (Pulteney). *J. Fish Dis.*, 37, 209-217.

[20] Ramón, M. Gracenea, M., Gonzalez-Moreno, O. (1999). *Bacciger bacciger* (Trematoda, Fellodistomidae) infection in commercial clams *Donax trunculus* (Bivalvia, Donacidae) from the sandy beaches of the Western Mediterranean. *Dis. Aquat. Org.*, 35, 37-46.

[21] Sandlund, N., Torkildsen, L., Magnesen, T., Mortensen, S., Bergh, Ø., 2006. Immunohistochemistry of great scallop *Pecten maximus* larvae experimentally challenged with pathogenic bacteria. *Dis. Aquat. Org.*, 69, 163-173.

[22] Soto, M., Pascual, S., Rodríguez, H., Gestal, C., Abollo, E., Arias, C., Estévez, J., 1996. *Nematopsis* spp. Schneider, 1892 (Apicomplexa: Gregarinida) in bivalve molluscs off Ría de Vigo (Galicia, NW Spain). *Bull. Eur. Assoc. Fish Pathol.*, 16, 157-160.

[23] Tuntiwaranuruk, C., Chalermwat, K., Pongsakchat, V., Meepool, A., Upatham, E. S., Kruatrachue, M., 2008. Infection of *Nematopsis* oocysts in different size classes of the farmed mussel *Perna viridis* in Thailand. *Aquaculture*, 281, 12-16.

In: Oysters and Clams
Editor: Jesús L. Romalde

ISBN: 978-1-63485-074-2

Chapter 6

SPATIAL AND TEMPORAL DISTRIBUTION OF *VIBRIO VULNIFICUS* IN THE DIFFERENT MOLLUSCS CULTIVATED IN THE EBRO DELTA

Carmen Lopez-Joven[1,*], Ignacio de Blas[2] and Ana Roque[1,†]

[1]Institute of Agriculture and Food Research and Technology (IRTA-SCR), Tarragona, Spain

[2]Department of Animal Pathology, Facultad de Veterinaria, Universidad de Zaragoza, Zaragoza, Spain

ABSTRACT

Prevalence and characteristics of *Vibrio vulnificus* in the environment are unclear, which limits the knowledge about this pathogen and the expression of its virulence. The diversity of *Vibrio* species associated with bivalves and their abundance in water is seasonally variable, reaching its maximum in summer when temperature is at its highest. Bivalve molluscs are considered an important source of food, but they can also transmit food-borne illnesses when they are eaten raw or undercooked, due to their filter-feeding mechanism where they can naturally accumulate marine biotoxins, viruses, chemicals and pathogenic bacteria such as *V. vulnificus*. To better understand the ecology of *V. vulnificus*, the distribution and prevalence of this bacterium was investigated over four campaigns in two bays of the Ebro delta, in four species of commercially important bivalve molluscs. Environmental factors such as temperature and salinity were taken into account. The results showed that the global prevalence of *V. vulnificus* in both bays was 10.9% (159/1459). *Vibrio vulnificus* was present every sampling years but not in all samples molluscs. Using multivariate analysis, a significant logistic regression model to explain the presence/absence (P/A) of *V. vulnificus* in bivalve molluscs was generated, and it allowed us to establish that the presence of *V. vulnificus* depends on molluscs, year of harvest, and temperature. Results of the present chapter contribute to the production of

* Present address: Instituto de Medicina Preventiva Veterinaria, Facultad de Ciencias Veterinarias, Universidad Austral de Chile, Edificio Saelzer 5º piso, Campus Isla Teja, CC567, Valdivia, Chile.

† Corresponding Author, Email: ana.roque@irta.es.

data body to be used in quantitative risk analysis for this important human, eel and fish pathogen.

Keywords: prevalence, *Vibrio vulnificus*, bivalve molluscs, ebro delta, water temperature, salinity

INTRODUCTION

Bivalve molluscs are considered an important source of food, especially in populations living near the sea. However, they represent a public health challenge since they are generally eaten raw or undercooked and their filter feeding nature contributes to them accumulating a number of microorganisms such as bacteria and viruses which enter from the surrounding water [1]. Pathogenic *Vibrios* species can represent one of the major marine-borne biological risks that reach human food web through shellfish consumption [2, 3]. In particular, *V. vulnificus,* Gram-negative halophilic bacterium, which is a natural inhabitant of temperate estuarine and marine waters all over the world [4-6], may cause three clinical presentations: gastroenteritis, severe wound infections and septicemia that can reach fatal consequences [7, 8]. Gastroenteritis or severe wound infections, leading to necrotizing cellulitis can happened in healthy individuals, while septicemia, is developed normally in immunocompromised individuals [9-13]. For the gastroenteritis, it is known that though no figure has been estimated in healthy individuals, an infectious dose of less than 100 bacteria is needed in predisposed individuals. Symptoms appear 16 to 38h after the ingestion of contaminated food, and the mortality rate is very high (60%) in people with liver disease or immunocompromised [14].

The literature published about classification of *V. vulnificus* populations has sometimes been controversial [15-17]. Strains of *V. vulnificus* can be phenotypically (biochemically) and genetically differentiated depending on their characteristics. Phenotypically, three biotypes have been described, all of which are pathogenic: biotype 1 was the first biotype described which is present in almost all human infections being the most common worldwide [18, 19], biotype 2 mainly affecting eels and causes rapidly fatal septicemia. On rare occasions, it has been isolated from human cases [20-22], and biotype 3, was the most recently described as a secondary human wound infections pathogen associated with tilapia aquaculture, which has been reported only in Israel [23, 24]. Genetically, two main clusters have been considered, one cluster more likely to cause clinical infections (C-genotype) and another cluster, less infectious, dominated by strains from environmental sources and shellfish (E-genotype) [25-31].

Although worldwide, the impact of outbreaks of diarrhea caused by eating bivalve molluscs is widely recognized, European monitoring programs on the quality of seafood, do not legislate for contamination by vibrios in these organisms. For this lack of legislation, several reasons have been pointed out and one of them is the lack of data on the frequency/presence of these pathogens in the water and food. At European Union level, the opinion issued by "The Scientific Committee on Veterinary Measures relating to Public Health" found that although *V. parahaemolyticus* and *V. vulnificus* are important pathogens for public health, there are several points that prevent the legislation, one being the lack of data on the prevalence of these species in the U. E [32].

Spain is considered an important producer and consumer of bivalve molluscs in Europe and is the second largest worldwide [33, 34]. More than 60 thousand tons of bivalves are consumed per year [35], raw or cooked. Moreover, Catalonia is one of the main areas of aquaculture production; and, farming of bivalves is concentrated in two bays (Alfacs and Fangar) in the delta of the Ebro River, where it represents one of the most important economic activities. The four main species harvested are oysters (*Crassostrea gigas)*, mussels (*Mytilus galloprovincialis*) and two species of clam: grooved carpet shell (*Ruditapes decussatus*), and Manila clam (*R. philippinarum*). Salinity, temperature and pH moderately alkaline of the two bays support the growth of vibrios [36]; annual temperature and salinity vary between 7 and 31°C and from 16 to 38 ppt in Alfacs Bay, respectively; and between 6 to 30°C and from 21 to 38 ppt in Fangar Bay.

Considering all such information, present chapter seeks to examine the spatial distribution and temporal changes of *V. vulnificus* in Ebro delta, Spain. In addition, it is one of the few studies which have focused on the examination of large numbers of oysters, mussels, and clams aiming to investigate the prevalence and spatial distribution and temporal change of *V. vulnificus* in bivalves in the two bays of Ebro delta, and to determine its relationship with environmental parameters (temperature and salinity) from the surrounding waters. Although, temperature effects were reduced due to the fact the study was conducted during the months of July and August (when temperature is higher), differences were still detected.

METHODS

Sampling Sites and Collection Samples

From 2006 to 2010, four surveillance campaigns for pathogenic *V. vulnificus* detection in commercial bivalves from the Ebro delta bays were performed.

Initially, sample sizes were calculated according to different criteria for each species, so minimum expected prevalence for clams (10%) was higher than oysters and mussels (5%), and confidence level (CL) was establish between 95 and 99%, taking into account the sample sizes, and considering that some bivalves are eaten raw or lightly cooked. Thus, CL for oysters was incremented to 99% to ensure detection of pathogens since normally this species is consumed raw. However, sample sizes were changed along the study and in 2010, we reduced confidence level for oysters due to economic reasons. Moreover, some deviations in sample sizes from surveillance scheme were originated as consequence of low number of adult specimens for collection (see Lopez-Joven et al. [37] for details).

A total of 1459 species of bivalves were sampled: 709 and 750 from Alfacs and Fangar bays, respectively. From Alfacs bay, 283 were oysters (*C. gigas*), 306 were mussels (*M. galloprovincialis*) and 120 clams (*Ruditapes* spp.) were processed. And, from Fangar, 330 oysters, 300 were mussels and 120 clams were processed. Similar batches for each species (*M. galloprovincialis, C. gigas,* and *Ruditapes* spp.) were harvested for each year of study. Oysters' samples were 180, 178, 180, and 120 in 2006, 2008, 2009 and 2010, respectively. Mussels samples were 127, 179, 180 and 120 in 2006, 2008, 2009 and 2010, respectively, and clams samples were 60 for each year.

Mussel and oysters were sampled from six sites of the culture area, three on each bay, at the beginning (N40º37'112"E0º37'092"-Alfacs; N40º46'723"E0º43'943"-Fangar), middle (N40º37'125"E0º38'570"-Alfacs; N40º46'666"E0º45'855"-Fangar) and end (N40º37' 309"E0º39'934"-Alfacs; N40º46'338"E0º44'941"-Fangar) of the culture polygons. Clams were sampled from only one site per bay, in the Alfacs bay from a natural bed of *R. decussatus* (N40º37'44"E0º38'0") and in the Fangar bay from an aquaculture bed of *R. philippinarum* (N40º47'3"E0º43'8") (Figure 1). All bivalves were collected in July and August of each sampling year. Bivalves were transported in cool conditions. Transport lasted less than 1 hour in all occasions.

Temperature and salinity data were recorded using a CTD datalogger (Seabird, USA) located in the middle point of the area of the bivalves culture raft, at a depth of 2 m and recording data every 60 min.

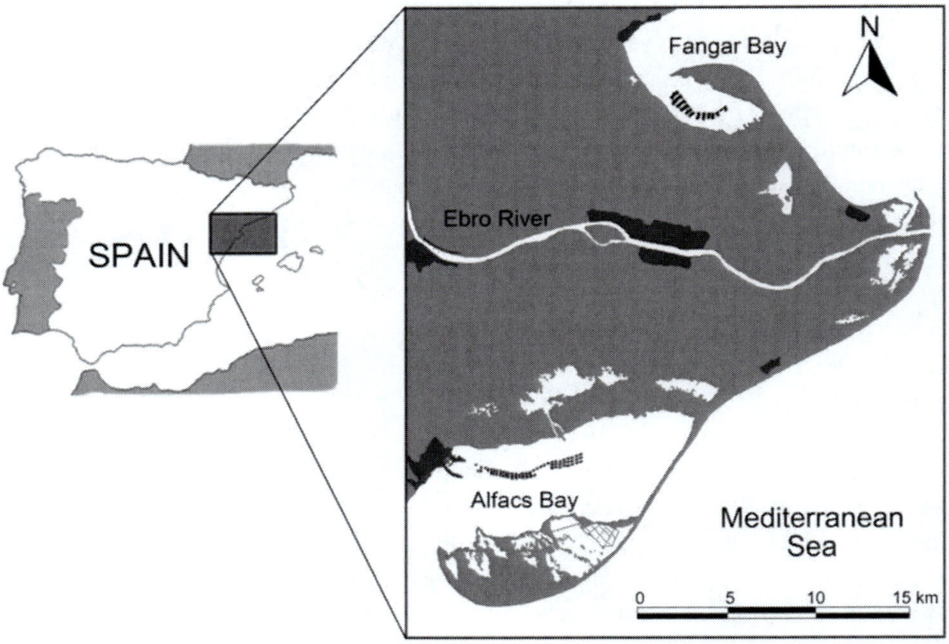

Figure 1. Map of the study area showing the location of the sampling sites in the Ebro Delta, Spain.

Microbiological and Molecular Analysis

Microbiological analysis was carried out as is described by Roque et al. [38], in order to obtain bacterial isolates to be tested by one-step PCR. All bivalves were individually processed and homogenized for 30 s using an ultra-turrax (IKA, T25 basic). Then, 1 ml of the homogenate was inoculated into 9 ml of alkaline peptone water (Scharlau, Spain). Following a 6-h incubation at 37°C, one loopful of the contents of each tube of alkaline peptone water was streaked onto CHROMagar vibrio plates (CHROMagar, France) and incubated for 18 h at 37°C. Blue turquoise colonies were purified, and each purified isolate was cryopreserved at -80°C.

Total DNA was extracted from each purified isolate using the Wizard genomic DNA purification kit (Promega), following the instructions of the manufacturer. DNA concentration was verified by spectrophotometry and the concentration of each DNA was adjusted to 50 ng μl^{-1}. A one-step PCR analysis was performed to identify/confirm *V. vulnificus* identity through the detection of a fragment of the gene citolysin-haemolysin. PCRs analyses were performed according to the protocol of Coleman et al. [39], which identify a portion of the 1,416-bp *cytolysin-hemolysin* gene that is unique to *V. vulnificus* (as species marker), with the following amplification conditions on the thermocycler (Eppendorf Mastercycler Personal, Hamburg, Germany): initial denaturation 3 min at 94ºC, 30 cycles of amplification, each one consisting of a denaturation at 94ºC for 45 sec, annealing of the primers at 55ºC for 45 sec, and extension at 72ºC for 45 sec, with a final extension of 2 min at 72ºC. Positive and negative controls were included in all reaction mixtures: positive control CAIM 610T (ATCC 27562 LMG 13545) [40, 41] [Collection of Aquatic Important Microorganisms (http://www.ciad.mx/caim/CAIM.html)], and negative control DNA-free molecular grade water (Sigma-Aldrich, Spain). Expected amplicons (344 bp) were visualized in 2% agarose gels stained with ethidium bromide.

Statistical Analysis

Pearson's Chi-Square test was carried out to assess the influence of host species of mollusc, location (Alfacs or Fangar bays) and year of sampling on prevalences (proportion of molluscs contaminated by *V. vulnificus*).

A logistic regression analysis was performed using *V. vulnificus* contamination as dependent variable and year, location, mollusc species, salinity and temperature as independent variables. Forward stepwise method was applied, and significance of the model was evaluated with omnibus test. Using coefficients of the model (Bx) Odds Ratio ($OR=e^{Bx}$) were calculated to identify significant risk factors.

Desired alpha error was established at 0.05. Statistical analysis was performed using SPSS 19.0 software (Chicago, IL, USA). All data collected represents data points and not a continuous variable.

RESULTS

A thousand four hundred and fifty nine bivalve molluscs were processed during four surveillance campaigns in the Alfacs and Fangar bays to assess the prevalence of *V. vulnificus*. Moreover, temperature and salinity conditions of the water in sampling sites during each campaign (Table 1) were analysed to identify potential significant risk factors.

Global Prevalence of *V. vulnificus*

Different trends in *V. vulnificus* prevalence and their relationship with water temperature (ºC) and salinity (ppt) from sampling sites are shown in Figure 2. Overall, in the two bays, *V.*

vulnificus was detected in 159 (10.9%) of the 1459 samples identified by a portion of the 1,416-bp *cytolysin-hemolysin* gene. As shown in Table 2, when statistical analysis was performed in order to compare prevalence of *V. vulnificus* among sampling sites (Alfacs and Fangar bays), they were found not significantly different (P = 0.902), being Alfacs' prevalence 11.0%, (78/709) and Fangar's prevalence 10.8%, (81/750). However, when this analysis was stratified by year, significant differences were observed between sampling bays: in 2006 (P = 0.003), the proportion of bivalves molluscs from Alfacs that contained *V. vulnificus* (9.6%, 187/709) was higher than in Fangar (2.2%, 180/750), and in 2010 (P = 0.001), the proportion of bivalves molluscs from Fangar that contained *V. vulnificus* (23.3%, 150/750) was higher than in Alfacs (8.1%, 135/709). However, no effect of sampling site was detected neither in 2008 (P = 0.313), nor in 2009 (P = 0.684). When prevalences of *V. vulnificus* were analysed in each bay over the sampling period, it was observed that the prevalences in Alfacs Bay were significantly different (P = 0.005) over the studied period due mainly to prevalence's increase in 2008: 2006 (9.6%), 2008 (17.4%), 2009 (7.2%), 2010 (8.1%) (Figure 2A). The prevalence of *V. vulnificus* in Fangar bay was also significantly different among years (P < 0.001), but fluctuating over time: 2006 (2.2%), 2008 (13.8%), 2009 (6.2%), and 2010 (23.3%) (Figure 2B).

Table 1. Temperature and salinity conditions of the sampling locations during each campaign (n = 3). Each value (measurement of a specific day) is the product of 24 measurements, expressed as average ± standard deviation

	Temperature (°C)		Salinity (ppt)	
Campaign	Alfacs	Fangar	Alfacs	Fangar
2006	27.67 ± 2.34	25.41 ± 1.57	35.50 ± 0.50	36.62 ± 0.66
2008	26.93 ± 0.68	26.37 ± 1.17	35.17 ± 1.76	36.50 ± 0.50
2009	27.54 ± 0.85	26.02 ± 0.17	36.01 ± 0.27	35.74 ± 0.90
2010	27.05 ± 1.07	25.74 ± 0.44	35.78 ± 0.42	37.04 ± 0.52

Prevalences of *V. vulnificus* stratified by years and by species of bivalves molluscs (without stratification by bay) are shown in Table 3. Statistical analysis indicated that significant differences were found among species of bivalve molluscs (P < 0.001) along the time: where prevalence in oysters (*C. gigas*) 9.0% (55/613) and in mussels (*M. galloprovincialis*) 8.4% (51/606) were lower than in both species of clams: *R. decussatus* 21.7% (26/120), and in *R. philippinarum* 22.5% (27/120). Regarding prevalences of *V. vulnificus* within each bivalve molluscs, independently, significant differences were detected within all bivalves along the time: oysters (P < 0.001), mussels (P = 0.006), and within both species of clams, *R. decussatus* (P = 0.032), and *R. philippinarum* (P = 0.002). In detail, higher prevalences of contamination by *V. vulnificus* were found in *R. decussatus* in 2006 (36.7%), and in *R. philippinarum*, in 2008 (33.3%), and in 2010 (40.0%). When stratified analysis was performed by year, significant differences among prevalences were found in 2006 ($P < 0.001$), in 2008 ($P = 0.018$), and in 2010 ($P = 0.002$), where higher prevalences of *V. vulnificus* were recorded in both species of clams, *R. decussatus* in 2006 and in *R. philippinarum* in 2008 and 2010, as it was showed above.

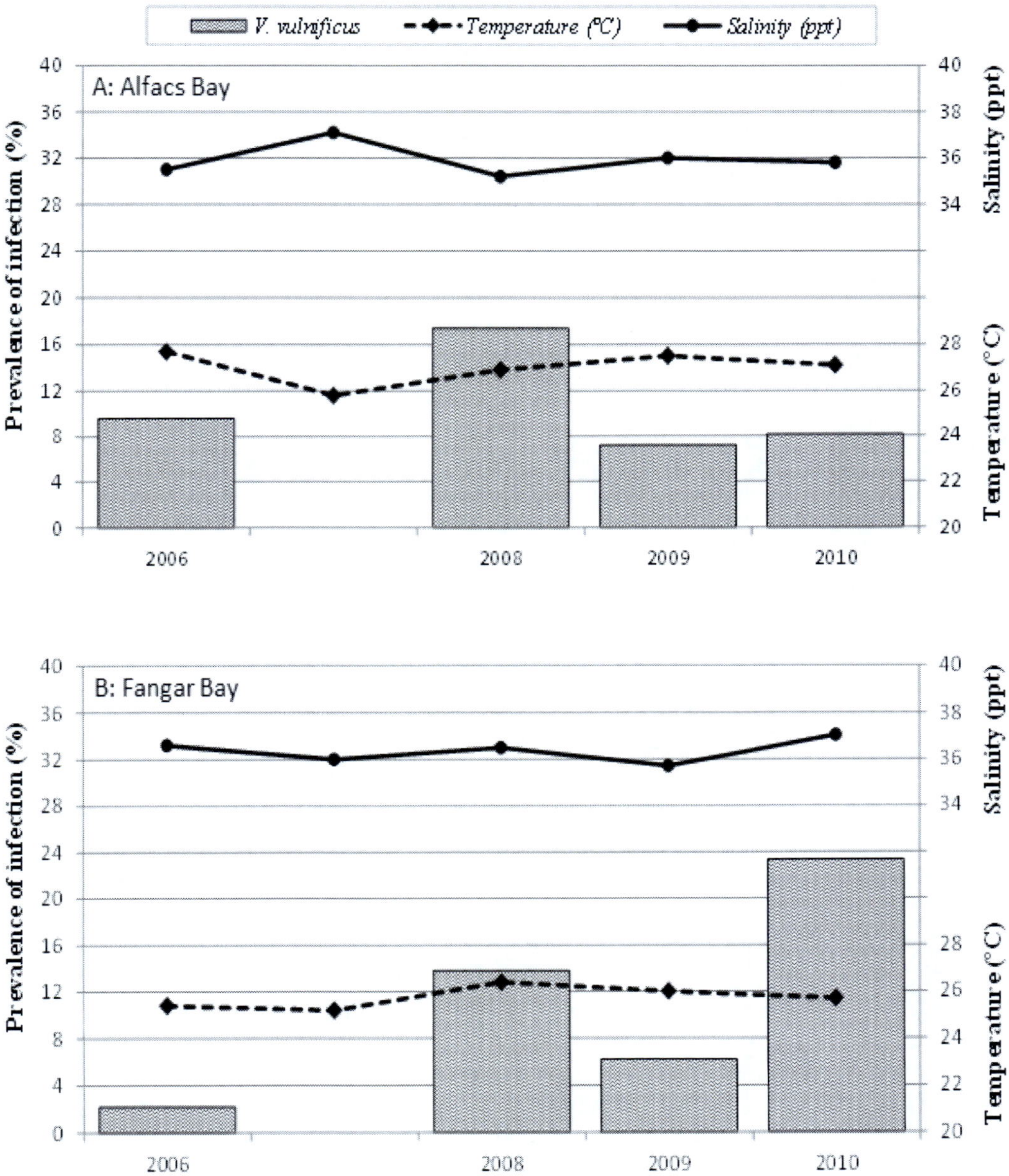

Figure 2. Prevalence of *Vibrio vulnificus* in (A) Alfacs Bay, and (B) Fangar Bay. Each bar represents the arithmetic mean of four species of bivalves averaged in sampling months. Solid line represents salinity (ppt); dashed line, water temperature (°C) during each campaign.

Table 2. Prevalence of *Vibrio vulnificus*, stratified by years and by bays

	2006		2008		2009		2010		molluscs	Prevalence *Vv*		
	n	%	n	%	n	%	n	%	n	n	***Vv* %**	*P*-value[2]
Alfacs	187	9.6%	207	17.4%	180	7.2%	135	8.1%	709	78	11.0%	**0.005**
Fangar	180	2.2%	210	13.8%	210	6.2%	150	23.3%	750	81	10.8%	**<0.001**
Total	**367**	**6.0%**	**417**	**15.6%**	**390**	**6.7%**	**285**	**16.1%**	1459	159	10.9%	**<0.001**
P-value[1]	**0.003**		**0.313**		**0.684**		**0.001**			**0.902**		

[1] P-value: Prevalence of *V. vulnificus* (*vvhA*) stratified by years.

[2] Prevalence of *V. vulnificus* (*vvhA*), stratified by bays.

Table 3. Prevalence of *Vibrio vulnificus* stratified by years and by species of bivalve molluscs

	2006		2008		2009		2010		Sampled molluscs	Prevalence *Vv*		
	n	%	n	%	n	%	n	%	n	n	%	*P-value*[2]
Crassostrea gigas	180	1.1%	178	12.4%	150	8.7%	105	17.1%	**613**	**55**	9.0%	**<0.001**
Mytilus galloprovincialis	127	3.9%	179	14.0%	180	5.6%	120	9.2%	**606**	**51**	8.4%	**0.006**
Ruditapes decussatus	30	36.7%	30	26.7%	30	6.7%	30	16.7%	**120**	**26**	21.7%	**0.032**
Ruditapes philippinarum	30	13.3%	30	33.3%	30	3.3%	30	40.0%	**120**	**27**	22.5%	**0.002**
Total	**367**	**6.0%**	**417**	**15.6%**	**390**	**6.7%**	**285**	**16.1%**	**1459**	**159**	10.9%	**<0.001**
P-value[1]	**<0.001**		**0.018**		**0.591**		**0.002**			**<0.001**		

[1] *P*-value: Prevalence of *V. vulnificus (vvhA)*, stratified by years.

[2] *P*-value: Prevalence of *V. vulnificus (vvhA)*, stratified by species of molluscs.

Table 4. Prevalence of *Vibrio vulnificus*, stratified by years, by bays, and by species of bivalve

	2006				2008				2009				2010			
	Alfacs		Fangar		Alfacs		Fangar		Alfacs		Fangar		Alfacs		Fangar	
	n	*Vp* %	n	*Vp* %	n	*Vp* %	n	*Vp* %	n	*Vp* %	n	%	n	%	n	%
Crassostrea gigas	90	2.2%	90	0.0%	88	15.9%	90	8.9%	60	10.0%	90	7.8%	45	11.1%	60	21.7%
Mytilus galloprovincialis	67	7.5%	60	0.0%	89	15.7%	90	12.2%	90	5.6%	90	5.6%	60	1.7%	60	16.7%
Ruditapes decussatus	30	36.7%	-	-	30	26.7%	-	-	30	6.7%	-	-	30	16.7%	-	-
Ruditapes philippinarum	-	-	30	13.3%	-	-	30	33.3%	-	-	30	3.3%	-	-	30	40.0%
Total	**187**	**9.6%**	**180**	**2.2%**	**207**	**17.4%**	**210**	**13.8%**	**180**	**7.2%**	**210**	**6.2%**	**135**	**8.1%**	**150**	**23.3%**
P-value [1]	**<0.001**		**0.001**		**0.350**		**0.003**		**0.595**		**0.646**		**0.022**		**0.044**	

[1] *P*-value: Prevalences of *V. vulnificus* (*vvhA*) stratified by years and by bays.

When statistical analysis was performed to study prevalences of *V. vulnificus*, stratified by years, by species of bivalves and by bays, results (see Table 4) indicated that significant differences were found among species of bivalves in Alfacs bay in 2006 ($P < 0.001$), and in 2010 ($P = 0.022$), where prevalence in *R. decussatus* was 36.7%, and 16.7%, respectively, higher than in *C. gigas* or *M. galloprovincialis*, but not in 2008 ($P = 0.350$) and 2009 ($P = 0.595$). In Fangar Bay, this analysis indicated that *V. vulnificus* was not present in all samples molluscs, e.g., *V. vulnificus* was neither detected in *C. gigas* nor in *M. galloprovincialis* in 2006. And, prevalences among species of bivalves in Fangar bay, were different in 2006 ($P = 0.001$), where only *V. vulnificus* was found in *R. philippinarum* (13.3%), and in 2008 ($P = 0.003$) and in 2010 ($P = 0.044$), where prevalences of *V. vulnificus* were higher in *R. philippinarum* (33.3% and 40.0%, respectively). No significant differences ($P = 0.646$) were found in prevalences of *V. vulnificus* in mollucs in 2009.

In summary, in Alfacs Bay the year 2008 was found to have higher prevalences of *V. vulnificus* than the other years studied (17.4%), while in Fangar Bay, higher prevalences of *V. vulnificus* were recorded in 2010 (23.3%).

Multivariate Analysis of Influence of Different Factors on the Presence of *V. vulnificus*

A logistic regression with Forward Stepwise method was used to analyse the relationship between the prevalence of *V. vulnificus* with species of bivalves, shell length (mm), year of harvest, location (sampling bay), water temperature, and salinity.

A significant logistic regression model that included bay to explain the presence/absence (P/A) of *V. vulnificus* in bivalves was generated ($P < 0.001$) (see Table 5 for details), and it established that the presence of *V. vulnificus* depends on species of molluscs, year of harvest and water temperature.

Results showed that the risk of finding *R. decussatus* and *R. philippinarum* containing *V. vulnificus* was higher than for other species of bivalves. The probability of finding *V. vulnificus* in *R. decussatus* and in *R. philippinarum* were 2.22 and 3.28 times higher than in *C. gigas,* respectively ($P = 0.006$, and $P < 0.001$, respectively).

Table 5. Description of the logistic regression model selected to explain the presence/absence (P/A) of (presence of *vvhA* gene) *Vibrio vulnificus* including location, species of bivalve, year of campaign, length, temperature and salinity. (Method: Forward Stepwise Conditional)

Variables included in the selected model	B	OR (e^B)	*P*-value
R. decussatus vs *C. gigas*	0.799	2.223	0.006
R. philippinarum vs *C. gigas*	1.187	3.279	<0.001
2008 vs 2006	1.097	2.996	<0.001
2010 vs 2006	1.130	3.096	<0.001
Temperature (ºC)	0.186	1.205	0.033
Constant	− 7.926	<0.001	0.001

Model summary: -2 Log likelihood = 934.598; Nagalkerke R Square = 0.094; Omnibus tests of model coefficients, $P < 0.001$.

Results also showed that year of sampling was significant using as reference year 2006. Therefore, the risk (expressed as Odds Ratio) to find *V. vulnificus* varied along time. Results obtained in 2008 and in 2010, respect to 2006 indicated that the risk to find *V. vulnificus* in bivalves was 3.00 and 3.10 times higher in 2008 and 2010, respectively, than 2006 ($P <$ 0.001, in both cases). Presence of *V. vulnificus* was significantly associated with water temperature ($P = 0.033$). The risk to find bivalves containing *V. vulnificus* increased 1.21 times with each unit (ºC) that water temperature increased.

On the other hand, results showed that shell length of molluscs ($P = 0.802$), location ($P =$ 0.545), and salinity ($P = 0.823$) had no significant effect in this logistic regression model (data not shown).

DISCUSSION

Present chapter includes new data about the prevalences of *V. vulnificus* in the two Ebro delta bays, in the Mediterranean coast. Results were obtained after examining intraseasonal relationships between selected environmental parameters (temperature and salinity) and the prevalences of *V. vulnificus* in four different species of bivalves (*C. gigas*, *M. galloprovincialis, R. decussatus* and *R. philippinarum*) cultured in the Ebro delta in four different years.

In this chapter temperature and salinity conditions of the bays during each campaign did not suffer big fluctuations since all campaigns took place during the summer (July and August), when vibrios are more frequently present and in higher numbers. Although no significant difference was found when both bays were global analysed, differences were found between both bays when the model was stratified by year, showing differences in 2006 and in 2010. Examination of data indicated that Alfacs Bay presented higher temperatures (ranged between 26.93 and 27.67ºC) than Fangar Bay (where, water temperature vary from 25.41 to 26.37°C), due to the basin volume in Alfacs is about ten times larger than in Fangar, needing more time to renew its water [36, 42]. The logistic regression model (which included location, species of bivalve, year of campaign, shell length, temperature and salinity) showed the risk of finding *V. vulnificus* is correlated with water temperature. This information is agreement with other studies which indicate that the highest abundances of *V. vulnificus*, both in the water column and in association with shellfish, occur in warm waters (typically exceeding 20°C) [43, 44]. Therefore, this parameter could explain the higher prevalences of *V. vulnificus* in Alfacs bay in 2006, taking into account that temperature in this bay is 1 or 2ºC higher than in Fangar bay. However, temperature could not explain the higher prevalences of *V. vulnificus* in the other bay (Fangar bay) in 2010, where temperature was lower than Alfacs. Then, correlations with other parameters are needed.

Salinity is also an important parameter to growth of human pathogenic *Vibrio* spp. and their proliferation in the environment. Moderate salinity (ranging from 5 to 25 ppt) [43, 44] or decrease in salinity caused by heavy rainfall and major flooding (ranged 10 and 15‰) favoured growth of *V. vulnificus* [45]. Data in this chapter showed Alfacs bay presented lower salinities than Fangar bay (around 1 or 2 ppt lower), caused by differential evaporation rates and freshwater inputs from nearby agricultural (rice) fields [42, 46]. However, salinities range in this summer period are far those optimal for growing *Vibrio* spp. ranging between 35.17 and 36.01 ppt in Alfacs and from 35.74 to 37.04 ppt. in Fagar bay.

Several studies have demonstrated that when salinities increased *V. vulnificus* levels decreased. Kaspar and Tamplin [43] in seawater microcosms demonstrated when salinities increased at 30, 35, and 38 ppt, *V. vulnificus* levels decreased by 58, 88, and 83%, respectively. And, the same adverse trend was reported by Motes et al. [44] in a study carried out in oysters collected at three northern Gulf Coast sites and two Atlantic Coast sites from July 1994 through September 1995, where they observed lower numbers of *V. vulnificus* at salinities above 28 ppt, which is typical of some Atlantic coastal sites in North and South Carolina. Wright et al. [47] in field surveys, showed that the abundance of *V. vulnificus* in the Chesapeake Bay was inversely correlated to salinity. Lipp et al. [48], found that the abundance of *V. vulnificus* in the Gulf of Mexico was positively correlated to salinity below 15 ppt and inversely correlated to salinity when the salinity was greater than 15 ppt. Randa et al. [49] reported elevated levels of *V. vulnificus* at 5-10 ppt, along with a positive correlation at higher salinities (20-25 ppt). And, Jacobs et al. [50] suggested an optimal salinity of 11.5 ppt, with abundance declining with distance from optimal. Finally, others studies did not reveal a significant relationship between *V. vulnificus* abundance and salinity [51, 52]. Thus, the relationship between salinity and *V. vulnificus* is not clear. While most studies support a mesohaline salinity distribution, *V. vulnificus* has been isolated from higher salinity environments [53, 54].

Data shown in this chapter indicate *V. vulnificus* has been isolated in bivalves harvested from waters of Ebro delta with high salinities above 35 ppt in summer season. So, salinities of >25 ppt do not suppress *V. vulnificus* levels totally even in warm waters, although it can decrease their numbers. In addition, the results show the risk of finding *V. vulnificus* in summer season is not correlated with salinity. However, more studies in different seasons should be carry out to give better understanding of this phenomenon in the Ebro delta.

Several studies have attempted to determine correlations of abundance or presence of *V. vulnificus* with water temperature, and to a lesser extent with salinity, though due to the weakness of the correlation of *V. vulnificus* concentrations with salinity this relationship is still unclear [43, 47, 51, 52, 55, 56]. Understanding the relationship between *V. vulnificus* and climatic parameters like temperature or salinity may contribute to forecasting the probable concentrations of the bacterium along coastal waters as well as among potential hosts [57].

While positive correlation with water temperature is a consistent finding in the literature (generally at > 20°C), the relationship of *V. vulnificus* with salinity has been more variable and model interpretation should be limited to studies areas. But, these are not the only parameters that influence the bacterium's abundance and distribution. Factors such as plancton composition, dissolved oxygen, particulate organic matter availability, and chlorophyll a, presence of fish and shellfish, as well as levels of freshwater inputs could influence the ecology of this bacterium. Vanoy et al. [58] reported *V. vulnificus* to range from <0.1 to >10^5 cfu/g of sediments taken from Galveston Bay in Texas. Wright et al. [47] found *V. vulnificus* in plancton in the Chesapeake Bay, suggesting that this bacterium may inhabit habitats similar to *V. cholerae* and *V. parahaemolyticus*. DePaola et al. [59] isolated *V. vulnificus* throughout the winter months from the intestines of estuarine fish from the Gulf of Mexico, at densities higher than those found in oysters, sediment, or seawater.

In addition, the results obtained in the stratified analysis by year and by species, where higher prevalences of *V. vulnificus* were recorded in *R. decussatus* (2006) and in *R.*

philippinarum (2008 and 2010), it would suggest a potential host effect. Similar results were found when *V. parahaemolyticus* prevalences were investigated and higher prevalences of *V. parahaemolyticus* in *R. decussatus* and in *R. philippinarum* were recorded [37].

Conclusion

In conclusion, this chapter evidences temporal and spatial variations in the prevalences of *V. vulnificus* in both bays of the Ebro Delta and in association with bivalve molluscs. Apparently, *V. vulnificus* populations in bivalves are controlled quantitatively and qualitatively by different factors. Higher prevalences of *V. vulnificus* were recorded in *R. decussatus* and in *R. philippinarum*, which in this area may indicate either a potential host effect or an effect due to culture systems, since oysters and mussels are grown in suspended ropes, clams are grown on the floor beds. It is also possible that the filtration rate of clams, under the sampling conditions, were closer to optimal when compared to mussels and oysters which would contribute to higher filtration rates in clams with consequent accumulation of bacteria.

Present chapter collected a considerable amount of data on the presence of *V. vulnificus* over a long period of time which makes these data robust. Countries like Australia and the USA have legislation for the safety of bivalves for consumption which includes pathogenic vibrios indicating that present data can be very useful when performing a risk analysis for assessing the consequences of *V. vulnificus* presence in commercial bivalves in the Mediterranean countries and others areas with similar environmental conditions of those of the Mediterranean coast like Central Chile and the western cape of South Africa.

Acknowledgments

The authors acknowledge the financial support of the projects RTA2007-00063-00-00 awarded to AR from the Instituto Nacional de Investigación y Tecnología Agraria y Alimentaria (INIA, Spain) with FEDER funds. Dr. C. Lopez-Joven acknowledges the FPI grant from INIA. Our sincere thanks to Beatriz Lacuesta for technical support and staff of the IRTA SMM for assistance in collecting field samples.

References

[1] Lopez-Joven, C., Roque, A., Furones, M. D., Giménez Papiol, G., (2014). Biological Risks Related to Shellfish Consumption. In: Hay, R. M. (Ed.) *Shellfish: Human Consumption, Health Implications and Conservation Concerns*. NOVA Science Publishers, Inc. Hauppauge, New York, pp: 211-242.

[2] Blake, P. A., Weaver, R. E., Hollis, D. G., (1980). Diseases of humans (other than cholera) caused by *Vibrios*. Annu. Rev. Microbiol. 34, 341-367.

[3] Austin, B., (2010). Vibrios as causal agents of zoonoses. *Vet. Microbiol.* 140, 310-317.

[4] Oliver, J.D., Warner, R.A., Cleland, D.R., (1982). Distribution and ecology of *Vibrio vulnificus* and other lactose-fermenting marine vibrios in coastal waters of the southeastern United States. *Appl. Environ. Microbiol.* 44, 1404-1414.

[5] Harwood, V.J., Gandhi, J.P., Wright, A.C., (2004). Methods for isolation and confirmation of *Vibrio vulnificus* from oysters and environmental sources: a review. *J. Microbiol. Methods.* 59, 301-316.

[6] Broza, Y.Y., Danin-Poleg, Y., Lerner, L., Valinsky, L., Broza, M., Kashi, Y., (2009). Epidemiologic study of *Vibrio vulnificus* infections by using variable number tandem repeats. *Emerg. Infect. Dis.* 15, 1282-1285.

[7] Oliver, J.D., Kaper, J.B., (1997). *Vibrio* species. In: Doyle, M.P., Beuchat, L.R., Montville, T.J. (Eds.) *Food Microbiology: Fundamentals and Frontiers.* ASM Press, Washington D.C., pp: 228-264.

[8] Jones, M.K., Oliver, J.D., (2009). *Vibrio vulnificus*: disease and pathogenesis. *Infect. Immun.* 77, 1723-1733.

[9] Hlady, W.G., Klontz, K.C., (1996). The epidemiology of *Vibrio* infections in Florida, 1981-1993. *J. Infect. Dis.* 173, 1176-1183.

[10] Shapiro, R.L., Altekruse, S., Hutwagner, L., Bishop, R., Hammond, R., Wilson, S., Ray, B., Thompson, S., Tauxe, R., Griffin, P.M., (1998). The role of Gulf Coast oysters harvested in warmer months in *Vibrio vulnificus* infections in the United States, 1988-1996. *J. Infect. Dis.* 178, 752-759.

[11] Tantillo, G.M., Fontanarosa, M., Di Pinto, A., Musti, M., (2004). Updated perspectives on emerging vibrios associated with human infections. *Lett. Appl. Microbiol.* 39, 117-126.

[12] Oliver, J.D., (2005). Wound infections caused by *Vibrio vulnificus* and other marine bacteria. *Epidemiol. Infect.* 133, 383-391.

[13] Horseman, M.A., Surani, S., (2011). A comprehensive review of *Vibrio vulnificus*: an important cause of severe sepsis and skin and soft-tissue infection. *Int. J. Infect. Dis.* 15, e157-166.

[14] Centers for Disease Control and Prevention (CDC). *Vibrio illness (Vibriosis).* Atlanta, GA. Last reviewed Oct 21, 2013. Available at: http://www.cdc.gov/vibrio/vibriov.html (accessed October 17, 2015).

[15] Strom, M.S., Paranjpye, R.N., (2000). Epidemiology and pathogenesis of *Vibrio vulnificus. Microbes Infect.* 2, 177-188.

[16] Chiang, S.R., Chuang, Y.C., (2003). *Vibrio vulnificus* infection: clinical manifestations, pathogenesis, and antimicrobial therapy. *J. Microbiol. Immunol. Infect.* 36, 81-88.

[17] Gulig P.A., Bourdage K.L., Starks A.M., (2005). Molecular pathogenesis of *Vibrio vulnificus. J. Microbiol.* 43, 118-131.

[18] Oliver, J.D., (1989). *Vibrio vulnificus*, In: Doyle M. P. (Eds.) *Food-Borne Bacterial Pathogens*. pp: 569-600. Marcel Dekker, Inc. New York, NY.

[19] Oliver, JD., (2015). The biology of *Vibrio vulnificus*, Microbiol Spectrum 3, VE-0001-2014.

[20] Tison, D.L., Nishibuchi, M., Greenwood, J.D., Seidler, R., (1982). *Vibrio vulnificus* biogroup 2: new biogroup pathogenic for eels. Appl. Environ. Microbiol. 44, 640-646.

[21] Amaro, C., Biosca, E.G., Fouz, B., Alcaide, E., Esteve, C., (1995). Evidence that water transmits *Vibrio vulnificus* biotype 2 infections to eels. *Appl. Environ. Microbiol.* 61, 1133-1137.

[22] Amaro, C., Biosca, E.G., (1996). *Vibrio vulnificus* biotype 2, pathogenic for eels, is also an opportunistic pathogen for humans. *Appl. Environ. Microbiol.* 62,1454-1457.

[23] Bisharat, N., Agmon, V., Finkelstein, R., Raz, R., Ben-Dror, G., Lerner, L., Soboh, S., Colodner, R., Cameron, D.N., Wykstra, D.L., Swerdlow, D.L., Farmer, J.J. 3rd., (1999). Clinical, epidemiological, and microbiological features of *Vibrio vulnificus* biogroup 3 causing outbreaks of wound infection and bacteraemia in Israel. Israel *Vibrio* study Group. *Lancet.* 354, 1421-1424.

[24] Efimov, V., Danin-Poleg, Y., Raz, N., Elgavish, S., Linetsky, A., Kashi, Y., (2013) Insight into the evolution of *Vibrio vulnificus* biotype 3's genome. *Front. Microbiol.* 4,393.

[25] Nisson, W.B., Paranjype, R.N., Depaola, A., Strom, M.S., (2003). Sequence polymorphism of the 16S rRNA gene of *Vibrio vulnificus* is a possible indicator of strain virulence. *J. Clin. Microbiol.* 41, 442-446.

[26] Bisharat, N., Cohen D.I., Harding R.M., Falush D., Crook D.W., Peto T., Maiden, M.C., (2005). Hybrid *Vibrio vulnificus*. *Emerg. Infect. Dis.* 11, 30-35.

[27] Rosche, T.M., Yano, Y., Oliver, J.D., (2005). A rapid and simple PCR analysis indicates there are two subgroups of *Vibrio vulnificus* which correlate with clinical or environmental isolation. *Microbiol. Immunol.* 49, 381-389.

[28] Cohen, A.L., Oliver, J.D., DePaola, A., Feil, E.J., Boyd, E.F., (2007). Emergence of a virulent clade of *Vibrio vulnificus* and correlation with the presence of a 33-kilobase genomic island. *Appl. Environ. Microbiol.* 73, 5553-5565.

[29] Gonzalez-Escalona, N., Jaykus, L.A., Depaola, A., (2007). Typing of *Vibrio vulnificus* strains by variability in their 16S-23S rRNA intergenic spacer regions. *Foodborne Pathog. Dis.* 4, 327-337.

[30] Vickery, M.C., Nilsson, W.B., Strom, M.S., Nordstrom, J.L., DePaola, A., (2007). A real-time PCR assay for the rapid determination of 16S rRNA genotype in *Vibrio vulnificus*. *J Microbiol. Methods* 68, 376-384.

[31] Morrison, S.S., Williams, T., Cain, A., Froelich, B., Taylor, C., Baker-Austin, C., Verner-Jeffreys, D., Hartnell, R., Oliver, J.D., Gibas, C.J., (2012). Pyrosequencing-based comparative genome analysis of *Vibrio vulnificus* environmental Isolates. PLoSONE 7, e37553.

[32] European Commission, (2001). Opinion of the Scientific Committee on Veterinary Measures Relating to Public Health on *Vibrio vulnificus* and *Vibrio parahaemolyticus* (in Raw and Undercooked Seafood). In: European Commission Health & Consumer Protection Directorate General. Brussels, 1-64.

[33] APROMAR, (2011). *La acuicultura marina de peces en España.* Edición especial 25 aniversario. Asociación empresarial de Productores de cutlivos Marinos, Cádiz, España, 77 pp.

[34] Eurostat, (2012). The Statistical Office of the European Communities. Total Production of Fish, Crustaceans, Mollusks and Other Aquatic Organisms from Aquaculture ("fish-farming"). Available at: http://epp.eurostat.ec.europa.eu/tgm/ table. do?tab=table&init=1&plugin=0&language=en&pcode=tag00075 (accessed February 21,2012).

[35] FAO, (2008). FishStat Plus - Departamento de pesca y acuicultura, Organización de las Naciones Unidas para la Agricultura y la Alimentación. http://www.fao.org/fishery/statistics/software/fishstat/es.

[36] Montilla, R., Palomar, J., Santmarti, M., Fuste, C., Viñas, M., (1994). Isolation and Characterization of Halophilic *Vibrio* from Bivalves Bred in Nurseries at the Ebre Delta. *J. Invertebr. Pathol.* 63, 178-181.

[37] Lopez-Joven, C., de Blas, I., Furones, M. D., Roque, A., (2015). Prevalences of pathogenic and non-pathogenic *Vibrio parahaemolyticus* in mollusks from the Spanish Mediterranean Coast. *Front. Microbiol.* 6, 736.

[38] Roque, A., Lopez-Joven, C., Lacuesta, B., Elandaloussi, L., Wagley, S., Furones, M.D., Ruiz-Zarzuela, I., de Blas, I., Rangdale, R., Gomez-Gil, B., (2009). Detection and identification of *tdh*- and *trh*-positive *Vibrio parahaemolyticus* strains from four species of cultured bivalve molluscs on the Spanish Mediterranean Coast. *Appl. Environ. Microbiol.* 75, 7574-7577.

[39] Coleman, S.S., Melanson, D.M., Biosca, E.G., Oliver, J.D., (1996). Detection of *Vibrio vulnificus* biotypes 1 and 2 in eels and oysters by PCR amplification. *Appl. Environ. Microbiol.* 62, 1378-1382.

[40] Reichelt, J.L., Baumann, P., Baumann, L., (1976). Study of genetic relationships among marine species of the genera *Beneckea* and *Photobacterium* by means of *in vitro* DNA/DNA hybridization. *Arch. Microbiol.* 110, 101-120.

[41] Farmer III, J.J., (1980). Revival of the name *Vibrio vulnificus*. *Int. J. Syst. Bacteriol.* 30, 6560.

[42] Camp, J., Delgado, M., (1987). Hidrografía de las bahías del delta del Ebro. Invest. Pesq. 51, 351-369.

[43] Kaspar, C.W., Tamplin, M.L. (1993). Effects of temperature and salinity on the survival of *Vibrio vulnificus* in seawater and shellfish. *Appl. Environ. Microbiol.* 59, 2425-2429.

[44] Motes, M.L., DePaola, A., Cook, D.W., Veazey, J.E., Hunsucker, J.C., Garthright, W.E., Blodgett, R.J., Chirtel, S.J., (1998). Influence of water temperature and salinity on *Vibrio vulnificus* in North Gulf and Atlantic Coast oysters (*Crassotrea virginica*). *Appl. Environ. Microbiol.* 64, 1459-1465.

[45] Esteves, K., Hervio-Heath, D., Mosser, T., Rodier, C., Tournoud, M-G., Jumas-Bilak, E., Colwell, R.R., Monfort, P. (2015). Rapid proliferation of *Vibrio parahaemolyticus*, *Vibrio vulnificus* and *Vibrio cholerae* during freshwater flash floods in French Mediterranean coastal lagoons. *Appl. Environ. Microbiol.* doi:10.1128/AEM.01848-15.

[46] Camp, J., (1994). Aproximaciones a la Dinámica Ecológica de una Bahía Estuárica Mediterránea. Ph.D. thesis, University of Barcelona, Barcelona.

[47] Wright, A.C., Hill R.T., Johnson J.A., Roghman M.C., Colwell R.R., Morris J.G., (1996). Distribution of *Vibrio vulnificus* in the Chesapeake Bay. *Appl. Environ. Microbiol.* 62,717-724.

[48] Lipp, E.K., Rodriquez-Palacios, C., Rose, J.B., (2001). Occurrence and distribution of the human pathogen *Vibrio vulnificus* in a subtropical Gulf of Mexico estuary. *Hydrobiologia* 460, 165-173.

[49] Randa, M.A., Polz, M.F., Lim, E., (2004). Effects of Temperature and Salinity on *Vibrio vulnificus* Population Dynamics as Assessed by Quantitative PCR. *Appl. Environ. Microbiol.* 70, 5469-5476.

[50] Jacobs, J.M., Rhodes, M., Brown, C.W., Hood, R.R., Leigh, A., Long, W., Wood R., (2010). Predicting the Distribution of *Vibrio vulnificus* in Chesapeake Bay. NOAA Technical Memorandum NOS NCCOS 112. NOAA National Centers for Coastal Ocean Science, Center for Coastal Environmental Health and Biomolecular Research, Cooperative Oxford Laboratory. Oxford, MD. 12 pp.

[51] Høi, L., Larsen, J.L., Dalsgaard, I., Dalsgaard, A., (1998). Occurrence of *Vibrio vulnificus* in Danish marine environments. *Appl. Environ. Microbiol.* 64,7-13.

[52] Pfeffer, C.S., Hite M.F., Oliver J.D., (2003). Ecology of *Vibrio vulnificus* in estuarine waters of eastern North Carolina. *Appl. Environ. Microbiol.* 69, 3526-3531.

[53] Tamplin, M., Rodrick, G.E., Blake, N.J. Cuba, T., (1982). Isolation and characterization of *Vibrio vulnificus* from two Florida estuaries. *Appl. Environ. Microbiol.* 44, 1466-1470.

[54] Oliver, J.D., Warner, R.A., Cleland, D.R., (1983). Distribution of *Vibrio vulnificus* and Other Lactose-Fermenting Vibrios in the Marine Environment. *Appl. Environ. Microbiol.* 45, 985-998.

[55] O'Neill, K.R., Jones, S.H., Grimes, D.J. (1992). Seasonal incidence of *Vibrio vulnificus* in the Great Bay Estuary of New Hampshire and Maine. *Appl. Environ. Microbiol.* 58, 3257-3262.

[56] Thompson, J.R., Randa, M.A., Marcelino, L.A., Tomita-Mitchell, A., Lim, E. and Polz, M.F., (2004). Diversity and dynamics of a north Atlantic coastal Vibrio community. *Appl. Environ. Microbiol.* 70, 4103-4110.

[57] Mahmud, M.Z., Neogi S.B., Kassu A., Huong B.T.M., Jahid I.K., Islam M.S., Ota F. (2008). Occurrence, seasonality and genetic diversity of *Vibrio vulnificus* in coastal seaweeds and water along the Kii Channel, Japan. *FEMS Microbiol. Ecol.* 64, 209-218.

[58] Vanoy, R.W., Tamplin, M.L., Schwarz, J.R., (1992). Ecology of *Vibrio vulnificus* in Galveston Bay oysters, suspended particulate matter, sediment and seawater: detection by monoclonal antibody-immunoassay - most probable number procedures. *J. Ind. Microbiol.* 9, 219-223.

[59] Depaola, A., Capers, G.M., Alexander, D., (1994). Densities of Vibrio vulnificus in the Intestines of Fish from the U.S. Gulf Coast. *Appl. Environ. Microbiol.* 60, 984-988.

In: Oysters and Clams
Editor: Jesús L. Romalde

ISBN: 978-1-63485-074-2

Chapter 7

An Ecological Approach to Understanding Host-Pathogen-Environment Interactions: The Case of Brown Ring Disease in Clams

Christine Paillard*
Laboratoire des Sciences de l'Environnement Marin (LEMAR), UMR 6539
UBO/CNRS/IRD/Ifremer, Université de Brest, Plouzané, France

Abstract

Brown Ring Disease (BRD) is used as a case study to describe an ecological approach to understanding host-pathogen interactions. BRD is caused by a bacterial microparasite, *Vibrio tapetis*, which elicits the inhibition of the calcification process. Epidemiological studies of BRD are facilitated by the use of characteristic clinic signs on the inner surface of the valves. The phenology and dynamics of BRD are then described by coupling epidemiological, ecological and physiological data. Spatial distribution of BRD prevalence in natural clam populations and *V. tapetis* load in the sediment allows the investigation of the transmission mode of this pathogen. Experimental testing of different transmission modes has demonstrated the predominant role of feces and pseudofeces in *V. tapetis* transmission and diffusion processes. Some other hosts can harbor *V. tapetis*, including reservoir species such as cockles but also vertebrate predators such as fishes, and cross-species transmission could potentially play a major role in dispersion of the pathogen during its life cycle. Environmental factors such as temperature, salinity and nutrients strongly modulate the coevolution between *Vibrio* and bivalve.

Keywords: microparasite, *Vibrio*, bivalve, transmission, clinical signs, shell disease, environment, temperature

* Corresponding Author address: Laboratoire des Sciences de l'Environnement Marin (LEMAR), UMR 6539 UBO/CNRS/IRD/Ifremer, Université de Brest, Technopôle Brest-Iroise, rue Dumont D'Urville, 29280 Plouzané, France. E-mail address: christine.paillard@univ-brest.fr.

INTRODUCTION

Taking an ecological approach to the investigation of host-pathogen interactions primarily requires a natural environment approach known as epidemiology. Investigating the distribution of the disease in space and time allows analysis of the factors (environmental, genetic...) concurrent with its onset and development. In combination with laboratory experiments, this approach can be used to assess the role of key factors on host-pathogen interactions. The development of a numerical model of Brown Ring Disease (BRD) allows knowledge to be synthesized in a simple manner (etiology, diagnostic tools, pathogen characterization, mechanisms of host-pathogen interaction, epidemiology), to prioritize different stages and to point to gaps and flaws of the three component system, Environment, Host and Pathogen (HPE) [23]. The HPE system can be reconsidered taking into account the entire microbial community: on one hand, the pathogen community and the co-infection processes, and on the other hand, the microbiota that the host harbors [10, 28, 29]. This ecological approach is highly interdisciplinary, drawing on techniques from fields including genetics, molecular biology, genomics, immunology, epidemiology and modeling (Figure 1).

EPIDEMIOLOGICAL APPROACH

Epidemiology is the study of diseases and health factors (environmental, genetic) in a population. The originality of epidemiology is its focus on the population rather than the individual. Quantification is paramount in epidemiology and quires to be precise and rigorous to determine the epidemiological indicators for each type of disease. These indicators will most often be expressed in relative terms compared to the population in which it is observed.

Brown Ring Disease provides a useful case study to describe the uses of epidemiological parameters, because the disease diagnostic is based mostly on macroscopic signs. In fact, BRD is due to a bacterial microparasite, *Vibrio tapetis*, which elicits shell growth abnormalities by inhibiting calcification process and inducing the deposit of brown conchiolin between the mantle and the inner surface of the valves [21]. This defense process, known as nacrezation or pearl formation [6] embeds parasites inside the shell matrix. This shell repair process is very powerful against macroparasites, such as trematodes, but in the case of microparasites such as bacteria with high generation rates, this mechanism is less efficient. However, depending on the HPE interaction modulation factors, it has been demonstrated that conditions favorable to the host such as nutrition and high temperature could enhance BRD recovery [14, 20, 24].

The development of clinical signs for BRD using a classification system to assess conchiolin deposition, shell repair and phase of the disease has already been established (Figure 2) [19, 21].

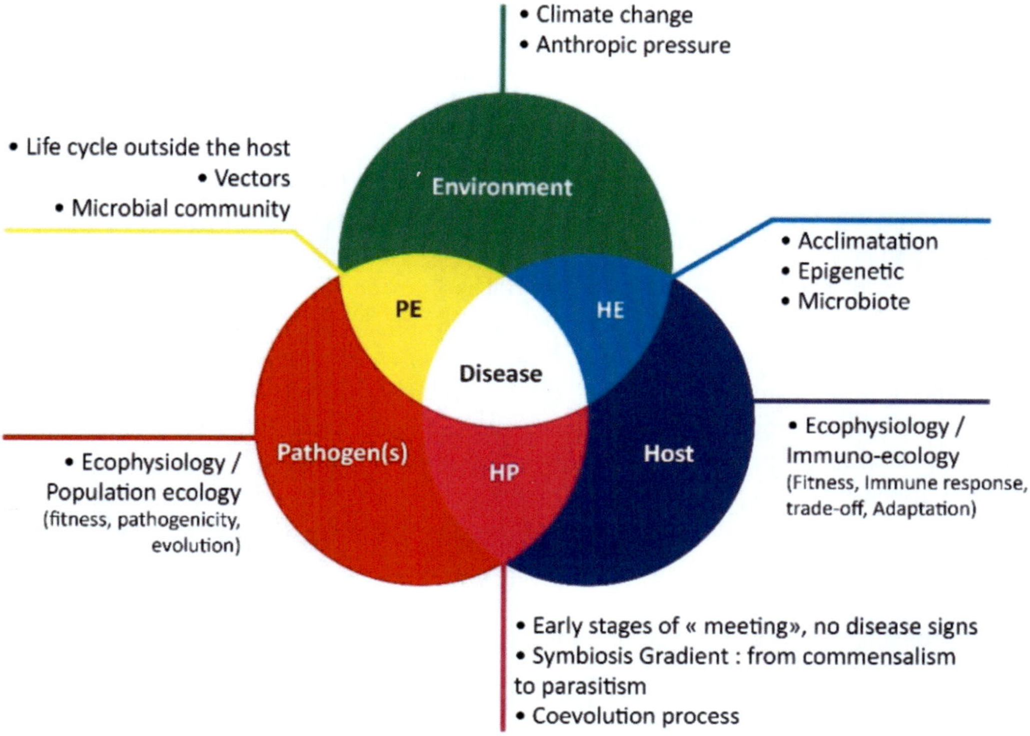

Figure 1. The ecological approach of the Host-Pathogen-Environment System.

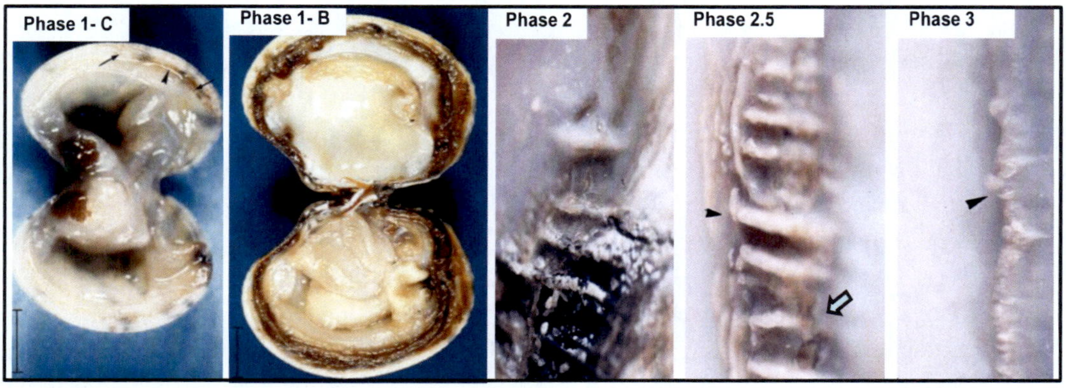

Figure 2. BRD clinical signs (BRD+) that could be observed macroscopically and microscopically from Phase 1 to Phase 3 of Brown Ring Disease. In Phase 1 (cases B and C), only conchiolin deposit is observed without shell repair. In the other phases (Phase 2, 2.5 and 3), the conchiolin deposit is recovered by shell repaired layers at different recovery steps, intermediate (phase 2), almost recovery (Phase 2.5) and completely recovery (Phase 3).

Epidemiological Indicators Can Be Used to Describe and Quantify Brown Ring Disease (BRD)

Prevalence of BRD (BRD%): total number of individuals infected with BRD clinical signs as a percentage of the total number of sampled individuals ($n \geq 100$), at a given moment. The prevalence provides static information about the frequency of the disease, but not its development (increase, decrease or increase of new cases, such as BRD intensity). It corresponds to the balance of a disease at a given time. The prevalence is not to be confused with the incidence, which is the rate at which newly infected BRD individuals appear in a specific population during a given period.

Prevalence of conchiolin disease stages (CDS%): number of individuals with a specific stage of brown ring relative to the total number of sick animals, during a specific time period or at a given time.

Prevalence of shell repair stages (SRS%): The number of individuals at shell specific stage of repair as a percentage of the total number of animals either currently sick or previously sick (and now recovered) during a specific time period or at a given time.

Another type of indicator is the pathogen load (number of cells or CFU/ml) in clams or other hosts or in the surrounding environment (sediment, seawater). Pathogen concentration is measured using different techniques previously described (bacteriological analysis, ELISA or SSP-PCR, qPCR) [5, 23]

Considering all epidemiological indicators together, BRD in clams can be considered in a phenological way, taking into account not only the changes in the clinical signs, but also the pathogen load and mortality of clams that is not necessarily related to the presence of BRD signs. Figure 3 summarizes these results and integrates them to form a more complete vision of the vibriosis, taking into account changes in pathogen load and BRD microscopic clinical signs.

Six Phases of Vibriosis in Clams Can Be Identified Over Time

Six phases of vibriosis in clams can be identified over time.

- Phase 0 is characterized by the absence of both BRD clinical signs (BRD–) and the pathogen after detection. It therefore represents either a healthy individual, or a clam beginning BRD incubation, with a pathogen load below the limit detection of *V. tapetis* by qPCR (1.125×10^1 bacteria/ml) [5].
- Phase 0.5 is characterized by an absence of BRD clinical signs (BRD–), but by obligatory positive detection *V. tapetis*. *V. tapetis* load can be measured in the extrapallialfluids (Figure 3, cases B or C), and/or in the hemolymph and tissues (Figure 3, cases A or B). It is a phase characterized by an absence of external BRD clinical signs and the presence of internal alterations such as increased hemocyte concentration a few hours after the inoculation, both in the mantle cavity and in the adductor muscle [3]. For clams which develop shell repair defense mechanisms, this phase is only temporary. For the others (cases A and B), invasion of *V. tapetis* within tissues will lead to mortality.

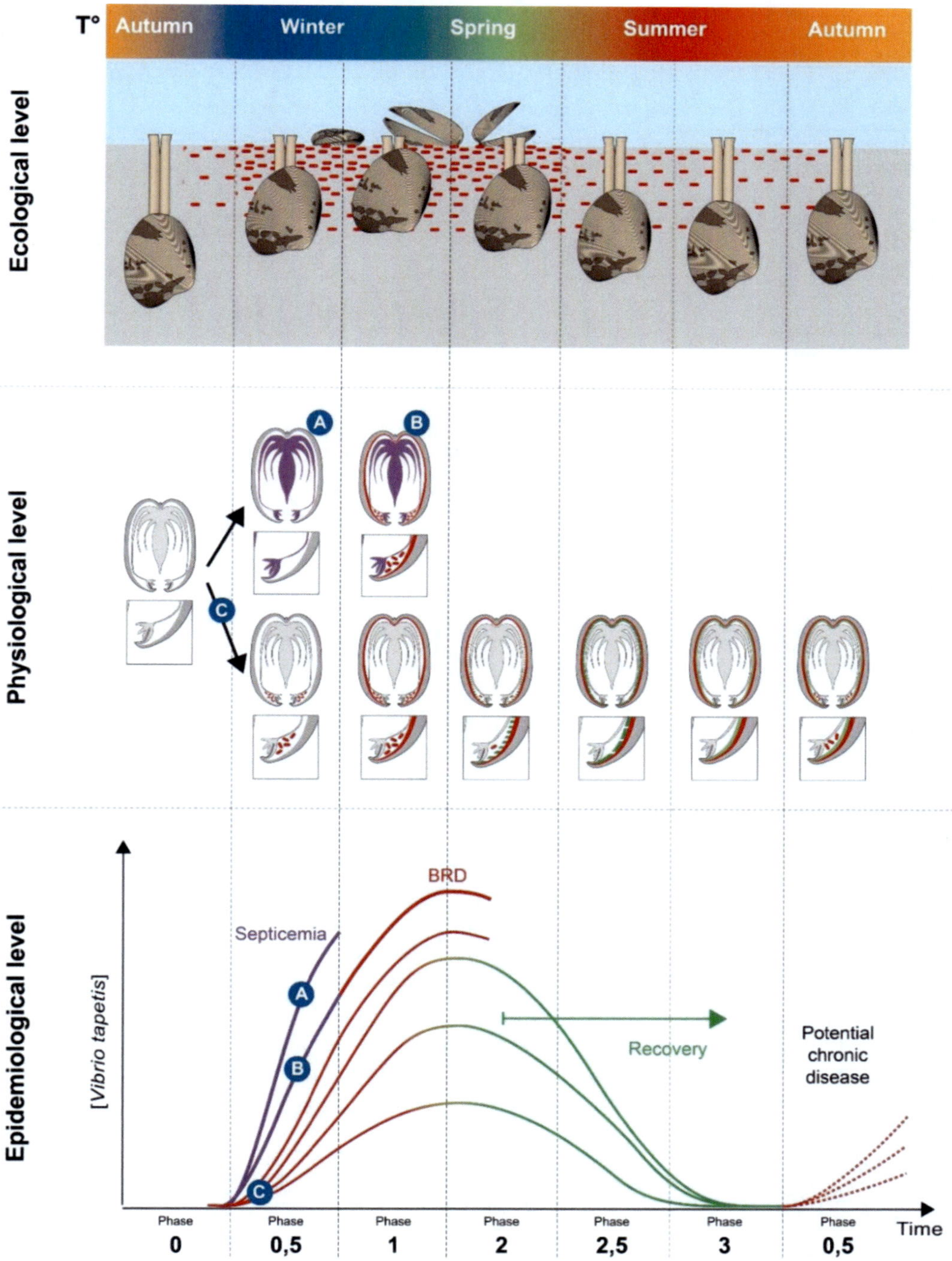

Figure 3. Synthesis of Brown Ring Disease (BRD) phenology, described at ecological, physiological and epidemiological levels. A: *Vibrio tapetis* only within tissues (purple); C: *V. tapetis* only in extrapallial and pallial fluids (red); B: *V. tapetis* in internal and external compartments (purple and red). Recovery process in green.

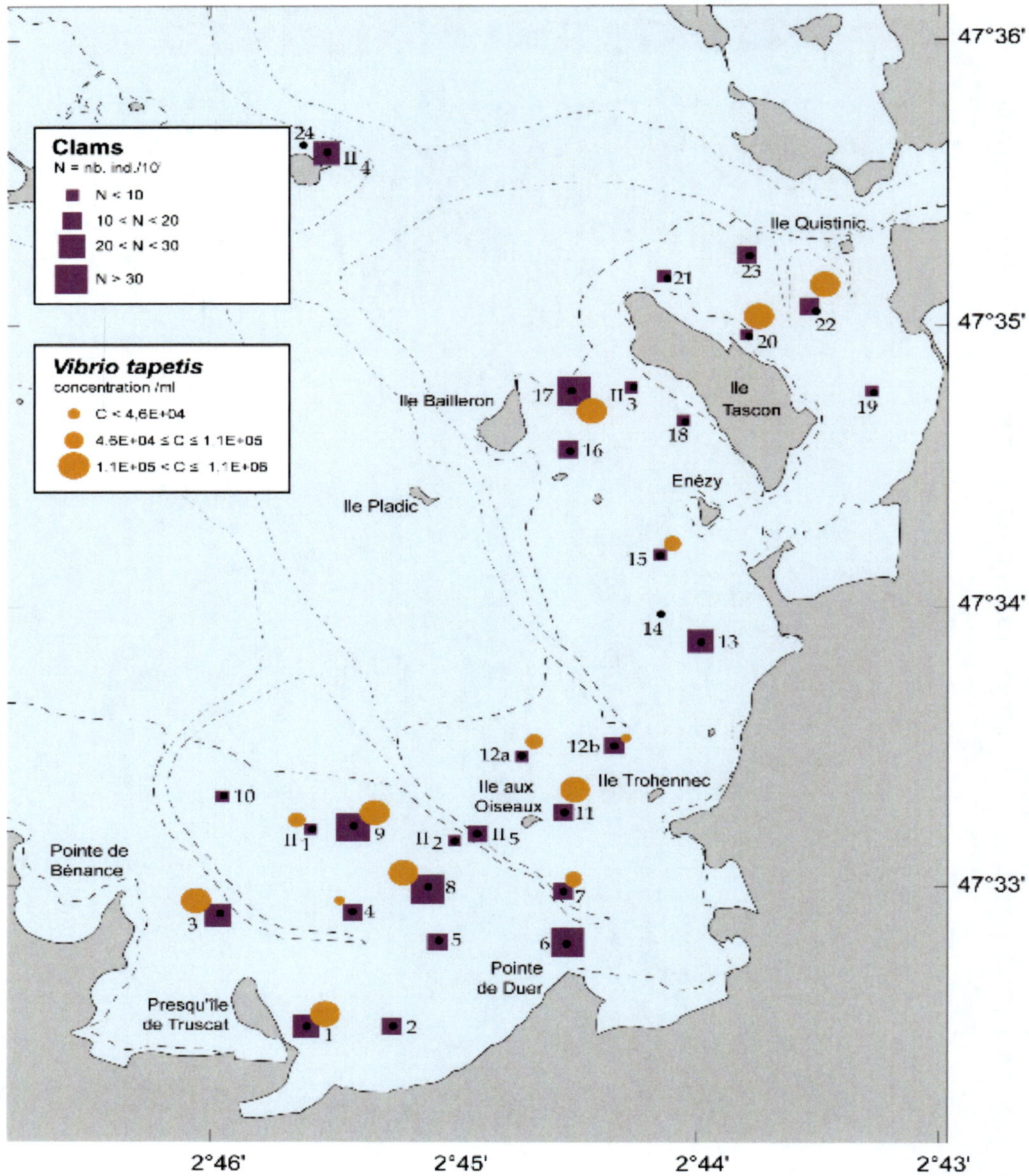

Figure 4a. Density of Manila clams (purple squares) and *V. tapetis* concentrations in the surface (0-2 cm) sediment (yellow circles) in the Gulf of Morbihan, France. March campaign 1999.

Phases 1 to 3 are all characterized by clinical signs (Figure 2).

- Phase 1 consists of the appearance and development of the conchiolin deposit (CD). *V. tapetis* concentration in extrapallial fluids tends to increase with CD stages.
- Phase 2 or the intermediate repair phase of BRD. The concentration of *V. tapetis* reaches maximum values.

- Phase 2.5 or the healing phase or "recovery." The concentration of *V. tapetis* starts to decrease slightly, reaching values similar to beginning of Phase 1. If there is no relapse, recovery will occur soon after.
- Phase 3 characterizes a recovered animal. A scar is visible on the shell and the concentration of *V. tapetis* is below the limit detection of *V. tapetis* [23]. The absence of *V. tapetis* must be confirmed using qPCR method [5] during this phase.

BRD associated mortalities occur generally in clams in phase 0.5 (cases A and B) when *V. tapetis* has colonized tissues, but also in phase 1 and 2, when the clams exhibit strong development of brown ring disease (phase 1 case B) or when the shell repair process is not completely finished (Phase 2, cases B and C) (Figure 2).

Using these indicators, Figure 3 represents a synthesis of BRD dynamics coupling prevalence of disease phase and *V. tapetis* load at individual level. It is interesting to note that mortalities can also occur without BRD signs (if the bacteria is able to penetrate within tissues and multiply in the hemolymph) [3] and at later stages of BRD development (in advanced stages of phase 2) [23].

The prevalence of BRD phase could represent a pertinent integrative indicator to determine health status of clam population.

Spatial Distribution of BRD and *V. tapetis* in the Field

Spatial distribution of disease (prevalence, pathogen load…) is also very useful in epidemiological studies. Spatial distribution of BRD prevalence was assessed according to the host density and density of pathogens in sediment (performed by ELISA) [18] during field sampling carried out during spring in the Gulf of Morbihan (Figures 4a and 4b). The distribution of BRD prevalence among burrowing and rising clams did not show any relationship with either host densities or pathogen densities in clams and in the sediment. However, a significant relationship between the density of the clams and the average concentration of *V. tapetis* in the sediment was demonstrated. Areas with very high densities of clams were also those that harbored the highest densities of *V. tapetis* in sediments. Simulations of Lagrangian trajectories for different numbers of particles or "larvae" using the hydrodynamic model of the Gulf of Morbihan [13] have confirmed that the areas with high densities of clams in the center of Truscat sandbank (cf. Figure 4a, purple squares) were those subject to low current swirling waters, favoring the accumulation and deposition of fine particles [11]. Larval transport and the spread of bacteria could also follow the same pattern of dispersion within the Truscat sandbank, resulting in the strong correlation between clam density and pathogen load.

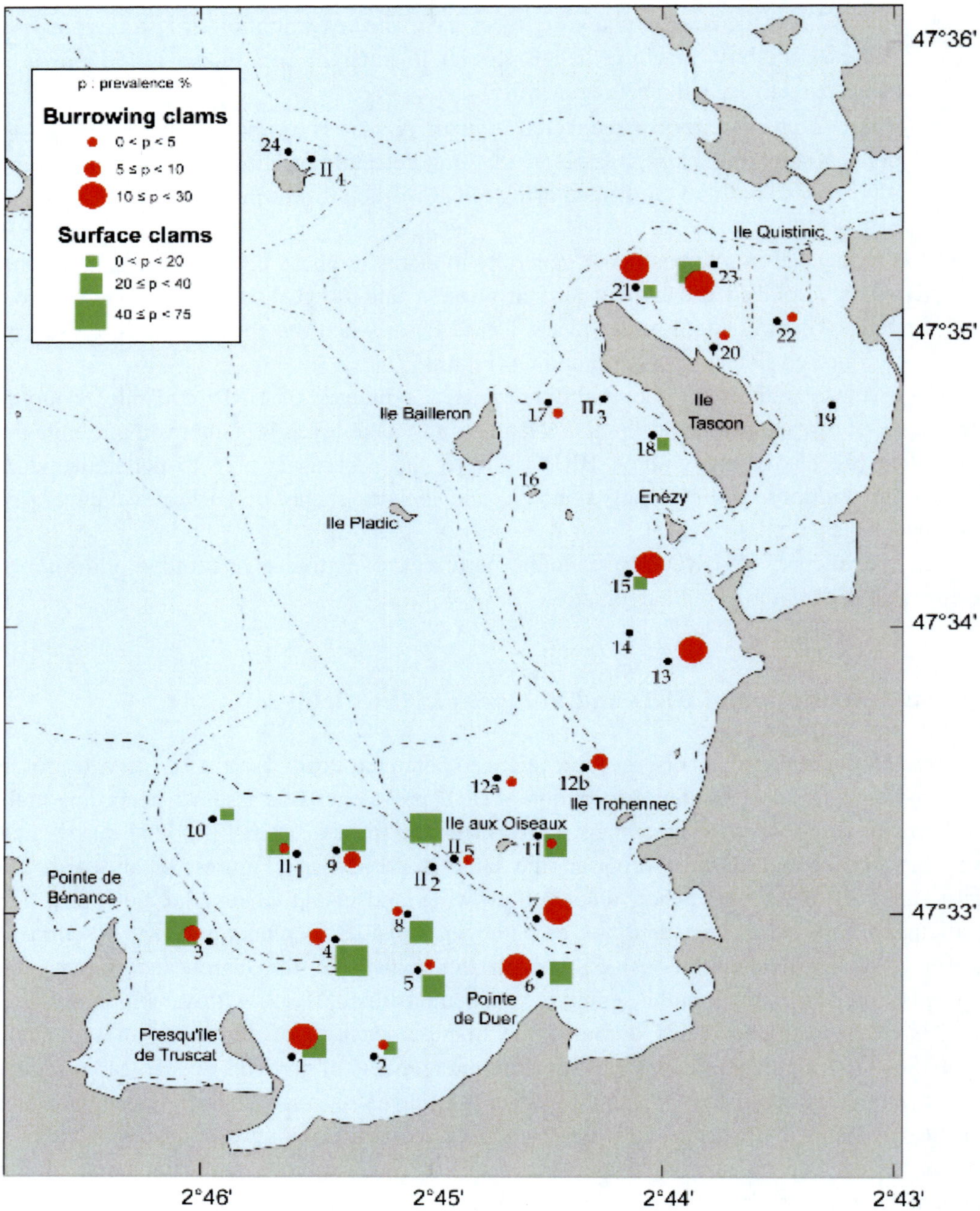

Figure 4b. BRD prevalence in burrowing (red circles) and in rising clams at the surface of the sediment (green squares) in the Gulf of Morbihan, France. March campaign 1999.

Disease Transmission Modes

Considering the results of spatial distribution of BRD and *V. tapetis* in the natural population of clams, an experiment was performed to investigate transmission of BRD. Three disease transmission modes were considered: by water, via contact, and by feces and pseudofeces. To determine the role of water in transmission, the animals were placed just under the surface, on top of a grid that allowed their feces to sink to the bottom of the tanks. Clams were not in contact with feces or with other clams. For testing the role of contact transmission, the bivalves were placed in the same conditions as above; however, they were kept in continuous contact with each other. To simulate the process of infection by feces and pseudo-feces, clams were fixed to the bottom of the tank, without contact with each other. Feces and pseudo-feces were resuspended by stirring every two days. *V. tapetis* contamination was performed by placing *V. tapetis* inoculated individuals (10% of the total number of clams) in each aquarium. Each transmission mode was investigated at two different densities: 100 and 400 clams/m^2. For each transmission mode and for each clam density, the effect of water change was also tested (no water change, or water change every two days). For this experiment, because disease-free clams were not available in the wild or from hatcheries, a treatment bath of chloramphenicol (8 mg/l for 48 hours) was performed on the clams prior to the start of the experiment. A subsequent thermic treatment (clams were subjected to a temperature change of 4°C every 24 hours, starting at the experimental temperature of 15°C, rising to a maximum of 27°C, and falling again to 15°C) and another antibiotic bath of 48 hours were applied to limit the development of opportunistic bacteria. The animals were fed every two days. After one month, the animals were sacrificed and the pathogenic load in the pallial and extrapallial fluids of clams determined by the ELISA method [18]. The results are summarized in Figure 5.

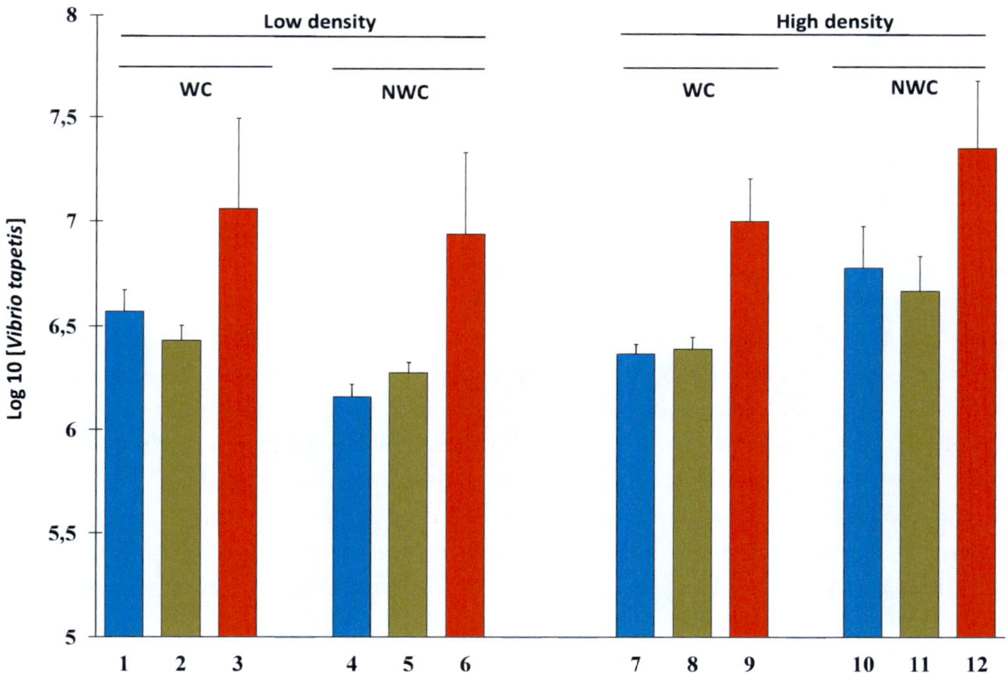

Figure 5. Log 10 [*V. tapetis*] in the pallial and extrapallial fluids of clams for each density condition and transmission mode at day 30. Blue: transmission by water; green: contact transmission; red: transmission through feces. WC: with water change every two days. NWC: No water change. Low density (100 clams/m^2); High density (400 clams/m^2).

Whether water was renewed or not, the concentration of *V. tapetis* in fluids of clams placed in tank bottom in contact with feces and pseudofeces was significantly higher than those placed in open water or in intimate contact between each other. We were able to demonstrate that the concentration of *V. tapetis* in diseased clam feces represented 80 to 90% of the total heterotrophic microflora. When clams were at high density, water exchange allowed a reduction in the pathogen load (factor 10). Transmission through the water only appeared to be effective if the density of the host was high (at least 400 clams/m^2) and if the water was not renewed. These are the conditions under which clams live in the Gulf of Morbihan (muddy sediments and densities of 100-600 clams/m2) (Figure 4a). However, within the trays on which the animals were placed in intimate contact, the distribution of hosts with high pathogen loads was independent of the location of the animals inoculated with *V. tapetis*. In conclusion, two possible transmission paths have been demonstrated: by water and by feces. Whatever the density of clams or water renewal treatment, transmission through the feces seems crucial.

Vertical transmission of *V. tapetis* was also considered, but to date, no experiments have been conducted to test this hypothesis.

LIFE CYCLE OF *V. TAPETIS*

Previously, we have shown that *V. tapetis* could be found in sediment, where high clam density occurs. Rodrigues et al. [27] have demonstrated the ability of *V. tapetis* to form a biofilm *in vitro* on abiotic surfaces. Within the clams, a biofilm of *V. tapetis* on shell secretions constitutes the first steps of the infectious process. Then, *V. tapetis* multiplies in different compartments, primarily in the brown deposit, the periostracal secretion, the external mantle epithelium and the pallial and extrapallial fluids [1,2]. In times of winter mortality or pre-spring time, *V. tapetis* can be also isolated in the hemolymph of moribund clams [Paillard et al., unpublished data]. *V. tapetis* can also be detected during the same period in the pore water of the sediment, the water surrounding moribund clams, and in the feces and pseudofeces of BRD clams after placing them in sterile vials [17].

Hosts other than *R. decussatus* and *R. philippinarum*, have also been identified. *V. tapetis* has been isolated from other species of veneridae that suffer from BRD (*Venerupis aurea*, *Tapes rhomboïdes*, *Dosynia exoleta)*, and the cockle, *Cerastoderma edule*, has also been identified as a potential vector species [16]. Indeed the ability to cross species barriers is a feature of *V. tapetis*, and can also occur from invertebrates to vertebrates. In fact, *V. tapetis* has been isolated from several species of fishes during their mortality events, such as flat fishes (*Hypoglossus hypoglossus*, *Dicologoglossa cuneata*, *Solea solea*) and the wrasse (*Symphodus melops*) [7, 11, 14]. All these fishes are likely to be predators of clams and could also potentially be considered as secondary hosts of *V. tapetis*. Figure 6 summarizes the life cycle of *V. tapetis* as a pathogen of clams.

VIBRIO BEHAVIOR MANIPULATION?

The two schematic figures presented (ecological level of Figure 3 and Figure 6) summarize our knowledge of how micro-habitats, transmission modes and seasonal change are able to influence the spread of *V. tapetis*. It thus appears that a predation/parasitism interaction could exacerbate mortalities; as weakened infected bivalves rise to the surface of the sediment and are more accessible to predators. In addition, the rising clams have a very high pathogenic load of vibrios, varying from 2 x 10^5 to 8 x 10^9 CFU/ml, which is on average 100 times higher than in a burrowing clam. The rising clams therefore also constitute a source of contamination and spread of the pathogen. Could this be similar to the favorisation process induced by trematode parasites in the clam *Tapes aureus*? [7]. These parasites modify the host behavior by promoting its rising in the sediment, making it more accessible to the secondary hosts (birds) [4]. Indeed, this same behavior was observed for heavily *V. tapetis* infected clams, and rising clams could even act like delayed explosive bombs in the predator! Thus, fish-eating clams, such as wrasse and flatfish may play a major role in the infectious cycle of *V. tapetis*, especially during periods of winter mortality.

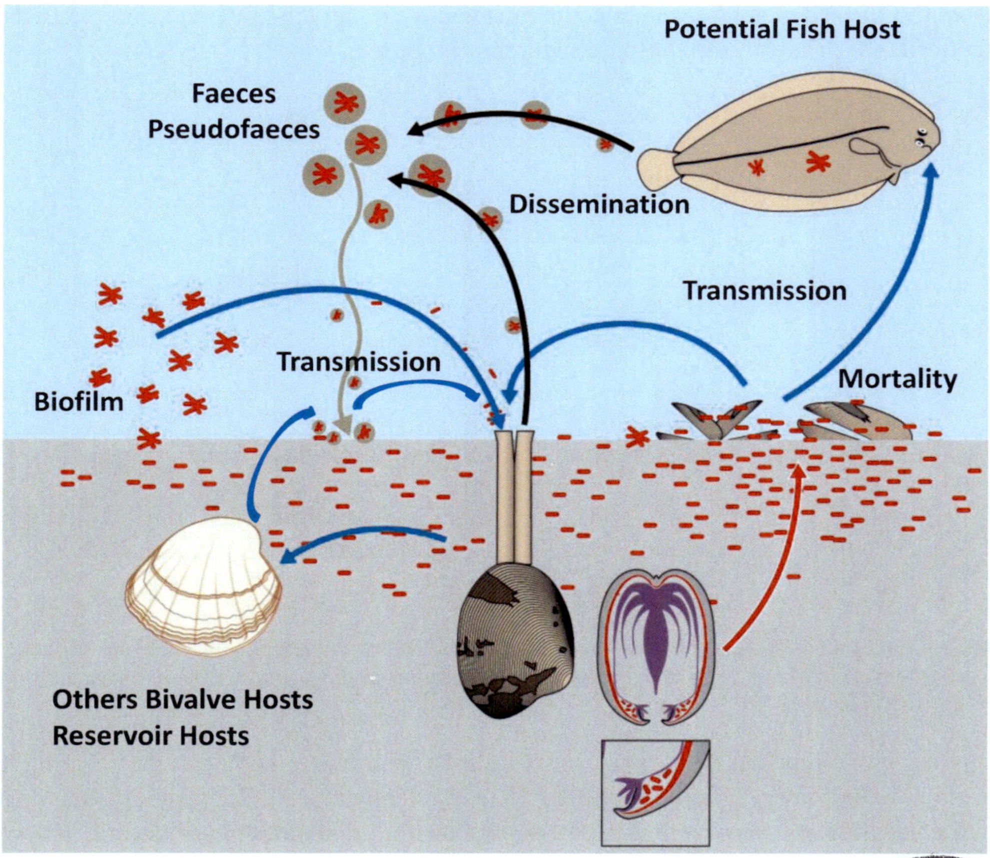

Figure 6. Life cycle of *V. tapetis* pathogenic to clams. The black arrows indicate dissemination, the blue arrows indicate transmission modes and the red arrows indicate mortality associated with the rising of the clams to the surface of the sediment.

Impact of Environmental Factors on the Host-Pathogen Interactions

In natural populations of Manila clam along the coasts of Brittany (Bay of Brest, North Finistere and Morbihan), we have shown previously that maximum prevalence occurs during the winter and beginning of spring periods, peaking in March. Animals from one natural population in the Gulf of Morbihan, ("Île aux oiseaux") were sampled during this spring period and maintained in aquaria at different temperature (4, 14, 18 and 21°C), allowing quantification of the BRD dynamic. At 14°C, the prevalence reached 90% after two months (Figure 7). At other temperatures (4, 18 and 21°C), the clams developed BRD but in much lower proportions. Environmental conditions strongly modulate the development of BRD. Experimental infections in laboratory with *V. tapetis* at various temperature have also shown that maximum BRD development in clams occurs at 14°C, and that at 21°C the prevalence is very low due to high recovery rates.

Temperature is also a major modulating factor impacting growth and survival of *V. tapetis.* Maximum growth occurs around 22°C but its survival decreases rapidly after 25°C and death is observed at 27°C [22]. The death of *V. tapetis* at a temperature generally measured in the sediment at summer conditions could explain the low prevalence of BRD in southern sites such as the Adriatic and Mediterranean Seas.

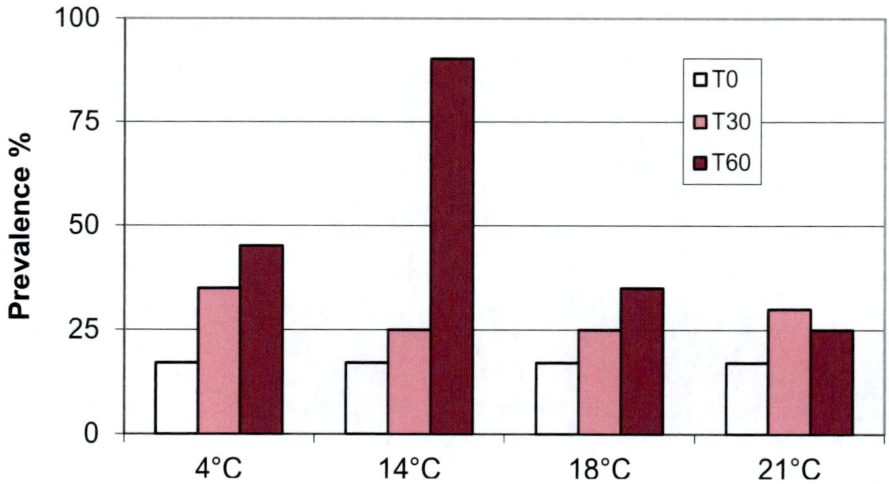

Figure 7. Evolution of BRD prevalence in clams harvested from natural populations (Île aux oiseaux, Gulf of Morbihan) in March 1998 and maintained in aquariums without food for two months at different temperatures (4, 14, 18 and 21°C).

Other environmental factors such as salinity can also modulate the *Vibrio*-clam interaction; in fact lower salinities enhance BRD development [25]. The quantity and quality of nutrients available also have a great impact on the condition index and on recovery processes of clam [9, 24]. Figure 8 integrates these different pressures on the host-pathogen environment system. The theoretical individual-based model of Brown Ring Disease in Manila clams suggested also that the outcome of the host–parasite interaction is dependent on food supply and temperature [23, 26]. Simulations with increase temperature (1 or 2°C) favor generally disease development, indicating that climate warming might favor the spread of BRD, although this disease is a cold water vibriosis.

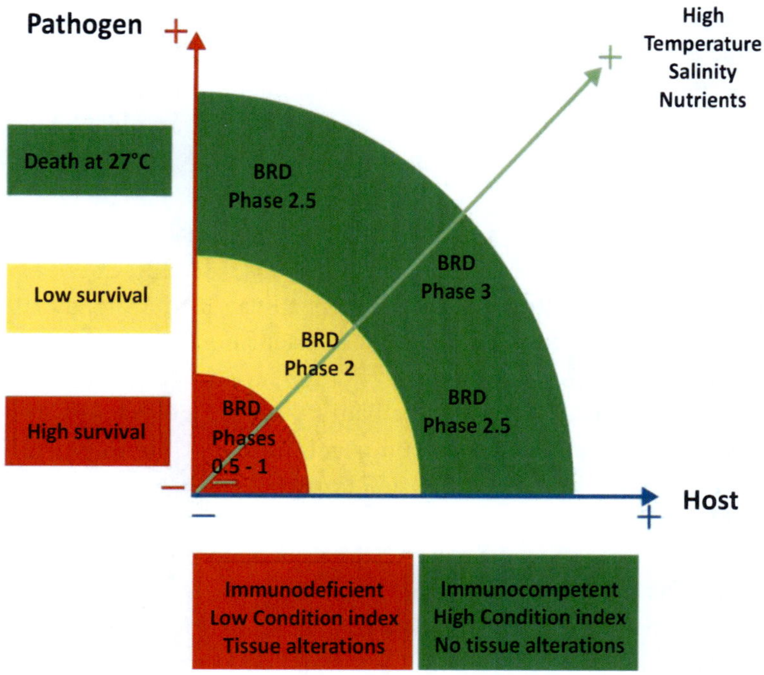

Figure 8. Synthesis of the Host-Pathogen-Environment system in the case of Brown Ring disease (BRD) in *R. philippinarum.*

CONCLUSION

Brown Ring Disease in clams, caused by *Vibrio tapetis*, represents a unique model to explore how a bacterium, acting as a microparasite, can intimately alter bivalve physiology and ecology. Adopting an ecological approach to studying BRD will allow the coupling of various phenotypes (characteristic external signs, pathogen load and other life traits such as physiological and immunological patterns) to better understand host-parasite coevolution; moreover, BRD phenology could also be followed along with its response to climatic phenomena. New investigations into the role of microbiota on the host-pathogen-environment interplay represent an intriguing possibility for future research into ecological approaches to mollusc diseases.

ACKNOWLEDGMENTS

I would like to thank, Romain Froquet, Susan Ford and Vanessa Thomann for the analysis of clams from field studies and laboratory experiments, Lionel Alano from Bailleron island for field survey and Sébastien Hervé for the illustrations and Ewan Harney for English corrections.

REFERENCES

[1] Allam, B., Ford, S.E., 2006. Effects of the pathogenic *Vibrio tapetis* on defence factors of susceptible and non-susceptible bivalve species: I. Haemocyte changes following in vitro challenge. *Fish Shellfish Immunol.* 20, 374–383.

[2] Allam, B., Paillard, C., Auffret, M., Ford, S.E., 2006. Effects of the pathogenic *Vibrio tapetis* on defence factors of susceptible and non-susceptible bivalve species: II. Cellular and biochemical changes following in vivo challenge. *Fish Shellfish Immunol.* 20, 384–397.

[3] Allam, B., Paillard, C., Ford, S.E., 2002. Pathogenicity of *Vibrio tapetis*, the etiological agent of brown ring disease in clams. *Dis. Aquat. Org.* 48, 221–231.

[4] Bartoli, P. 1974. Recherches sur les Gymnophallidae F.N. Morozov, 1955 (Digenea), parasites d'oiseaux des côtes de Camargue: systématique, biologie et écologie. Thèse, Univ. Aix-Marseille.

[5] Bidault A., Richard G., Le Bris C., Paillard C. 2015. Development of a Taqman real time PCR assay for rapid detection and quantification of *Vibrio tapetis* in extrapallial fluids of clams. *PeerJ* 3:e1484; DOI 10.7717/peerj.1484.

[6] Cheng T.C. 1967. Internal defense mechanisms. *Adv. Mar. Biol.* 5, 60-80.

[7] Combes, C. (Ed.) 2001. Parasitism. The Ecology and Evolution of Intimate Interactions. The University of Chicago Press. 728 pp.

[8] Declercq, A. M., K. Chiers, M. Soetaert, A. Lasa, J. L. Romalde, H. Polet, F. Haesebrouck, and A. Decostere. 2015. *Vibrio tapetis* isolated from vesicular skin lesions in Dover sole *Solea solea*. *Dis. Aquat. Org.* 115, 81-86.

[9] Delaporte M, Soudant P., Moal J., Lambert C., Miner P., Quere C., Choquet G., Paillard C. & Samain J.F. 2003. Effect of a mono-specific algal diet on immune functions in two bivalve species *Crassostre gigas* and *Ruditapes philippinarum*. *J. Mar. Exp. Biol.* 206, 3053-3064.

[10] Desriac, F., Le Chevalier, PA.; Brillet, B.; Leguerinel, I., Thuillier, B., Paillard, C., Fleury, Y. 2014. Exploring the hologenome concept in marine bivalvia: hemolymph microbiota as a pertinent source of probiotics for aquaculture. *FEMS Microbiol. Lett.* 350, 107-116.

[11] Guillou, J., F. Dumas, V. Mariette, and M. Le Pennec. 2002. Simulations de la dérive lagrangienne des larves. *Rapport PNEC.* 11pp.

[12] Jensen S., Samuelsen O. B, Andersen K., Torkildsen L., Lambert C., Choquet G., Paillard C., Bergh O. 2003. Characterization of strains of *Vibrio splendidus* and *Vibrio tapetis* isolated from corkwing wrasse (*symphodus melops*) suffering vibriosis. *Dis. Aquat. Org.* 53, 25-31.

[13] Janin J-M, Marcos F. 1996. Qualité des eaux du Golfe de Morbihan - Utilisation d'un modèle hydrodynamique. Journées Nationales Génie Côtier - Génie Civil, Dinard, pp 297-305. DOI:10.5150/ jngcgc.1996.031.

[14] Le Bris C., G. Richard, C. Paillard, C. Lambert, C. Seguineau, O. Gauthier, F. Pernet, F. Guérard, 2015. Immune responses of phenoloxidase and superoxide dismutase in the Manila clam *Venerupis philippinarum* challenged with *Vibrio tapetis* - Part I: Spatio-temporal evolution of enzymes' activities post-infection. *Fish Shellfish Immunol.* 42, 16-24.

[15] Lopez, J. R., S. Balboa, S. Nunez, E. de la Roca, R. de la Herran, J. I. Navas, A. E. Toranzo, and J. L. Romalde. 2011. Characterization of *Vibrio tapetis* strains isolated from diseased cultured Wedge sole (*Dicologoglossa cuneata* Moreau). *Res. Vet. Sci.* 90, 189-195.

[16] Maes P., Paillard C. 1992. Effet de *Vibrio* P1, pathogène de *Ruditapes philippinarum* sur d'autres espèces de bivalves. *In*: Les mollusques marins, biologie et aquaculture, Ifremer, Actes de colloques, no 14: 141-148.

[17] Maes, P. 1992. Pathologie bactérienne chez deux invertébrés marins d'interêt commercial, *Ruditapes philippinarum* et *Paracentrotus lividus*. University of West Brittan, Brest. 217 pp.

[18] Nöel D., Nicolas J.-L., Boulo V., Mialhe E., Roch P. 1996. Development of a colony-blot ELISA assay using monoclonal antibodies to identify Vibrio P1 responsible for "brown ring disease" in the clam *Tapes philippinarum*. *Aquaculture* 146, 171-178.

[19] Paillard C. 2004. Review of brown ring disease, a vibriosis affecting clams. *Aquat. Living Res.* 17, 467-475.

[20] Paillard C., Allam B., Oubella R. 2004. Temperature effects on Brown Ring Disease susceptibility and defense-related activities in clams *Ruditapes philippinarum*. *Dis. Aquat. Org*. 59, 249-262.

[21] Paillard C., Maes P. 1994. The brown ring disease symptom in the Manila clam, *Ruditapes philippinarum*: establishment of a classification system. *Dis. Aquat. Org.* 19, 137-146.

[22] Paillard C., Maes P., Mazurie J., Claude S., Marhic A., Le Pennec M. 1997. Epidemiological survey of the brown ring disease in clams of Atlantic coast: role of temperature in variations of prevalence. Proccedings of VIIIe Symposium of the International Society for Veterinary Epidemiology and Economics. Paris. AEEMA publications. n° 31-32 (2): pp 14.03.1-14.03.3.

[23] Paillard, C., F. Jean, S. E. Ford, E. N. Powell, J. M. Klinck, E. E. Hofmann, J. Flye-Sainte-Marie. 2014. A theoretical individual-based model of Brown Ring Disease in Manila clams, *Venerupis philippinarum*. *J. Sea Res.* 91, 15-34.

[244] Plana S. Sinquin G. Maes P., Paillard C., Le Pennec M. 1996. Variations in the biochimical composition of the juveniles of Manila clam, *Ruditapes philippinarum*, infected by a *Vibrio sp. Dis. Aquat. Org.* 24, 205-213.

[25] Reid H. Soudant P., Lambert C, Choquet G., Paillard C., H. Birkbeck C. 2003. Salinity effect on the vibriosis model. *Dis. Aquat. Org.* 56, 249-258.

[26] Richard G., Le Bris C., Lambert C., Guérard F., C. Paillard 2015 - Immune responses of phenoloxidase and superoxide dismutase in the Manila clam *Venerupis philippinarum* challenged with *Vibrio tapetis* - part II: combined effect of temperature and two *Vibrio tapetis* strains. *Fish Shellfish Immunol*.44, 79-87.

[27] Rodrigues S, Paillard C, Le Pennec G, Dufour A, Bazire A. 2015. *Vibrio tapetis*, the causative agent of Brown Ring Disease, forms biofilms with spherical components. *Front. Microbiol.* 6:1384. doi: 10.3389/fmicb.2015.01384.

[28] Rosenberg, E., O. Koren, L. Reshef, R. Efrony, and I. Zilber-Rosenberg. 2007. The role of microorganisms in coral health, disease and evolution. *Nat. Rev. Microbiol.* 5, 355-362.

[29] Seabloom, E. W., E. T. Borer, K. Gross, A. E. Kendig, C. Lacroix, C. E. Mitchell, E. A. Mordecai, and A. G. Power. 2015. The community ecology of pathogens: coinfection, coexistence and community composition. *Ecol. Lett.* 18, 401-415.

In: Oysters and Clams
Editor: Jesús L. Romalde
ISBN: 978-1-63485-074-2

Chapter 8

ALTERATIONS IN THE PHYSIOLOGICAL, METABOLIC AND ONTOGENETIC PROCESSES OF OYSTERS AND CLAMS AFFECTED BY ENVIRONMENTAL STRESSORS AND PATHOGENS

Celia Vazquez Boucard[1,2,*], Cristina Escobedo Fregoso[1,3], Larisa Lee[1,2], Emilie Bigaud[1,2] and Gerardo Anguiano[4,5]

[1]Centro de Investigaciones Biológicas del Noroeste (CIBNOR), La Paz, Baja California Sur, México

[2]Laboratory of Proteomic and Genetic Toxicology,

[3]Genomic Laboratory,

[4]Universidad Juárez de Durango. Facultad de Ciencias Químicas, Durango, México

[5]Laboratory of Molecular Biomedicine

ABSTRACT

In recent decades the massive development of human activities and the demographic pressures caused by steady population growth coupled with events of natural origin have combined to generate a substantial intensification of environmental stressors, be they chemical, biotic or abiotic in nature. In the coastal marine environments that harbour mollusc populations the effects of this problematic are reflected in the development of diseases capable of producing massive mortalities in both wild and cultivated organisms, as well as alterations at different levels of biological organization. In order to detect modifications at the molecular, biochemical, cellular, physiological or behavioural level caused by one or more factors of environmental stress, biological indicators have been developed and sentinel organisms utilised. Numerous biomarkers have been employed, some in international biomonitoring programs that regulate, and provide information on, the environmental health of habitats and/or sites where organisms are cultivated. Numerous studies of oysters and clams exposed to environmental stressors have been

[*]Corresponding author: Laboratory of Proteomic and Genetic Toxicology. Centro de Investigaciones Biológicas del Noroeste (CIBNOR), Instituto Politecnico Nacional 190, Colonia Playa Palo Alto de Santa Rita, 23096, La Paz, Baja California Sur, México. Email: cboucard@cibnor.mx.

conducted in relation to ecotoxicology, genetics and ecology. These organisms manifest molecular and cellular responses to environmental stressors in: a) greater DNA damage; b) toxicogenomic damage determined by studies of the transcriptome; c) variations in gene expression and enzymatic activity; d) alterations of such physiological functions as respiration, osmoregulation, detoxification, the immune system, reproduction, energy metabolism, etc. and, e) genetic variability and diversity. Meanwhile, at the level of populations, research has centred on studies of parameters associated with the population dynamics of age, fecundity, abundance, and biometry, among others, that measure total abundance and indexes of biotic and biomass diversity, etc. Evaluating the effects of environmental stressors allows us to better understand the response and adaptive mechanisms of individual organisms and populations exposed to stressful conditions as means of regulating the environmental conditions of cultivation systems, estimating levels of environmental degradation, and opening up the possibility of restoring ecosystems that suffer pressures induced by human activity.

Keywords: bivalves, biomarkers, environmental stress, diseases, genetic, physiology

INTRODUCTION

The massive development of human activities and demographic pressures has caused a considerable intensification of environmental stress factors in areas of great biological and economic importance, such as coastal zones. Tourist, industrial, aquaculture and shipping activities, among others, that increase with the development of urban zones and intensive agriculture in river basins are sources of intense stress for organisms that live, grow, feed and reproduce in coastal environs. The molluscs that inhabit these niches are permanently exposed to foreign chemical particles that can produce toxic effects and even alter molecular and biochemical mechanisms necessary for their genomic and epigenetic integrity. The presence of contaminants in aquatic environments affects the tissues and health of molluscs and likely of their consumers as well, including humans, due to the bioaccumulation and biomagnification processes characteristic of these toxic compounds. Numerous studies have evaluated the effects of exposure to such compounds using biomarkers in both experimental trials and projects involving sentinel organisms. Sentinel species are biological models that permit a broad, detailed identification and evaluation of physiological, metabolic and ontogenetic processes that may be altered by exposure to environmental contaminants. Because of their characteristics as sedentary, filtering organisms, oysters, mussels and clams are considered excellent bioindicators for use in international biomonitoring programs designed to regulate and provide information on the ecological health of the environments where these organisms live and/or are cultivated. In general, responses to these environmental alterations consist of a series of metabolic and genetic events that the organisms utilize to maintain their integrity. During the detoxification phase, some molluscs are unable to eliminate all bioaccumulated contaminants (organic and inorganic), resulting in lesions at different levels of their biological organization. Processes potentially affected include: reproduction (sterility, decreased fecundity rates, low production of viable gametes); digestion (alterations due to oxidizing stress); cell structure (altered DNA); transcription (modified genetic expression); and the immune system (altered immune activity of haemocytes). Determining the effects of this stress will allow us to better understand the adaptive mechanisms and individual responses of populations to such conditions. Current knowledge in biology allows us to propose and apply an ecophysiological focus related to the vital functions of organisms and their responses at the genomic and cellular levels. This strategy enables researchers to determine the adaptive processes of organisms exposed to environmental alterations.

Epidemics produced by the presence of pathogens in aquatic environments cause significant economic losses in international aquacultural production. In molluscs, the incidence of disease outbreaks and the appearance of "emerging" illnesses have increased in recent years, probably due to significant anthropogenic activity. Because bivalve molluscs feed on suspended materials obtained by filtering water, while feeding they can accumulate virus and bacteria at concentrations higher than those actually present in the aquatic environment. Because some mollusc species are not susceptible to certain pathogens, they may act occasionally as vectors or passive carriers.

The objective of this review is to bring together existing information on two main topics. First, it discusses the toxic effect of contaminants detected in aquatic environments and the biological responses that molluscs are capable of presenting when exposed. We also evaluate different biomarkers used to measure the effects of exposure to xenobiotics, emphasizing the importance of recognizing genetic expression in order to identify the relations of coordinated groups of genes that respond to stress factors. Second, it examines the responses produced by pathogens-hosts interactions at the transcriptional level, as well as immunogenic responses in resistant and susceptible species of molluscs caused by pathogens capable of causing massive mortality. Finally, this review offers a general panorama of the current situation of diseases declared by the World Organization for Animal Health (OIE) as causing mortality in oysters and clams.

ENVIRONMENTAL STRESSORS

In coastal marine zones, environmental stress can be chemical, biotic or abiotic in nature. Chemical contaminants include primarily polycyclic aromatic hydrocarbons (PAHs), a broad series of compounds generated as secondary products of combustion processes: e.g., polychlorinated biphenyls (PCBs) applied in commercial and industrial sectors; pesticides (insecticides, herbicides, fungicides, etc.); heavy metals (cadmium, lead, mercury, arsenic, etc.); fertilizers, and drug derivatives, among others. Sources of abiotic stress vary and correspond to fluctuations in numerous physicochemical parameters of the environment, including changes in temperature, salinity, turbidity, pH, UV rays, and dissolved oxygen. These physicochemical parameters can vary as an indirect consequence of the organic or inorganic contaminants present in the environment. Other causes of environmental stress include global climate change, higher sea levels, and soil erosion, as well as anthropogenic pressures that result from industrial activity, demographic growth, over-fishing, deforestation, toxic wastes, etc. [108]. He factors of a natural or anthropogenic character such as those mentioned here can promote the adaptation, acclimatization and evolution of natural populations, including man [28]. The scientific community, authorities and the general public are all becoming increasingly aware of the kinds of damage that these stressors can cause to both the viability of ecosystems and human health.

Utilization of Molluscs as Sentinel Organisms for Environmental Alterations

It is difficult to evaluate contaminants in the marine environment due to factors such as diversity and seasonal and spatial variation. Also, contaminants may be present in the environment in trace amounts, which complicates their detection and assessment difficulty. But the greatest difficulty lies in the nature of these contaminants and the multitude of new molecules introduced day-by-day by human activities, many of which have not yet been identified in their totality by current analytical techniques.

Sentinel species are particularly important biological models in environmental toxicology and ecotoxicology. By obtaining comprehensive and detailed biological information on the physiological, metabolic and ontogenic processes of these species, it becomes possible to identify and evaluate possible biological characteristics or functions that can be utilized as biomarkers of environmental alteration. Another key aspect is characterizing the levels of activity or natural condition of these markers in the sentinel species and the correlation that exists between them and the presence of contaminants in organisms [123, 142]. Bioindicators or sentinel organisms provide indirect indications of the presence, effects and/or sensitivity of contaminants and/or pathogens in the environment they inhabit [142, 152]. Organisms are chosen on the basis of several relevant characteristics, including sampling feasibility, special biological conditions, and the ecological or commercial importance of the species, among others [86].

Molluscs are members of a particularly important taxonomic group due to the role they perform in the structure and functions of ecosystems. Bivalve molluscs (especially mussels and oysters) have been used by international institutions for several decades [88, 91, 107] to monitor the integrity of coastal bodies based on the concept of "mussel watch." In addition, certain species have great economic value in aquaculture and the fishing industry. The main advantages that these species offer include their broad distribution in most aquatic environments, high conservation of biological and metabolic processes, and homology with those of vertebrates. Also, certain features of their biology may favour evaluations of bioaccumulation and responses to toxic compounds, as in the case of filtering benthic organisms with a sessile or sedentary lifestyle in their adult phase, among others [123]. Because of these qualities, when these organisms inhabit contaminated sites where they suffer chronic exposure over several generations, the concentrations of xenobiotic compounds they may accumulate in their soft tissues can be greater than those that exist in the external environment [78].

Molluscs may also be vectors or passive carriers of numerous pathogens, bacteria, parasites and viruses that can impact humans [105]. Vazquez Boucard et al. [150], for example, reported the presence of the White Spot Syndrome Virus (WSSV), crustacean pathogen, in the gills and digestive glands of *C. gigas* oysters placed in water supply canals at a shrimp farm where a WSSV outbreak had occurred. Later studies demonstrated that *C. gigas* oysters can be used as bioindicators of this virus (WSSV) because when they cohabit with infected shrimp or are exposed to seawater containing WSSV, they can accumulate the virus in their gills and digestive glands without becoming infected. The use of a WSSV-nested PCR to screen oysters placed in water canals at the entrance to a shrimp farm allowed detection of WSSV 16 days prior to disease outbreak. The finding that *C. gigas* can concentrate small amounts of WSSV present in seawater without being harmed makes it an ideal sentinel species for use on shrimp farms [152].

Significant efforts have been made to characterize and describe the genic and protein sequences of bivalve molluscs that have ecological or economic importance, thus increasing our knowledge of the molecular character of these species and their potential use as bioindicators of environmental contamination [45, 103, 147].

BIOMARKERS OF EXPOSURE TO ENVIRONMENTAL STRESSORS

Environmental stress has often been evaluated by quantifying the chemical contaminants present in the environment; however, these measures are judged incomplete for assessing the health conditions of an ecosystem and, above all, for estimating the biological and ecological conditions of aquatic systems [1]. Also, abiotic factors like oxygen availability, temperature, salinity, ultraviolet rays and pH can vary under the effects of contaminants and induce additional stress for exposed organisms. Climate change, soil erosion, deforestation, toxic wastes, mass urbanization and industrialization, as well as infections by pathogenic agents may also produce collateral effects that further damage environmental conditions and the health status of human and wild populations [108].

Nowadays, biomolecules are increasingly used to gain information regarding the development of biological processes. Biomarkers reflect the severity of the effects and/or mechanisms of toxic action involved. Ideally, these biomarkers could establish causal relations between the presence of toxins and their adverse effects, give "early warning" signals of the presence and effect of the toxins and/or pathogens present in aquatic environments, and predict effects at higher levels of biological organization [154]. The integral evaluation of biomarkers has been used successfully in conjunction with the quantification of associated toxins in the environment and sentinel organisms to assess the ecological impacts produced by environmental toxins, thus allowing identification of contaminants responsible for environmental deterioration and the sites at greatest risk, while prioritizing their management and monitoring remedial actions [18, 153].

However, success in the search for biomarkers has been only partial and subject to controversy. While a certain number of biomarkers are being used regularly to detect the presence of toxic compounds in the environment –such as pesticides, hydrocarbons and heavy metals– responses can be unclear or ambiguous, which may complicate interpreting results or reduce their value as indicators of alterations in natural environments [64, 129]. According to Sarkar et al. [129], the main advantages of biomarkers of contamination are: a) recognition of the interaction between toxin and organism and evaluations of the effects of exposure at sub-lethal levels; b) detection of the presence of contaminants both known and unknown; c) early sub-lethal detection of the effects of exposure that support decision-making to remedy or prevent ecological disasters; d) spatial and temporal evaluation of bioavailable contaminants; e) indications of the danger or environmental risk due to the presence of a contaminating compound; f) their function in relation to chemically-related toxins; g) contribution to establishing routes of exposure to different trophic levels and prioritize monitoring schemes and remedial strategies; and g) allowing the integration of toxicological interactions that lead to the expression of a cumulative effect at different levels of biological organization.

Causal relations have been established between enzymatic levels of Cytochrome P450, of esterase's (AChE, carboxylesterase) and the activity of ethoxyresorufin-O-deethylase (EROD), concentrations of polycyclic aromatic hydrocarbons and organophosphate pesticides, and the induction of metallothionein in relation to heavy metal levels in the environment [67, 68]. Aquatic organisms have developed antioxidant defence mechanisms both enzymatic, such as superoxide dismutase (SOD) or catalase, and non-enzymatic, such as glutathione, uric acid and bilirubin [5, 8]. The enzyme AChE is present in several tissues, but it is an important target for pesticides in the nervous system. This enzyme is related to the hydrolysis of acetylcholine in the cholinergic synapses [66]. Inhibition of AChE produces an accumulation of neurotransmitters and a disruption of neurotransmission [93]. AChE activity in *C. gigas* has been demonstrated in gill tissue [22]. This study reported two types of AChE, one of which is not sensitive to organophosphate and carbamate pesticides. There are reports of its use as a biomarker of exposure to organophosphate pesticides in some species of bivalve molluscs exposed in experimental media, in biomonitoring programs of wild organisms, and in transplanted sentinel organisms [21, 99]. With respect to the effects of organochlorine pesticides, there are reports of AChE inhibition in *Rudatipes decussatus* clams [84] and *C. gigas* oysters [11] exposed to lindane. However, we cannot discard the possibility that inhibition of the activity of the enzyme AChE in the presence of organochlorine compounds is due to the oxidizing effect of this pesticide on that protein, and not because it presents the characteristics of a specific enzymatic inhibitor, as in the case of organophosphate pesticides and carbamates [119]. Anguiano et al. [10] have shown the effect on AChE inhibition in *C. gigas* exposed to organophosphate, carbamate and organochlorine pesticides using dichlorvos, carbofuran, oxamil and lindane. Another biomarker of exposure to xenobiotics is induction of cytochrome P450 proteins, which act by metabolizing endogenous and exogenous substances including environmental contaminants. P450 cytochromes play a primordial role in detoxification mechanisms, so exposure to non-polar xenobiotics like polychlorinated dibenzo-p-dioxins (PCDD), polychlorinated biphenyls (PCBs) or polycyclic aromatic hydrocarbons (PAHs), leads to the induction of enzymes including 7-ethoxyresorufin-O-deethylase (EROD). EROD activity has been proposed as a biomarker for contaminated aquatic environments, but it has shortcomings, such as variations in its activity as a function of uncontrollable parameters –like water temperature– that produce seasonal variation in its physiological activity. Another problem with P450 activity is the lack of information on predictions of the dose-response relation; a datum that is required before this enzyme can be used as a biomarker for regulatory purposes. Aspects of aquatic invertebrate cytochrome P450 enzymes have been studied, but so far little progress has been made in identifying P450 gene diversity in aquatic invertebrates. The tools of molecular biology have aided greatly in this search, and are likely to identify as much diversity in this protein superfamily in molluscs as occurs in higher marine and terrestrial organisms [138].

Edge et al. [54] examined the relationship between biomarker responses by oysters in the presence of metals and PAHs, and determined that lysosomal membrane destabilization proved to be the most useful indicator. We suggest that this may be an especially useful biomarker for incorporation into local environmental monitoring programs.

The genetic damage caused by exposure to xenobiotics varies widely. The cell cycle may be affected by lesions or molecular damage to its DNA strands that may induce blockages of phases of the cycle until the lesions are repaired. If the damage is not corrected, the cell will be not eliminated by apoptosis and was duplicated, which may trigger problems such as mutagenesis, clastogenesis, teratogenesis or carcinogenesis. The compounds that induce this type of damage are called genotoxins, and include benzopyrene, atrazine and some organochloride pesticides [25, 127, 136].

One of the tests most commonly used to evaluate genotoxicity in molluscs is the micronucleus assay, conducted with haemocytes and gill cells (e.g., [82]). This assay consists in estimating the proportion of the smallest nuclei (i.e., micronuclei) formed by the fragmentation of chromosomes during cell division and as a result of a deficient spindle apparatus [130]. A larger proportion of micronuclei has been observed in molluscs from sites exposed to contamination by petroleum, persistent organic contaminants, and metals [16, 60, 79, 94]. Other aberrations of the nuclei may also appear, such as binucleated cells and nuclear buds, which have been associated with toxic substances [16, 94]. Despite the widespread application of this test in the biomonitoring of molluscs, it is important to consider that physiological or environmental factors may occur that induce the formation of micronuclei without necessarily being associated with exposure to contaminants. These factors include water temperature, the cell type used, salinity, the organism's size and, possibly, diet [15].

The comet assay (i.e., single-cell gel electrophoresis assay) can also be used to evaluate genotoxic damage in organisms since it can detect breaks in the DNA chains of individual cells and estimate the proportion of damage to the DNA in the cells of the tissue studied [43]. In molluscs, this test has been conducted with gill cells, digestive gland cells, and haemolymph [11, 44, 122, 151].

Cells from different tissues may have distinct sensitivities to some xenobiotic substances. For example, in *Corbicula fluminea* a direct relation was found between the concentration of methyl methanesulfonate, an alkylating agent and carcinogen, and the number of cells with genetic damage in haemolymph and digestive gland cells. The gill cells, however, presented an exponential increase in damage as well as more basal damage than the cells from other tissues [122]. In addition, different mollusc species may vary in their sensitivity to certain toxic agents. For example, the haemocytes of the common cockle (*Cerastoderma edule*) are more sensitive to damage by peroxide than those of mussels (*Mytilus edulis)* [38], which are those most commonly used in ecotoxicological monitoring. In a similar study, mussels and clams (*Mya arenaria*) were exposed to soluble fractions of crude petroleum. The haemocytes from the exposed organisms showed no significant increase in damage in either species. However, an increase in genetic damage was observed in the gill cells only in mussels, not in clams [83]. These studies indicate the need to evaluate the sensitivity of different organisms and tissues to contaminants before choosing one species or tissue type as a biomarker.

The micronuclei and comet tests have also been used jointly to evaluate genotoxicity in diverse mollusc species, and results from some of those studies are in agreement. For example, zebra mussels (*Dreissena polymorpha*) transplanted to sites with distinct levels of contamination showed greater genetic damage in the most contaminated sites compared to reference sites using both tests [92]. In common mussels (*M. edulis*) exposed to tributyltin the genetic damage determined in haemolymph cells by the comet test agreed with that detected by the micronuclei test [82]. Similar agreement in results using these two tests has been found in golden mussels (*Limnosperna* fortunei), Mediterranean mussels (*M. galloprovincialis*), and green mussels (*Perna viridis*) [46, 136, 155, 156]. However, results from both tests are not always in agreement. For example, the comet assay found more severe genetic damage in zebra mussel haemocytes exposed to decabromodiphenyl ether (BDE-209) than in controls, but the micronuclei test failed to detect any difference between haemocytes from the control and exposed mussels [125]. A study conducted with Asian clams (*Corbicula fluminea*) also showed less sensitivity using the micronuclei test in terms of detecting damage caused by aromatic hydrocarbons present in gasoline compared to the comet test [61]. Distinct results from these two tests could be due to differential responses to the substance of exposure; for example, if the substance is capable of inducing primary damage that can be repaired, but not chromosomal damage, which is detected by analysing micronuclei [124]. It is also possible that the exposure time necessary to detect damage caused by one substance differs between these two tests. Finally, the comet assay can detect damage after short exposure periods, but this is only possible under certain circumstances for the micronuclei test. Depending on the specific xenobiotic involved, longer exposure times may be required to detect damage [61, 116].

In addition, studies have shown that genetic damage can fluctuate over time. Shaw et al. [135], for example, found seasonal differences in DNA damage measured using the comet assay in mussels that were not associated with levels of PAHs found in the mussels' tissues. Those authors emphasize, on the one hand, that these differences may be associated with fluctuations in contaminant levels over time, or perhaps with variations in other factors, such as temperature, precipitation, or the composition of the organisms themselves. On the other hand, genetic damage may reflect the efficiency of the organisms' mechanisms of genetic repair. For these reasons, they suggest that the comet assay is useful as a final biomarker of genotoxicity, but not as a biomarker for one specific compound [135].

The literature on biomarkers for genotoxicity in molluscs is extensive [23]. In light of the usefulness of certain organisms as sentinels capable of detecting contamination in aquatic environments, and the need for continuous evaluation of ecosystem status, we can expect that more of these studies will be conducted in the future. However, biomarkers of genotoxicity are not specific; that is, different compounds and factors –exogenous and endogenous– can influence the damage that an organism's genetic material suffers. Therefore, it is necessary to develop and implement monitoring techniques that do not depend on just one biomarker, but that include a series of physiological biomarkers capable of detecting a broader panorama of the effects that xenobiotics have on organisms and ecosystems in general.

More recently, the search for novel biomarkers has focused on the application of OMICS methodologies. Environmental proteomics focuses on the analysis of an organism's proteome and the detection of changes in the level of individual proteins/peptides in response to environmental stressors. Proteomics can provide a more robust approach for the assessment of environmental stress and therefore exposure to pollutants [34].

RESPONSES OF OYSTERS AND CLAMS TO ENVIRONMENTAL STRESSORS

Cellular and Molecular Alterations

The various biological systems integrated in living beings may respond differently to aggression by a toxin. The first level of integration of responses to xenobiotics is that of the molecule and cell. Molecular responses can be studied on the basis of damage to genetic material and its subsequent repair, by considering variation in genes or proteins, or by examining variation in activity and structural damage to proteins. Some metabolic pathways may also be stimulated when xenobiotic compounds are detected at the cellular level. Cellular responses are triggered by xenobiotic receptors that detect such compounds and then produce a series of enzymatic activities that allow them to intervene in detoxification processes. The main physiological and cellular processes affected by xenobiotic compounds include: transmission of nerve impulses, the cell cycle, and digestive, reproductive and excretory functions, as well as the immune defence system. Various biological factors influence sensitivity to the lethal effects of xenobiotics, such as the lipid content of the exposed organisms, their metabolic condition, and the physiological processes or states that exist at any one time, as well as stress and general health condition of the exposed organism We know, for instance, that lethal concentrations of pesticides and exposure times in molluscs are higher than CL_{50} levels for fish species [76]; likely due to bivalves' capacity to control the flow of water through their valves, which becomes their first response to the presence of stressor agents like pesticides, heavy metals, or radioactive elements [53]. Added to the reduction in exposure gained by closing their valves, bivalve molluscs also have the ability to modify their metabolic and physiological activity such that they can subsist for prolonged periods without ingesting food [137]. Moreover, these molluscs tend to present greater accumulation of lipophilic toxic compounds in their fatty tissues due to differences in the capacity of their digestive glands to metabolize exogenous compounds [102]. In bivalve organisms, water filtration activity has been used as an indicator of environmental stress [47]. Anguiano et al. [11], for example, demonstrated significant decreases in the percentage of feeding activity in *C. gigas* oysters exposed to the pesticide lindane, quantified as the filtration of micro-algae over time (i.e., filtration rate). Nevertheless, the reduction in filtration rates does not necessarily reflect toxic damage by lindane, though it is likely to be a sign of stress that, if maintained over a sufficient period of time, could prove detrimental to the organisms' health. Moreover, these bivalve species are a source of contamination for their predators due to the latter's low capacity for biotransformation and high efficiency in accumulating certain toxic agents, which ultimately alter the functioning of ecosystems. Compounds with neurotoxic properties induce alterations in neuronal communication processes such as the inhibition of AChE activity in *C. gigas* oysters produced by organophosphate pesticides and carbamates [22], and, in *Ruditapes decussatus* clams, alterations in the activity of ionic canals in neuronal membranes due to exposure to the organochloride pesticide lindane [84].

Regarding the respiratory process, the presence of xenobiotics can modify breathing rates and generate cellular alterations in tissues due to the peroxidation of membranes and the oxidation of macromolecules caused by the action of reactive oxygen species (ROS) [19, 156]. The production of ROS is prevalent in the world's oceans, and oxidative stress is an important component of stress responses in marine organisms exposed to a variety of insults that result from changes in environmental conditions, such as thermal stress, ultraviolet radiation and pollution [97].

Research has described alterations in the digestive process of bivalve molluscs due to oxidative stress and the accumulation of liposoluble toxic organic substances in tissues with high lipid content. These compounds induce cellular responses to ROS, enzyme biotransformation, and the elimination of hydrophobic molecules such as the monooxygenase system (cytochrome P450, reduction of glutathione) and the synthesis of chelating proteins from such metals as the metallothioneins [17, 69, 139]. In bivalve molluscs, the main excretory organs are the pericardial glands and kidneys. The latter play an important role in eliminating hydrocarbons and toxic metals stored in lysosomes and eventually eliminated through the urine. Accumulation of these metals in renal tissues has been reported to increase in the presence of kidney damage [140].

The effect of toxic compounds on reproductive processes is the most widely-documented aspect due to the magnitude of the damage that may occur in wild populations, since this directly impacts the ability of species to subsist over time. Organic pollutants that exhibit endocrine-disrupting activity (Endocrine aquatic Disruptors-EDs) are considered ubiquitous contaminants that occur frequently in a broad range of aquatic ecosystems and are resistant to degradation in the natural environment. Although their potential is low compared to that of natural hormones, their concentrations in the environment may be sufficiently high to produce adverse biological effects [121]. Compounds like polychlorinated biphenyls, organochloride pesticides, plasticizers and surfactants all have the ability to induce, among other alterations, the feminization of male and masculinization of female *C. gigas* oysters [7], and decreased fecundity rates in *C. virginica* oysters [40]. Exposure to estradiol can also cause male-to-female sex-reversal in *C. gigas* oysters. Nonylphenol, a common component in personal care products, affects larval development [115], sex ratios, gamete viability [114], and sperm motility [113] in *C. gigas*, as well as delayed gametogenesis in *Mya arenaria* clams [73]. Due to the persistent nature of most xenobiotic estrogens and their metabolites, many can bioaccumulate and biomagnify in different environmental compartments, including molluscs. The cellular responses of *C. virginica* to titanium dioxide nanoparticles were studied by Johnson et al. [90], who proved that at low concentrations they can cause sub-lethal toxicity which can improve under conditions of natural light. In estuary environments where it is probable that these nanomaterials will accumulate, agglomeration rates, interaction with organic materials, and responses to sunlight are fundamental for determining the extent of their bio-reactivity and biological impacts.

Induction and Inhibition of Proteins

The effects induced by the pesticides metconazole and isoproturon in *C. gigas* were studied by Epelboin et al. [57], who found an over-activation of the sensing-kinase AMP-activated protein kinase α (AMPKα), a key enzyme involved in energy metabolism and, more specifically, glycolysis, together with a decrease in hexokinase and pyruvate kinase activity; whereas 2-DE proteomic revealed that fructose-1, 6-bisphosphatase (F-1, 6-BP), a key enzyme in gluconeogenesis, was up-regulated. Both pesticides enhanced the superoxide dismutase activity of oysters. Isoproturon enhanced catalase activity, and metconazole enhanced peroxiredoxin activity. The effects of a mixture of pesticides containing alachlor, metolachlor, atrazine, terbuthylazine, diuron, fosetyl aluminium, carbaryl, and glyphosate on the physiology of *C. gigas* was analysed by Geret et al. [75]. The concentration of the mixture of these pesticides corresponded to pesticide concentrations in the shellfish cultivation area of the Marennes-Oleron basin in France. Changes in metabolic activities characterized by increased glutathione synthetase (GS), catalase (*CAT*), and glutathione 5-transferase (*GST*) levels were found, together with the formation of DNA adducts and alterations in haemocyte parameters. Anguiano et al. [11] demonstrated the inhibition of AChE in *C. gigas* as a response to exposure to dichlorvos, carbofuran, oxamyl and lindane.

The sensitivity of immune systems to diverse xenobiotics like PBC, PAHs or pesticides has been studied in diverse organisms. While oysters have no adaptive immunity, they can prosper as filterers in estuaries rich in pathogens. Due to these unique adaptations, they provide interesting models for the study of the evolution of host-defence systems. Studies with bivalve molluscs have associated increases in the immune activity of haemocytes to the presence of xenobiotics in tissues [118]. Results reveal a depression of phagocytosis and several subcellular and physiological changes in oysters exposed to cadmium alone, or to cadmium and copper, suggesting an overall alteration of the phagocytic function [13]. Lacoste et al. [96] provided valuable information on the influence of stress on immune functions in oysters by studying the number of circulating haemocytes, migratory and phagocytic activities, and the production of ROS in haemocytes, measured before, during, and after, application of the stressor.

Induction of vitellogenin (VTG), the precursor molecule of yolk proteins, has been used extensively as a biomarker in both laboratory and field studies with numerous species (101). Single cDNA sequence encoding of two vitellogenins in *C. gigas* was isolated and characterized *(Cg-Vtg1(100)*;*Cg-Vtg2)* [98], providing the opportunity to investigate regulation of vitellogenesis and the interactions of endocrine modifiers in this process. It is well-known that the development of gametes and vitellogenesis in marine bivalve molluscs are targets of endocrine modifiers; however, additional knowledge of the endocrine function in molluscs and its role in reproduction is required. We need to determine the concentrations of endocrine modifiers in a realistic environmental range, and the base levels of vitellogenins with their seasonal variation, before we can begin to utilize VTG as a biological marker of exposure to endocrine disrupters [121].

Genetic Alterations

The DNA of organisms exposed to various factors of environmental stress can suffer several different genetic alterations. Genetic lesions of the point mutation type can occur, threatening to alter not only the gene sequence that is affected directly, but also its expression and the function of its expressed product (i.e., proteins). Genotoxic agents can cause chromosomic lesions that modify the number (aneuploidy or polyploidy) or structure of chromosomes (inversions, breakage, etc.). Damage to the double strand of DNA can be irreversible on organisms (carcinogenesis, teratogenesis) and may also affect the reproduction and "fitness" of individuals (i.e., their capacity to transmit their genes to subsequent generations [149]. Barranger et al. [14] determined the genotoxic effects of diuron exposure on oyster genitors, the possible transmission of damaged DNA to offspring, and its repercussions for oyster fitness. Results obtained by the Comet assay clearly showed a higher level of DNA strand breaks in both the haemocytes and spermatozoa of diuron-exposed genitors. The transmission of damaged genetic material to gamete cells could be responsible for the genetic damage measured in offspring. Anguiano et al. [11] evaluated organismal toxicity, cytotoxicity, genotoxicity and filtration rates in response to different concentrations of sub-chronic lindane exposure (gamma-hexachlorocyclohexane) in *C. gigas*. Their results revealed a sensitive effect categorized as filtration rate < genotoxicity < cytotoxicity < mortality, and revealed that a toxicant can have toxic effects at different concentrations depending on the effect level evaluated. Molecular and cellular biomarkers were more sensitive to lindane than organism mortality, and filtration rates were more sensitive than either the molecular or cellular endpoints examined in that study. The embryotoxic and genotoxic effects of diuron, glyphosate and Roundup in oysters were studied by Akcha et al. [3]. Only diuron affected embryo-larval development from the lowest concentration of 0.05 μg L−1. Compared to the other two substances, diuron was also genotoxic for spermatozoa from 0.05 μg L−1 upwards. Conversely, it did not affect sperm mitochondrial function or acrosomal membrane integrity. Hamoutene et al. [83] determined DNA alterations in clams exposed to water contaminated with petroleum hydrocarbons, while Vazquez Boucard et al. [151] studied *C. gigas* oysters exposed to high concentrations of pesticides on the coasts of Sonora and Sinaloa, Mexico, determining that they are capable of significantly damaging DNA. Considering that important aquaculture and fishing activities are conducted in that study zone, human health is equally at risk if these foods are consumed in high amounts.

Bivalve molluscs are key model species for monitoring aquatic pollution by engineered nanomaterials (ENMs). In general, the main organ for ENM accumulation is the digestive gland, but its fate at the cellular level differs according to nano-specific properties, experimental conditions, and the particular species involved. The endosomal-lysosomal system and mitochondria are the main cellular targets of ENMs. The mode of action of metal-based ENMs is related primarily to the dissolution and/or release of the chemical component of the particle that induces immunotoxicity, oxidative stress and cellular injury to proteins or membranes, as well as DNA damage [126].

Anthropogenic environmental toxins can affect genetic diversity at both the individual and ecosystemic levels, as well as the genetic structure of entire populations. Fernandez Tajes et al. [62] examined the temporal pattern of genetic diversity among populations of bivalve pod razors (*Siliqua ensis*) in two localities in Galicia in the aftermath of an oil spill that dumped 63,000 tons of heavy fuel into a marine environment. They observed a significant reduction in genetic diversity.

In 1997, Moraga and Tanguy pioneered the search for genetic markers in *C. gigas* exposed to contaminants of anthropogenic origin [109]. They demonstrated a difference in the allozymatic genotypes of their study populations that varied according to the specific contaminant analysed. The six enzymatic genes determined (Aat-2, Ak, Pgdh, Pgi, Cap-l and Pgm) to be involved in physiological processes were considered as potential genetic markers. Later, those authors examined the effects of atrazine, isoproturon, alachloro, metolachlor and diuron on five enzymatic systems (AK, IGP, Cap, PGM, and MDH) implicated in physiological processes, and identified 6 alleles and 5 genotypes (Ak, IGP and PGM) as markers of resistance (110). More recently, analysis of the genetic structure of *Ruditapes decussatus* and *Ruditapes philippinarum* clam populations revealed that organisms inhabiting environments contaminated by heavy metals exhibit higher allelic diversity and possess alleles at PGM loci that could be selected for the presence of heavy metals [111].

Alterations of the Transcriptome

Exposure to xenobiotics, and other types of stress, can produce the inhibition or over-expression of genes that are sensitive to such stress. The mechanisms that regulate transcription and translation are complex, so any modification of gene expression is a precursor event that may precede other manifestations of toxicity, such as biological effects at higher levels of organization [157]. Changes in gene expression are considered important compounds in adaptation to the environment [133]. The great diversity in immune and stress response genes exhibited by expanded gene families, together with high-sequence and structural polymorphisms, may be central to oysters' adaptation to highly-stressful and widely-changing environments [81]. Chapman et al. [35] evaluated changes in gene expression in *C. virginica* oysters in relation to physicochemical conditions and the levels of metals and other contaminants in their environment. The transcriptomic data obtained provided a view of physiological response mechanisms to temperature and pH that were consistent with the known effects of these factors on the physiological functions of ectotherms, and indicated important links between transcriptomic and physiological results. The low levels of contaminants found played a relatively small role in the organisms' responses. At the molecular level, a high number of proteins and genes related to the response of *C. gigas* to xenobiotic compounds have been described: metallothioneins [144]; heat shock proteins 70 (HSP70) [26]; multi-xenobiotic response protein [106], and phosphoglucomutase (PGM) [143]. The transcriptional effects of exposure events to herbicides in *C. gigas* were examined by Tanguy et al. [145] and Boutet et al. [27], while Medeiros et al. [104] identified the influence of domestic sewage on gene expression in this species. The expression of genes associated with biotransformation pathways (CYP356A1, GSTO, MDR, FABP and ALAS) was also analysed, observing a significant induction in all genes examined in the sewage-exposed oysters. These genes are related to phases I (CYP356A1), II (GSTO), and III (MDR) of biotransformation systems, to the uptake and transport of hydrophobic ligands (FABP), and to the synthesis of the prosthetic group heme (ALAS). Those organisms were able to survive in contaminated conditions because their protective mechanisms were properly stimulated. Differential gene expression after exposure to heavy metals in *C.gigas* has been examined by Zapata et al. [160], Choi et al. [39] and Ivanina et al. [89].

Working with *C. gigas*, Moreau et al. [112] observed that the effects of metaldehyde on oysters may differ from one individual to another in relation to genetic diversity. Many microarray and suppression subtractive hybridization (SSH) studies have analysed the effects of environmental stress on gene transcription in marine species; however, there have been no unifying analyses of these data to identify common stress response pathways. Anderson et al. [9] conducted a meta-analysis of 14 studies that probed the effects of different environmental stressors on gene expression in oysters, and suggested a consensus model of sub-cellular stress responses. Exposure of *C. gigas* to *Gimnodinium catenatum*, a paralyzing toxin producer, caused genotoxic damage, especially to the *cyp356A1* gene that is involved in inflammatory processes, in regulating the cell cycle, and in detoxification. In addition, changes in the cytoskeleton arrangement could diminish the health of these oysters and thus increase their susceptibility to diseases [70]. David et al. [48] identified a new physiological and genetic indicator of multiple environmental stress response in *C. gigas*; namely, the peroxiredoxin 6 gene. Pharmaceuticals such as anti-inflammatory non-steroidal drugs are frequently detected in aquatic ecosystems. Serrano et al. [134] evaluated the effects of Ibuprofen (IBU) on antioxidant status in *C. gigas* oysters. The transcription of cytochrome P450 genes (CYP2AU2, CYP356A1, CYP3071A1, CYP30C1), glutathione S-transferase isoforms (GST-ω-like and GST-π-like), cyclooxygenase-like (COX-like), fatty acid binding protein-like (FABP-like), caspase-like, heat shock protein-like (HSP70-like), catalase-like (CAT-like) and catalase activity (CAT), glutathione peroxidase (GPx), glutathione reductase (GR), and glutathione S-transferase (GST) were also evaluated in the gills of oysters. A significant upregulation of CYP2AU1, CYP356A1, CYP3071A1, GST-ω-like, GST-π-like, COX-like, and FABP-like was observed in oysters exposed to IBU under different experimental conditions. *C. gigas* exposed to environmentally-relevant concentrations of IBU exhibited an increased transcription of certain genes and alterations of antioxidant and auxiliary enzymes that could, in the long term, cause damage to exposed organisms.

Contamination by hydrocarbons perturbs the metabolism of marine bivalve *C. gigas* oysters. To better understand the response of this organism to this type of contaminant, Boutet et al. [27] studied the gene expression implicated in the disintoxication of these compounds (cytochrome p4501A1-like protein, cytochrome b(5), flavin-containing monooxygenase 2, and glutathione S-transferase omega class), protection against oxidative stress (copper/zinc superoxide dismutase) and cell protection (heat shock protein 70 family). They found an increase in the level of mRNA of all the genes studied. David et al. [49] studied the molecular response to hypoxia stress and identified 10 genes that codified for the proteins and enzymes implicated in different physiological pathways, such as respiration, carbohydrate metabolism, lipid metabolism, oxidative metabolism and the immune system, protein regulation, nucleic acid regulation, metal sequestration and stress response. These genes can be utilized as biomarkers of environmental stress in molluscs. Tanguy et al. [146], meanwhile, performed a molecular characterization of the glutamine synthetase gene (CgGSII) in *C. gigas* and determined the expression in response to exposure to xenobiotic hydrocarbons, glyphosate and a mixture of atrazine, diuron and isoproturon, and developmental stage. CgGSII expression was highly-regulated in xenobiotic-exposed oysters according to developmental stage. Collin et al. [42] studied the response of *C. gigas* to a pesticide mixture consisting of lindane, metolachlor and carbofuran, and examined the expression of three genes implicated in iron metabolism and oxidative stress, as well as two ubiquitous stress proteins. However, because of the low dosage of the contaminants studied they found no significant response in the expression of the genes and proteins examined, though they did identify one strong, site- and organ-specific response to those pesticides. This study thus, (1) provides insight into bivalve responses to pesticide pollution at the level of the transcriptome, the first level of response for organisms facing pollution; and, (2) raises interesting questions concerning the importance of the sites and organs studied in the toxicogenomic field. Anguiano Vega [12] identified the main processes affected by exposure to lindane in *C. gigas*: translational and transcriptional processes; metabolism and electronic transport; motility and cellular structure; cellular cycle, and stress responses.

The main caveat of genomic techniques is that although they provide important information about changes at the mRNA level, these might not correlate with the protein content. Any mRNA produced can be degraded rapidly, translated inefficiently or edited by attachment of lipid, carbohydrate or phosphate residues [71]. The proteome, in contrast to the genome, is more complex and dynamic, changing in response to time, cell type and environmental stressors, subjected to translational and PTM regulation For instance, combining transcriptomic tools with functional genomics was utilized to study the interactive effects of a pesticide and heavy metal mixture in marine bivalves [52].

Epigenetic mechanisms induce changes in gene activity without altering the DNA sequence and these changes can be transferred to multiple generations, even if these are no longer exposed to the external factors that originally induced the epigenetic change. In ecotoxicology epigenomic applications are very limited although epigenetic changes can help to elucidate the effects of environmental stress caused by metal, persistent organic pollutants or endocrine disrupting chemicals. Extremely few epigenetic studies in oysters or clams have been performed to date. Gavery and Roberts [74], showed, for the first time, how DNA methylation has regulatory functions in the Pacific oyster C. gigas, especially in gene families involved in stress and environmental response [34].

Physiological Alterations

Global climate change represents a threat to biota because it increases atmospheric CO_2 which raises the temperatures of the world's oceans and causes acidification and hypercapnia, while decreasing carbonate saturation. Together, these factors have harmful effects on invertebrate development. Byrne [32] has described the ecotoxicology of global climate change, identified early life history bottlenecks in marine invertebrates, and variable species' responses, and presented variable experimental approaches that emphasize the sensitivity of marine organisms to changes in the ocean and the importance of this for the conservation of species by assuring that all ontogenic states are completely successful. Among the multi-stressor impacts of the heating and acidification of the oceans on marine invertebrates, Byrne and Przeslawski [31] have shown that fertilization can be robust to near-future warming and acidification, depending on the male-female mating pair. Although the larvae and juveniles of some species tolerate near-future levels of warming and acidification (+2°C/pH 7.8), projected far-future conditions (ca. $\geq$ 4°C/ $\leq$ pH 7.6) are severely deleterious, as they will cause reductions in the size and survival of larvae. These authors affirm that some species will show greater resistance than others to climate change and so may be able to tolerate, in the near future, oceanic changes through acclimatization or adaptations. Despite the abundant studies of the responses of organisms to multiple stress factors, none have yet adapted to the predictive model of global climatic warming that implies warmer oceans with reduced salinity. Cheng et al. [36] tested the interactive effects of diel-cycling hypoxia under conditions of warming and decreased salinities using ecologically-realistic exposures. They found no evidence of negative synergistic effects on oyster growth, but only additive and opposed effects of hypoxia (detrimental) and warming (beneficial). They concluded that studies of multiple environmental stressors can be greatly improved by integrating ecologically-realistic exposures and the timing of stressors found in nature with organismal life-history traits.

Based on a study of physiological parameters (condition index, valve growth, respiration and excretion) together with molecular analyses, Clark et al. [41] presented a first study designed to understand the energetic consequences of cumulative thermal stress on intertidal *C. gigas*, which could significantly impact coastal ecosystem biodiversity and functions in the future. Low salinity is one of the main factors that limits the distribution and survival of marine species. The *C. gigas* oyster is categorized as tolerant to relatively low salinity. Zhao et al. [161] monitored the genes that regulate responses to osmotic stress using next-generation sequencing, and identified a number of candidate genes as potential markers of tolerance to hypoosmotic stress in oysters.

In the past 50 years, important episodes of mortality during the cultivation of Pacific *Crassostrea gigas* oysters have been registered in different areas of the world [55]. Up to a certain point, these mortalities could be considered normal due to the high rate of reproduction (up to 40 million eggs per spawn) characteristic of this species. However, mortality rates among natural Pacific oyster populations remain low, as long as environmental conditions allow [20, 51, 131, 132]. Also, mortality among oysters during the summer season is a well-known, recurrent phenomenon on oyster farms that can affect both spawn and adults, and diploids and triploids [37]. In most cases, the cause of mortality is multifactorial and includes adverse physiological conditions, cultivation techniques, environmental conditions, genetic predisposition, and the presence of specific pathogens [128].

PATHOGENS IN OYSTERS AND CLAMS

Widely-dispersed pathogens that cause high mortality are regulated by the World Organisation for Animal Health (OIE). The pathogens that the OIE lists for molluscs include: the protozoans *Perkinsus marinus*, *Perkinsus olseni*, *Bonamia exitiosa*, *Bonamia ostreae*, and *Marteilia refringens*; the bacterium *Xenohaliotis californiensis*, and the abalone herpes-type virus. One virus not on the OIE's list but that also causes high mortality is the herpes-like microvariant virus (OsHv-1 μvar). In Mexico, the presence of the protozoan *P. marinus* has been detected in *Crassostrea gigas* [56], *Crassostrea virginica* [29, 80], and *Crassostrea corteziensis* oysters [33, 58]; while the OsHv-1 virus has been identified in *C. gigas* oysters and *Nodipecten subnodosus* clams [6], and the bacterium *Xenohaliotis californiensis* has been recorded in several species of *Haliotis* abalone. These are all important species in aquaculture, indeed *C. gigas* is the most widely-cultivated species of bivalve mollusc in Mexico. With regard to fishing, the *N. subnodosus* clam was of great importance because it has the largest abductor muscle of all bivalve species. It is one of the most important maritime species exploited in Mexico, followed by abalone. In recent years, however, catches of *N. subnodosus* have declined drastically, though the causes of high mortality rates in this species remain unknown. Unpublished studies suggest they could be due to a series of adverse environmental conditions compounded by the presence of pathogens. At present, the only pathogen on the OIE's list recorded in *N. subnodosus* is the herpes-type virus; however, the low number of viral copies suggests this could be a reservoir species for the virus, or that sampling has not been conducted during the period when this virus replicates [6].

Significant increases of mortality events involving Pacific *C. gigas* oysters were detected in 2008 and 2009 in mollusc-breeding zones along the entire French littoral that affected very different ecosystems, and in three such zones (bays) on the coast of Ireland. With respect to the etiological agent involved in those events, they were attributed to a combination of factors: the presence of herpesvirus-1 (OsHV-1), a species of *Vibrio*, and environmental factors. The scientific report issued by the EFSA AHAW Panel [55] regarding these problems cited the following: 1) the role of herpes virus ostreid (1-OsHV) and *Vibrio aestuarianus* in oyster mortality; 2) the efficacy of water treatment to inactivate OsHV-1 and *V. aestuarianus*; 3) the viability, availability and efficacy of preventive and control measures implemented to guard against diseases; 4) growth, water temperatures above 16°C, and water treatments (free of pathogens) in the effluents of seawater during the culling of organisms; and 5) the installations at the cultivation centres. Wild populations of *C. gigas* also contribute to propagating OsHV-1, and only a few zones in Europe are still free of this infectious agent. Fleury and Huvet [63] have demonstrated the importance of aging on the innate immunity of

C. gigas during episodes of massive mortality in the summer season; hypothesizing that this is the main determinant of their differential survival in the field.

Perkinsus marinus in Crassostrea

The first indications of the presence of this protozoan that causes high mortality were detected in 1948 in *C. virginica* oysters on the coast of Louisiana, U.S.A. Later, it was also found in oysters on the eastern coast of the U.S. and as far south as the state of Tabasco on the Gulf of Mexico [2]. This broad dispersion has allowed the genetic diversification of isolates such that pathogenicity now varies among genotypes. The *virginica* strain is the most virulent one [30]. Studies of oysters in northwestern Mexico describe that the genotype is phylogenetically-related to low-pathogenicity genotypes [58], according to Bushek and Allen's classification [30].

As far as we know, the most susceptible host for *P. marinus* is *C. virginica*, although it has been detected as well in cultivated *C. corteziensis* and *C. gigas* oysters in northwestern Mexico, with no reported cases of mortality due to its presence. This is because of the hosts' different inter-specific immune response mechanisms and genetic variability, differences in the pathogenicity of *P. marinus,* and variations in the environmental conditions at distinct cultivation sites.

High mortality caused by this disease has occurred in *C. virginica* on the eastern coast of the U.S., where it has been determined that the disease develops in relation to the advancing age of organisms. Greatest proliferation occurs among 1-to-2-year-old hosts, beginning with infections of the stomach and intestinal epithelium that later may spread into other tissues. *Perkinsus* is transmitted among hosts through faeces, and can remain in water for long periods until it finds suitable conditions for proliferation [120]. The optimal proliferation temperature is 28-to-32°C [72], but in conditions of low salinity proliferation decreases.

Immune Response of Oysters to Prevent Infection by *P. marinus*

Recognition and Phagocytosis

P. marinus is an intracellular protozoan recognised by the CvGal lectin of *C. virginica* haemocytes [148] that phagocyte and destroys it. However, the low percentage of phagocyted haemocytes indicates that *P. marinus* can survive phagocytosis in susceptible organisms. Observations of *P. marinus*-infected *C. virginica* under mechanical stress show that the percentage of phagocytosis is lower than in non-infected organisms, as is the expression of the high-mobility group protein gene associated with phagocytosis [125], given that phagocytosis is an early defence mechanism in non-stressed organisms. The fact that *P. marinus* can survive phagocytosis in *C. virginica* and *C. ariakiensis* [4], which are considered resistant species, indicates that this pathogen evades phagocytosis in order to proliferate. This suggests that the host turns to other mechanisms, perhaps apoptosis, to destroy the cell and prevent proliferation of *P. marinus*.

APOPTOSIS AS A SYSTEM OF DEFENCE AGAINST PATHOGENS

Cell death by apoptosis can occur when there is cell damage, or in the presence of intracellular pathogens, to prevent the infection of neighbouring cells. In this process the chromatin is condensed, the nucleus fragments, apoptotic bodies form and, finally, cell death occurs, but without causing inflammation. The extrinsic pathway for apoptosis can be activated by pathogens through signalling for a cascade of caspases that triggers DNA fragmentation and so prevents proliferation.

In species susceptible to *P. marinus*, like *C. virginica*, apoptosis is regulated by the pathogen that, upon infecting the host, causes a decrease in the number of apoptotic haemocytes, similar to infection by the protozoan *Haplosporidium nelsoni* [141]. This prevents cell destruction and allows the pathogen to become established. Induction of apoptosis via caspase-3 in *C. virginica* in response to *P. marinus* has been observed at 4 hrs post-infection

[87], but caspase-3 can be regulated by apoptosis inhibitors, as has been seen in *C. corteziensis* [58]. Under conditions of prolonged infection, it has been shown that caspase-3 activity has no relation to the infection, suggesting that other caspases may be involved, or that apoptosis may occur via the apoptosis-inducing factor [87].

Oysters induce apoptosis in the early stage to prevent infection by *P. marinus*, but the most virulent isolates of *P. marinus* can inhibit apoptosis for prolonged periods (8-24 hrs) compared to the less virulent ones (2-4 hrs) [87]. This early response to prevent proliferation also occurs in *C. gigas* oysters, a species resistant to *P. marinus*. There, apoptosis is greatest at three days, in contrast to the susceptible species, *C. virginica*, where it does not occur until the seventh day [77]. Therefore, responses to *P. marinus* vary according to the virulence of the isolate and the differential resistance of hosts. This suggests that apoptosis may play an important role in the progression and transmission of diseases in oysters [77, 141].

MOLECULAR MARKERS FOR OYSTERS ASSOCIATED WITH DISEASE RESISTANCE

Genetic Improvement

The diseases that affect *C. virginica*, such as MSX caused by *Haplosporidium nelsoni* and Dermo by *P. marinus*, have produced high mortalities that cause dramatic decreases in oyster production. To combat this, strategies have been developed that include improvement programs based on selecting groups of oysters that survived –or presented low mortality– during disease events at cultivation sites. These groups have been crossbred to produce progeny that are potentially resistant to endemic diseases. One such selection program generated families that were disease-resistant using 10 lineages [117]; however, the time required to prove that succeeding generations will, in fact, be resistant is substantial; hence, these programs are complemented by the search for molecular markers that will permit identification of disease-resistant progenitors.

For this same purpose, selection programs have been conducted with three genetic groups that generated lines which were subsequently compared to a line from the natural environment to evaluate survival at different sites. This showed that site-to-site variations in salinity constitute a variable that affects survival, as the survival response of the different families varied from one site to another. However, survival in general was always greater in the line from the natural environment, suggesting that this is likely the result of interaction between genotype and environmental conditions at the cultivation site [65]. Obviously, this increases the difficulty of selecting disease-resistant families.

Resistance to *P. marinus* has been attributed primarily to the presence of serine protease inhibitors, which inhibit the extracellular proteases subtilisin and perkinsus to prevent propagation of the pathogen and the resulting degradation of the host's tissues [158]. Therefore, in the presence of *P. marinus*, the F-3 generation of *C. virginica* oysters selected for its resistance to *P. marinus* presented a higher expression of the cvS1-1 gene, which codifies for the serine protease inhibitor, compared to organisms non-selected as resistant [95]. This defence mechanism has also been observed via inhibition of the proteases produced by pathogens like *P. marinus* and *Vibrio vulnificus*. Higher production of protease inhibitors in *C. gigas* resistant to *P. marinus* was seen when compared to C. *virginica* susceptible to this pathogen [59].

In light of these results, the expression of transcribed cvS1-1 has been proposed as a marker for selecting organisms resistant to *P. marinus* infection [95] for the purpose of performing genetic improvement programs. The disadvantage of this marker is that it requires conducting infection challenges with the reproducers to be used in generating families. Nonetheless, it has recently been shown that the cvS1-1 gene presents polymorphisms in oysters that survived an infection event by *P. marinus* [159]. This SNP is a mutation in the promotor region related to the increase in the transcription of cvS1-1, and the organisms that present this deletion confer resistance to *P. marinus* [85].

The phylogenetic comparison performed to evaluate the evolution of the species revealed that *C. gigas*, *C. virginica* and *C. corteziensis* –species endemic to north-western Mexico– share the same phylogenetic clade, though the sequence of the serine protease inhibitor of *C. virginica* diverges from that of *C. corteziensis* and *C. gigas*, the latter considered a *P. marinus*-resistant species. Thus, we suggest that the low mortality seen in *C. corteziensis* is because this inhibitor is similar to that of the resistant species, *C. gigas*, which diverges from *C. virginica's* inhibitor [58]. Up to now, the serine protease inhibitor is the marker that can be used as an indicator of oysters resistant to *P. marinus*. It can be identified by the polymorphism that occurs in resistant families [85].

EFFECTS OF PLOIDY ON RESPONSES TO PATHOGENS

The intraspecific differential resistance of *C. virginica* is also seen in *C. gigas* oysters whose juvenile organisms suffer high mortality caused by the ostreid herpesvirus type 1 OsHv-1, principally in Europe, Australia and New Zealand. Using organisms that survived mortality events, lines for generating OsHv-1-resistant families have been developed [128]. Also, studies have evaluated resistance to OsHv-1 in triploid *C. gigas*. Triploid oysters use energy during growth and so avoid the energy expenditures associated with reproduction, added to a decreased response capacity to pathogens. However, results show similar mortality rates for diploid and triploid oysters [50]. This may be due to the fact that OsHv-1 infections occur in early stages of growth, in contrast to *P. marinus* infections which occur in adult *C. virginica* oysters.

This leads to the conclusion that the triploid nature of some oysters is not associated with resistance to pathogens, though it does accelerate oyster growth and so reduce the period in which the organisms remain in cultivation sites before reaching commercial size. This decreases exposure time to pathogens like *P. marinus* that develop in adult organisms. The combination of selection programs with triploidy is an option for production in terms of improving oyster growth and survival.

Conclusion

The environmental impacts generated by natural events and anthropogenic activities are undeniable facts. The harmful consequences produced by such problematics on the economy and the health of both entire ecosystems and humans has been a concern of environmental regulatory authorities and human populations for decades because, for example, of the drastic reductions in fishing and the aquaculture of clams and oysters they have caused worldwide. The diversity of stressors to which organisms are exposed that damage the environment and physical parameters can, through complex synergistic interactions, produce effects on health, but determining the impacts of these interactions directly, either in natural settings or through experimental bioassays, is a complex, laborious process. Therefore, ongoing efforts to validate biomarkers and their application at the individual, population and community levels is necessary if we are to improve our understanding of toxicities and the causes of the appearance of new, emerging diseases, especially since the integrated use of biomarkers with direct analyses of contaminants and sentinel animals can provide information on the health status of individuals and populations based on relatively small samples. With respect to determining the status of ecosystems or environmental health, biomarkers can be used to relate processes of cell and molecular damage to higher levels of contamination (predictive capacity) that could cause pathologies or reduce the physiological and/or reproductive performance of organisms. Scant information exists on numerical modelling; a tool that would allow us to predict future environmental impacts and evaluate the risks of exposure to environmental stressors, so it is necessary to develop this further. One strategy of environmental management designed to ensure the sustainability of resources requires an integrated focus capable of both evaluating risks and making predictions. However, before such a strategy can influence decision-making regarding environmental policies, it is crucial to demonstrate the scientific solidity of the long-term predictions it generates.

Responses to environmental stressors can be detected in individual organisms or entire populations, manifested at times by modifications of behaviour, physiology or genetics, or simply through tolerance or intolerance to pollutants. At the molecular level, the main genomic sequences affected by environmental stressors such as hydrocarbons, pesticides and hypoxic events, among others, are those involved in mechanisms that regulate energy metabolism, the biosynthesis pathways of macromolecules (primarily transcription and translation), responses to oxidative stress, and ion- and electron-transporting molecules. The availability of powerful, vanguard analytical tools will allow us to evaluate the activity or expression of a group of genes from a specific tissue or of a group of cells subjected to a certain event. Used jointly with other technologies, such as proteomic analysis, these methods strengthen research and allow the identification of relations –direct or indirect– among the genes damaged by environmental stressors. Especially important in this regard are the genetic expression libraries already developed for *C. virginica* and *C. gigas* oysters. Adopting this focus will make it possible to identify new molecular biomarkers of exposure and determine the adverse effects of exposure or susceptibility to environmental stressors, in addition to verifying the biological processes affected by the presence of such stressors. The relation between the processes that govern responses at the individual level and those that determine consequences at the level of population are rarely evident. Measures effectuated at the molecular, cellular and biochemical levels provide valuable information on the health status of individual organisms whose responses are usually rapid, specific and sensitive, while also presenting strong toxicological significance. However, it is difficult to extrapolate individual results to higher levels of organization, where other factors intervene. In contrast, responses to stress observed at the population or community level tend to be slower and when detected may have already become irreversible. For these reasons it is necessary to conduct toxicologically- and ecologically-significant research that will provide answers to questions concerning ecotoxicology. The use of biomarkers and bioindicators can potentially improve our understanding of the mechanisms involved and allow us to relate different response levels to actual conditions of environmental stress.

REFERENCES

[1] Adams, S. M. (2002). Biological indicators of aquatic ecosystem stress. Introduction and overview. In: Adams, S. M. (Ed.) *Biological Indicators of Aquatic Ecosystems Stress.* pp 1-12. American Fisheries Society. Bethesda, Maryland.

[2] Aguirre-Macedo, L. M., Simá-Alvarez, R. A., Román-Magaña, M. K. & Güemez-Ricalde, J. I. (2007). Parasite survey of the eastern oyster *Crassostrea virginica* in coastal lagoons of the southern Gulf of Mexico. *J. Aquat. Anim. Health., 19*, 270–79.

[3] Akcha, F., Spagnol, C. & Rouxel, J. (2012). Genotoxicity of diuron and glyphosate in oyster spermatozoa and embryos. *Aquat. Toxicol.*, 106–107, 104–113.

[4] Alavi, M. R., Fernández-Robledo, J. A. & Vasta, G. R. (2009). Development of an *in vitro* assay to examine intracellular survival of *Perkinsus marinus* trophozoites upon phagocytosis by oyster (*Crassostrea Virginica* and *Crassostrea Ariakensis*) haemocytes. *J. Parasitol., 95*, 900–907.

[5] Almeida, C., Pereira, C., Gomes, T., Bebianno, M. J. & Cravo, A. (2011). DNA damages as a biomarker of genotoxic contamination in *Mytilus galloprovincialis* from the south coast of Portugal. *J. Environ. Monitor., 13*, 2559–2567.

[6] Alvarado-Arellano, A. (2013). Estudio Sobre La Diversidad Genética Del Herpesvirus de Los Ostreidos-1. Centro de Investigaciones Biológicas del Noroeste.

[7] Alzieu, C. (1998). Tributyltin: case of study of a chronic contaminant in the coastal environment. *Ocean Coast Manage., 40*, 23-36.

[8] Amiard, J. C., Amiard-Triquet, C., Barka, S., Pellerin, J. & Rainbow, P. S. (2006). Metallothioneins in aquatic invertebrates: Their role in metal detoxification and their use as biomarkers. *Aquat. Toxicol., 76*, 160-202.

[9] Anderson, K., Taylor, D. A., Thompson, E. L., Melwani, A. R., Nair, S. V. & Raftos, D. A. (2015). Meta-analysis of studies using suppression subtractive hybridization and microarrays to investigate the effects of environmental stress on gene transcription in oysters. *PLOS One*, *10*, e0118839. DOI: 10.1371/journal.pone. 0118839.
[10] Anguiano, G., Amador, A., Moreno Legoreta, M., Arcos Ortega, F. & Vazquez Boucard, C. (2009). Effects of exposure to oxamil, carbofuran, dichlorvos and lindano on acetylcholinesterase activity in the gills of the pacific oyster *Crassostrea gigas*. *Environ. Toxicol.*, *28*, 337-332.
[11] Anguiano, G., Llera Herrera, R., Rojas, E. & Vazquez Boucard, C. (2007). Subchronic organismal toxicity, cytotoxicity, genotoxicity and feeding response of pacific oyster *Crassostrea gigas* to lindane. Exposure under experimental conditions. *Environ. Toxicol. Chem.*, *26*, 2192-2197.
[12] Anguiano Vega, G. A. (2008). Desarrollo de biomarcadores de exposición al plaguicida organoclorado lindano en el ostión del Pacifico Crassostrea gigas. Ph.D. Thesis. *CIBNOR*, 157 pp.
[13] Auffret, M., Mujdzic, N., Corporeau, C. & Moraga, D. (2002). Xenobiotic-induced immunomodulation in the European flat oyster, *Ostrea edulis*. *Mar. Environ. Res.*, *54*, 585-9.
[14] Barranger, A., Akcha, F., Rouxel, J., Brizard, R., Maurouard, E., Pallud, M., Menard, D., Tapie, N., Budzinski, H., Burgeot, T. & Benabdelmouna, A. (2014). Study of genetic damage in the Japanese oyster induced by an environmentally-relevant exposure to diuron: evidence of vertical transmission of DNA damage. *Aquat. Toxicol.*, *146*, 93-104.
[15] Baršienė, J., Lyons, B., Rybakovas, A., Martinez-Gómez, C., Andreikenaite, L., Brooks, S. & Maes, T. (2012). Background document: micronucleus assay as a tool for assessing cytogenetic/DNA damage in marine organisms. *Report of the study group on Integrated Monitoring of Contaminants and Biological Effects.* Report number: ICES CM 2011/ACOM: 30. International Council for the Exploration of the Sea. Copenhagen, Denmark.
[16] Baršienė, J., Šyvokienė, J. & Bjornstad, A. (2006). Induction of micronuclei and other nuclear abnormalities in mussels exposed to bisphenol A, diallyl phthalate and tetrabromodiphenyl ether-47. *Aquat. Toxicol.*, *78*, S105-S108.
[17] Bebianno, M. J., Serafim, M. A. & Simes, D. (2000). Metallothioneins in the clam *Ruditapes decussatus*: an overview. *Analusis*, *28*, 386-390.
[18] Belabed, S. & Soltani, N. (2013). Acute toxicity of cadmium on *Donax trunculus*: acetylcholinesterase, glutathione S-transferase activities and pattern of recovery. *Eur. J. Exp. Biol.*, *3*, 54-61.
[19] Bellas, J., Beiras, R., Marino-Balsa, J. C. & Fernandez, N. (2005). Toxicity of organic compounds to marine invertebrate embryos and larvae: A comparison between the sea urchin embryogenesis bioassay and alternative test species. *Ecotoxicology*, *14*, 337-353.
[20] Beukema, J. J. & Decker, R. (2010). Extraordinarily low mortality rates reported in juvenile Pacific oysters in the German Wadden Sea: a com ment. *Helgol Mar Res.*, *64*, 69-70.
[21] Binelli, A., Ricciardi, F., Riva, C. & Provini, A. (2006). Integrated use of biomarkers and bioaccumulation data in Zebra mussel (*Dreissena polymorpha*) for site-specific quality assessment. *Biomarkers*, *11*, 428-448.

[22] Bocquene, G., Roig, A. & Fournier, D. (1997). Cholinesterase from the common oyster *Crassostrea gigas*. Evidence for the presence of a soluble acetylcholinesterase insensitive to organophosphate and carbamate inhibitors. *FEBS Lett.*, *407*, 261-266.

[23] Bolognesi, C. & Cirillo, S. (2014). Genotoxicity biomarkers in aquatic bioindicators. *Curr. Zool.*, *60*, 273-284.

[24] Bolognesi, C., Frenzilli, G., Lasagna, C., Perrone, E. & Roggieri, P. (2004). Genotoxicity biomarkers in *Mytilus galloprovincialis:* wild versus caged mussels. *Muta Res./Fund. Mol. Mech. Muta*, *552*, 153-162.

[25] Bouilly, K., Leitão, A., McCombie, H. & Lapègue, S. (2003). Impact of atrazine on aneuploidy in pacific oysters, *Crassostrea gigas*. *Environ. Toxicol. Chem.*, *22*, 219-223.

[26] Boutet, I., Tanguy, A., Rousseau, S., Auffret, M. & Moraga, D. (2003). Molecular identification and expression of heat shock cognate 70 and heat shock protein 70 genes in the Pacific oyster Crassostrea gigas. *Cell stress*, *8*(1), 76-85.

[27] Boutet, I., Tanguy, A. & Moraga, D. (2004). Response of the Pacific oyster *Crassostrea gigas* to hydrocarbon contamination under experimental conditions. *Gene Dev.*, *329*, 147-157.

[28] Burdon, J. J., Gibson, A. H., Searle, S. D., Woods, M. J. & Brockwell, J. (1999). Variation in the effectiveness of symbiotic associations between native rhizobia and temperate Australian *Acacia*: within-species interactions. *J. Applied Ecol.*, *36*, 398–408.

[29] Burreson, E. M., Sima-Alvarez, R., Vidal-Martinez, V. & Aguirre-Macedo, L. (1994). *Perkinsus marinus* (Apicomplexa) as a potential source of oyster *Crassostrea Virginica* mortality in coastal lagoons of Tabasco, Mexico. *Dis. Aquat. Org.*, *20*, 77–82.

[30] Bushek, D. & Allen, S. K. (1996). Host-parasite interactions among broadly distributed populations of the eastern oyster *Crassostrea Virginica* and the protozoan *Perkinsus marinus*. *Mar. Ecol. Prog. Ser.*, *139*, 127–41.

[31] Byrne, M. & Przeslawski, R. (2013). Multistressor impacts of warming and acidification of the ocean on marine invertebrates' life histories. *Integr. Comp. Biol.*, *53*, 582-96.

[32] Byrne, M. (2012). Global change ecotoxicology: Identification of early life history bottlenecks in marine invertebrates, variable species responses and variable experimental approaches. *Mar. Environ. Res.*, *76*, 3-15.

[33] Cáceres-Martínez, J., Vásquez-Yeomans, R., Padilla-Lardizábal, G. & Del Río Portilla, M. A. (2008). *Perkinsus marinus* in pleasure oyster *Crassostrea Corteziensis* from Nayarit, Pacific Coast of México. *J. Invertebr. Pathol.*, *99* (1), 66–73.

[34] Campos, A., Tedesco, S., Vasconcelos, V. & Cristobal, S. (2012). Proteomic research in bivalves: towards the identification of molecular markers of aquatic pollution. *J. Proteomics*, *75*, 4346-59.

[35] Chapman, R. W., Manciam A., Beal, M., Veloso, A., Rathburn, C., Blair, A., Holland, A. F., Warr, G. W., Didinato, G., Sokolova, I. M., Wirth, E. F., Duffy, E. & Sanger, D. (2011). The transcriptomic responses of the eastern oyster, Crassostrea virginica, to environmental conditions. *Mol. Ecol.*, *20*, 1431-49.

[36] Cheng, B. S., Bible, J. M., Chang, A. L., Ferner, M. C., Wasson, K., Zabin, C. J., Latta, M., Deck, A., Todgham, A. E. & Grosholz, E. D. (2015). Testing local and global stressor impacts on a coastal foundation species using an ecologically realistic framework. *Glob. Chang. Biol.*, doi: 10.1111/gcb.12895.

[37] Cheney, D. P., Elston, R. A., MacDonald, B. F., Cherr, G. N., Hamdoun, A. M. & Jacobsen, J. L. (2000). An update on the ongoing oyster summer mortality study: Mortality of the pacific oyster, *Crassostrea gigas*: Health screening, environmental links and management options. *J. Shellfish Res*, *19*, 353-359.

[38] Cheung, V. V., Depledge, M. H. & Jha, A. N. (2006). An evaluation of the relative sensitivity of two marine bivalve mollusc species using the Comet assay. *Mar. Environ. Res. Supplement*, *1*, S301-S305.

[39] Choi, Y. K., Jo, P. G. & Choi, C. Y. (2008). Cadmium effects the expression of heat shock protein 90 and metallothionein mRNA in the Pacific oyster *Crassostrea gigas*. *Comp. Biochem. Physiol. C, Pharmacol. Toxicol. Endocrinol.*, *147*, 286–292.

[40] Chu, F. L. E., Soudant, P. & Hale, R. C. (2003). Relationship between PCB accumulation and reproductive output in conditioned oysters *Crassostrea virginica* fed contaminated algal diet. *Aquat. Toxicol.*, *65*(3), 293-307.

[41] Clark, M. S., Thorne, M. A., Amaral, A., Vieira, F., Batista, F. M., Reis, J. & Power, D. M. (2013). Identification of molecular and physiological responses to chronic environmental challenge in an invasive species: the Pacific oyster, *Crassostrea gigas*. *Ecol. Evol.*, *3*, 3283-97.

[42] Collin, H., Meistertzheim, A. L., David, E., Moraga, D. & Boutet, I. (2010). Response of the Pacific oyster *Crassostrea gigas*, Thunberg 1793, to pesticide exposure under experimental conditions. *J. Exp. Biol.*, *213*, 4010-7.

[43] Collins, A. R., Azqueta Oscoz, A., Brunborg, G., Gaivão, I., Giovannelli, L., Kruszewki, M., Smith, C. C. & Štětina, R. (2008). The comet assay: topical issues. *Mutagenesis*, *23*, 143-151.

[44] Cotelle, S. & Ferard, J. F. (1999). Comet assay in genetic ecotoxicology. A review. *Environ Mol Mutagen*, *34*, 246-255.

[45] Cunningham, C., Hikima, J. I., Jenny, J. I., Chapman, R. W., Fang, G. C., Saski, C., Lundqvist, M. L., Wing, R. A., Cupit, P. M., Gross, P. S., Warr, G. W. & Tomkins, J. P. (2006). New resources for marine genomics: Bacterial artificial chromosome libraries for the eastern and Pacific oysters (*Crassostrea virginica* and *C. gigas*). *Mar. Biotechnol.*, *8*, 521-533.

[46] Danellakis, D., Ntaikou, I., Kornaros, M. & Dailianis, S. (2011). Olive oil mill wastewater toxicity in the marine environment: Alterations of stress indices in tissues of mussel *Mytilus galloprovincialis*. *Aquat. Toxicol.*, *101*, 358-366.

[47] Das, S. & Jana, B. B. (2003). Oxigen uptake and filtration rate as animal health biomarker in *Lamellidens marginalis*. *Indian J. Exp. Biol.*, *41*, 1306-1310.

[48] David, E., Tanguy, A. & Moraga, D. (2007). Peroxiredoxin 6 gene: a new physiological and genetic indicator of multiple environmental stress response in Pacific oyster *Crassostrea gigas*. *Aquat. Toxicol.*, *84*, 389-398.

[49] David, E., Tanguy, A., Pichavant, K. & Moraga, D. (2005). Response of the Pacific oyster *Crassostrea gigas* to hypoxia exposure under experimental conditions. *FEBS J.*, *272*, 5635-5652.

[50] Dégremont, L., Christophe, L., Maurouard, E., Nourry, M. & Benabdelmouna, A. (2014). Effect of ploidy on the mortality of *Crassostrea gigas* spat caused by OsHV-1 in France using unselected and selected OsHV-1 resistant oysters. *Aquat. Res.* DOI: 10.1111/are. 12536.

[51] Diederich, S. (2006). High survival and growth rates of introduced Pacific oysters may cause restrictions on habitat use by native mussels in the Wadden Sea. *J Exp Mar Biol Ecol*, 328, 2, 24, 211–227.

[52] Dondero, F., Banni, M., Negri, A., Boatti, L., Dagnino, A. & Viarengo, A. (2011). Interactions of a pesticide/heavy metal mixture in marine bivalves: a transcriptomic assessment. *BMC Genomics*, *12*, 195.

[53] Duquesne, S., Liess, M. & Bird, D. J. (2004). Sub-lethal effects of metal exposure: physiological and behavioral responses of the estuarine bivalve *Macoma balthica. Mar. Environ. Res.*, *58*, 245-250.

[54] Edge, K. J., Johnston, E. L., Roach, A. C. & Ringwood, A. H. (2012). Indicators of environmental stress: cellular biomarkers and reproductive responses in the Sydney rock oyster (*Saccostrea glomerata*). *Ecotoxicology*, *21*, 1415-25.

[55] EFSA AHAW Panel (EFSA Panel on Animal Health and Welfare). (2015). Scientific opinion on oyster mortality. *EFSA J.*, *13*, 4122, 59 pp.

[56] Enríquez-Espinoza, T. L, Grijalva-Chon, J. M., Castro-Longoria, R. & Ramos-Paredes, J. (2010). *Perkinsus marinus* in *Crassostrea gigas* in the Gulf of California. *Dis. Aquat. Organ.*, *89*, 269–73.

[57] Epelboin, Y., Quéré, C., Pernet, F., Pichereau, V. & Corporeau, C. (2015). Energy and antioxidant responses of pacific oyster exposed to trace levels of pesticides. *Chem. Res. Toxicol.*, *28*, 1831–1841.

[58] Escobedo-Fregoso, C., Arzul, I., Carrasco, N., Gutiérrez-Rivera, J. N., Llera-Herrera, R. & Vázquez-Juárez, R. (2015). Polymorphism at the ITS and NTS loci of *Perkinsus marinus* isolated from cultivated oyster *Crassostrea corteziensis* in Nayarit, Mexico and phylogenetic relationship to *P. marinus* along the Atlantic Coast. *Transbound. Emerg. Dis.*, *62*, 137–47.

[59] Faisal, M., MacIntyre, E. A., Adham, K. G., Tall, B. D., Kothary, M. H. & La Peyre, J. F. (1998). Evidence for the presence of protease inhibitors in eastern (*Crassostrea virginica*) and pacific (*Crassostrea gigas*) oysters. *Comp. Biochem. Physiol. Part B*, *121*, 161–68.

[60] Farhadi, A., Farahmand, H., Mirvaghefi, A. & Khalili, B. (2011). A genotoxicological study in Persian Gulf on rock oyster (*Soccostrea cucullata*) using micronuclei and RAPID assays. *Inter. J. Environ. Res.*, *5*, 567-572.

[61] Fedato, R. P., Simonato, J. D., Martinez, C. B. R. & Sofia, S. H. (2010). Genetic damage in the bivalve mollusk *Corbicula fluminea* induced by the water-soluble fraction of gasoline. *Muta Res./Gen. Toxicol. Environ. Muta*, *700*, 80-85.

[62] Fernández-Tajes, J., Arias-Pérez, A., Fernández-Moreno, M. & Méndez, J. (2012). Sharp decrease of genetic variation in two Spanish localities of razor clam *Ensis siliqua*: natural fluctuation or Prestige oil spill effects? E*cotoxicology*, *21*, 225-33.

[63] Fleury, E. & Huvet, A. (2012). Microarray analysis highlights immune response of pacific oysters as a determinant of resistance to summer mortality. *Mar. Biotechnol.*, *14*, 203-17.

[64] Forbes, V. E., Palmqvist, A. & Bach, L. (2006). The use and misuse of biomarkers in ecotoxicology. *Environ. Toxicol. Chem.*, *25*, 272-280.

[65] Frank-Lawale, A., Standish, K. A. & Dégremont, L. (2014). Breeding and domestication of eastern oyster (*Crassostrea Virginica)* lines for culture in the Mid-

Atlantic, USA: line development and mass selection for disease resistance. *J. Shellfish Res.*, *33*, 153–65.

[66] Fukuto, R. (1990). Mechanism of action of organophosphorus and carbamate insecticides. *Environ. Health Persp.*, *87*, 245-354.

[67] Galloway, T. S., Millward, N., Browne, M. A. & Depledge, M. H. (2002a). Rapid assessment of organophosphorus/carbamate exposure in the bivalve mollusk *Mytillus edulis* using combined esterase activities as biomarkers. *Aquat. Toxicol.*, *61*, 169-180.

[68] Galloway, T. S., Sanger, R., Smith, K., Fillmann, G., Readman, J., Ford, T. & Depledge, M. (2002b). Rapid assessment of marine pollution using multiple biomarkers and chemical immunoassays. *Environ. Sci. Technol.*, *36*, 2219-2226.

[69] Gamble, S. C., Golfard, P. S., Porte, C. & Livingstone, D. R. (1995). Glutathione peroxidase and other antioxidant enzyme function in marine invertebrates (*Mytilus edulis. Pecten Maximus, Carcinus maenas and Asterias rubens*). *Mar. Environ. Res.*, *39*, 191-195.

[70] Garcia Lagunas, N., Romero Geraldo, R. & Hernández Saavedra, N. Y. (2013). Genomics study of the exposure effect of *Gymnodiniun catenatum* a paralyzing toxin producer, on *Crassostrea gigas*. Defense systems detoxification genes. *PLOS One*, *8*, 1-9.

[71] Garcia-Reyero, N. & Perkins, E. J. (2011). Systems biology: leading the revolution in ecotoxicology. *Environ. Toxicol. Chem.*, *30*, 265–73.

[72] Gauthier, J. D. & Vasta, G. R. (1995). *In vitro* culture of the Eastern Oyster parasite *Perkinsus marinus*: Optimization of the methodology. *J. Inverter. Pathol.*, *66*, 156–68.

[73] Gauthier-Clerc, S., Pellerin, J., Blaise, C. & Gangé, F. (2002). Delayed gametogenesis of *Mya arenaria* in the Saguenay fjord (Canada): A consequence of endocrine disruptors? *Comp. Biochem. Physiol. Part C.*, *131*, 457-467.

[74] Gavery, M. R. & Roberts, S. B. (2010). DNA methylation patterns provide insight into epigenetic regulation in the Pacific oyster (*Crassostrea gigas). BMC Genomics*, *11*, 483.

[75] Geret, F., Burgeot, T., Haure, J., Gagnaire, B., Renault, T., Communal, P. Y. & Samain, J. F. (2011). Effects of low-dose exposure to pesticide mixture on physiological responses of the pacific oyster *Crassostrea gigas*. *Environ. Toxicol.*, *28*, 689-699.

[76] Geyer, H. J., Steinberg, C. E., Scheunert, I., Bruggemann, R., Schutz, W., Ketrup, A. & Rozman, K. (1993). A review of the relationship between acute toxicity of gamma-hexachlorocyclohexane and total lipid content of different fish species. *Toxicology*, *83*, 169-179.

[77] Goedken, M., Morsey, B., Sunila, I. & De Guise, S. (2005). Immunomodulation of *Crassostrea Gigas* and *Crassostrea Virginica* cellular defense mechanism by *Perkinsus marinus*. *J. Shellfish Res.*, *24*, 487–96.

[78] Goldberg, E. D. (1975). The mussel Watch a first step in global marine monitoring. *Mar. Pollut. Bull.*, *6*, 7, 111-111.

[79] Guidi, P., Frenzilli, G., Benedetti, M., Bernardeschi, M., Falleni, A., Fattorini, D., Regoli, F., Scarcelli, V. & Nigro, M. (2010). Antioxidant, genotoxic and lysosomal biomarkers in the freshwater bivalve (*Unio pictorum*) transplanted in a metal polluted river basin. *Aquat. Toxicol.*, *100*, 75-83.

[80] Gullian-Klanian, M., Herrera-Silveira, J. A., Rodriguez-Canul, R. & Aguirre-Macedo, L. (2008). Factors associated with the prevalence of *Perkinsus marinus* in *Crassostrea Virginica* from the Southern Gulf of Mexico. *Dis. Aquat. Org.*, *79*, 237–47.

[81] Guo, X., He, Y., Zhang, L., Lelong, C. & Jouaux, A. (2015). Immune and stress responses in oysters with insights on adaptation. *Fish Shellfish Immunol.*, *46*, 107-19.

[82] Hagger, J. A., Depledge, M. H. & Galloway, T. S. (2005). Toxicity of tributyltin in the marine mollusc *Mytilus edulis*. *Mar. Pollut. Bull.*, *51*, 811-816.

[83] Hamoutene, D., Payne, J. F., Rahimtula, A. & Lee, K. (2002). Use of the Comet assay to assess DNA damage in hemocytes and digestive gland cells of mussels and clams exposed to water contaminated with petroleum hydrocarbons. *Mar. Environ. Res.*, *54*, 471-474.

[84] Hamza-Chaffai, A., Romeo, M., Gnassia-Barelli, M. & El Abed, A. (1998). Effect of copper and lindane on some biomarkers measured in the clam *Ruditapes decussates*. *Bull. Environ. Contam. Toxicol.*, *61*, 397–404.

[85] He, Y., Yu, H., Bao, Z., Zhang, Q. & Guo, X. (2012). Mutation in promoter region of a serine protease inhibitor confers *Perkinsus marinus* resistance in the eastern oyster (*Crassostrea Virginica*). *Fish Shellfish Immunol.*, *33*, 411–17.

[86] Hodkinson, I. D. & Jackson, J. K. (2005). Terrestrial and aquatic invertebrates as bioindicators for environmental monitoring, with particular reference to mountain ecosystems. *Environ. Manage.*, *35*, 649-666.

[87] Hughes, F. M., Foster, B., Grewal, S. & Sokolova, I. M. (2010). Apoptosis as a host defense mechanism in *Crassostrea virginica* and its modulation by *Perkinsus marinus*. *Fish Shellfish Immun.*, *29*, 247–57.

[88] IFREMER, (2015), Qualité du Milieu Marin Littoral Synthèse Nationale de la Surveillance 2013. ODE/DYNECO/VIGIES/1507. http://envlit.ifremer.fr/content/download/82798/598028/version/6/file/ SyntheseNationaleBullSurvED2015.pdf.

[89] Ivanina, A. V., Cherkasov, A. S. & Sokolova, I. M. (2008). Effects of cadmium on cellular protein and glutathione synthesis and expression of stress proteins in eastern oysters, *Crassostrea virginica*. *J. Exp. Biol.*, *211*, 577-586.

[90] Johnson, B. D., Gilbert, S. L., Khan, B., Carroll, D. L. & Ringwood, A. H. (2015). Cellular responses of eastern oysters, *Crassostrea virginica*, to titanium dioxide nanoparticles. *Mar. Environ. Res.*, *111*, 135-143.

[91] Kimbrough, K. L., Lauenstein, G. G., Christensen, J. D. & Apeti, D. A. (2008). An assessment of two decades of contaminant monitoring in the Nation's Coastal Zone. *Silver Spring,* MD, NOAA/National Centers for Coastal Ocean Science, (NOAA Technical Memorandum NOS NCCOS, 74).

[92] Klobučar, G. I. V., Pavlica, M., Erben, R. & Papeš, D. (2003). Application of the micronucleus and comet assays to mussel *Dreissena polymorpha* haemocytes for genotoxicity monitoring of freshwater environments. *Aquat. Toxicol.*, *64*, 15-23.

[93] Koelle, G. B. (1994). Pharmacology of organophosphates. *J. Appl. Toxicol.*, *14*, 105-109.

[94] Koukouzika, N. & Dimitriadis, V. K. (2008). Aspects of the usefulness of five marine pollution biomarkers, with emphasis on MN and lipid content. *Mar. Pollut. Bull.*, *56*, 941-949.

[95] La Peyre, J. F., Xue, Q. G., Itoh, N., Li, Y. & Cooper, R. K. (2010). Serine protease Inhibitor cvSI-1 potential role in the eastern oyster host defence against the protozoan parasite *Perkinsus marinus*. *Dev. Comp. Immunol.*, 34, 84–92.

[96] Lacoste, A., Malhamb, S. K., Gelebarta, F., Cueffa, A. & Pouleta, S. A. (2002). Stress-induced immune changes in the oyster *Crassostrea gigas*. *Dev. Comp. Immunol.*, *26*, 1–9.

[97] Lesser, M. P. (2006). Oxidative Stress in Marine Environments. *Biochem. Physiol. Ecol. Ann. Rev. Physiol.*, *68*, 253-278.

[98] Llera Herrera, R., Vazquez Boucard, C., García Gasca, A. & Huvet, A. (2012). Co-expression and regulation of ovarian vitellogenesis in the pacific oyster *Crassostrea gigas*. *Aquac. Res.*, *45*, 448-459.

[99] Matozzo, V., Tomei, A. & Marin, M. G. (2005). Acetylcholinesterase as a biomarker of exposure to neurotoxic compounds in the clam *Tapes philippinarum* from the lagoon of Venise. *Mar. Pollut. Bull.*, *50*, 1686-1693.

[100] Matsumoto T., Nakamura, A. M., Mori, K. & Kayano, T. (2003). Molecular characterization of a cDNA encoding vitellogenin from the Pacific oyster *Crassostrea gigas*. *Zool. Sci.*, *20*, 37-42.

[101] Matthiessen, P. & Sumpter, J. P. (1998). Effects of estrogenic substances in the aquatic environment. In. Fish ecotoxicology. (Eds Braunbeck T, Hinton D, Streit B) 1 ed, Birkhausser Verlag Basel/Boston/Berlin, 119-335.

[102] McDowell, J. (1996). The bioaccumulation and biological effects of lipophilic organic contaminants. In Kennedy, V. S., Newell, R. I. E., Eblem A. F. (Eds.) *The eastern oyster Crassostrea virginica*. pp, 539-557. Maryland sea grant book. Maryland, USA.

[103] McKillen, D. J., Chen, Y. A., Chen, C., Jenny, M. J., Trent, H. F., Robalino, J., McLean, D. C., Gross, P. S., Chapman, R. W., Warr, G. W. & Almeida, J. S. (2005). Marine Genomics: A clearing-house for genomic and transcriptomic data of marine organisms. *BMC Genomics*, *6*, 34.

[104] Medeiros, I. D., Siebert, M. N., De Toledo e Silva, G., Moraes, M. O., Freire Marques, M. R. & Días Bainy, A. C. (2008). Differential gene expression in oyster exposed to sewage, *Mar. Environ. Res.*, *66* (1), pp. 156.

[105] Miller W. A., Atwill, E. R., Gardner, I. A., Miller, M. A., Fritz, H. M., Hedrick, R. P., Melli, A. C., Barnes, N. M. & Conrad, P. A. (2005). Clams (*Corbicula fluminea*) as bioindicators of fecal contamination with *Cryptosporidium* and *Giardia* spp. in freshwater ecosystems in California. *Int. J. Parasitol.*, *35*, 673-684.

[106] Minier, C., Lelong, C., Djemel, N., Rodet, F., Tutundjian, R., Favrel, P., Mathieu, M. & Leboulenger, F. (2002). Expression and activity of a multixenobiotic resistance system in the pacific oyster *Crassostrea gigas*. *Mar. Environ. Res.*, *54*, 455-459.

[107] Monirith, D. U., Takahashi, S., Nakata, H., Sudaryanto, A., Subramanian, A., Karuppiah, S., Ismail, A., Muchtar, M., Zheng, J., Richardson, B. J., Prudente, M., Duc Hue, N., Tana, T. S., Tkalin, A. V. & Tanabe, S. (2003). Asia-Pacific mussel watch: monitoring contamination of persistent organochlorine compounds in coastal waters of Asian countries. *Mar. Pollut. Bull.*, *46*, 281-300.

[108] Moore, M. N., Depledge, M. H., Readman, J. W. & Leonard, D. R. P. (2004). An integral biomarker based strategy for ecotoxicological evaluation of risk in environmental management. *Muta Res./Gen. Toxicol. Environ. Muta*, *552*, 247-268.

[109] Moraga, D. & Tanguy, A. (1997). Effects of anthropogenic factors on genetic diversity in the marine bivalve *Crassostrea gigas*: search for genetic markers. *Vie milieu*, *47*, 355-365.

[110] Moraga, D. & Tanguy, A. (2000). Genetic indicators of herbicide stress in the Pacific oyster *Crassostrea gigas* under experimental conditions. *Environ Toxicol Chem*, *19*(3), 706 – 711.

[111] Moraga, D., Mdelgi-Lasram, E., Romdhane, M. S., El Abed, A., Boutet, I., Tanguy, A. & Auffret, M. (2002). Genetic responses to metal contamination in two clams: *Ruditapes decussatus* and *Ruditapes philippinarum*. *Mar. Environ. Res.*, *54*, 521-5.

[112] Moreau, P., Burgeot, T. & Renault, T. (2015). *In vivo* effects of metaldehyde on Pacific oyster, *Crassostrea gigas*: comparing haemocyte parameters in two oyster families. *Environ. Sci. Pollut. Res. Int.*, *22*, 8003-9.

[113] Nice, H. E. (2005). Sperm motility in the Pacific oyster (*Crassostrea gigas*) is affected by nonyphenol. *Mar. Pollut. Bull.*, *50*, 1668-1674.

[114] Nice, H. E., Morritt, D., Crane, M. & Thorndyke, M. (2003). Long-term and transgenerational effects of nonylphenol exposure at a key stage in the development of *Crassostrea gigas*. Possible endocrine disruption? *Mar. Ecol. Prog. Ser.*, *256*, 293-300.

[115] Nice, H. E., Thorndyke, M. C., Morritt, D., Steele, S. & Crane, M. (2000). Development of *Crassostrea gigas* larvae is affected by 4-nonylphenol. *Mar. Pollut. Bull.*, *40*, 491-496.

[116] Nigro, M., Falleni, A., Barga, I. D., Scarcelli, V., Lucchesi, P., Regoli, F. & Frenzilli, G. (2006) Cellular biomarkers for monitoring estuarine environments: Transplanted versus native mussels. *Aquat. Toxicol.*, *77*, 339-347.

[117] Oliver, J. L., Gaffney, P. M., Allen Jr. S. K., Faisal, M. & Kaattari, S. L. (2000). Protease Inhibitory Activity in Selectively Bred Families of Eastern Oysters. *J. Aquat. Anim. Health*, *12*, 136–45.

[118] Oliver, M. N., Fisher, W. S., Winstead, J. T., Hemmer, B. L. & Long, E. R. (2001). Relationship between tissue contaminants and defense related characteristic of oysters Crassostrea virginica from live Florida bays. Aquat. *Toxicol.*, *55*, 203-222.

[119] Ozmen, M., Sener, S., Mete, A. & Kucukbay, H. (1999). *In vitro* and *in vivo* acetylcholinesterase-inhibiting effect of new classes of organophosphorus compounds. *Environ. Toxicol. Chem.*, *18*, 241- 246.

[120] Park, K. I., Yang, H. S., Kang, H. S., Cho, M., Park, K. J. & Choi, K. S. (2010). Isolation and identification of Perkinsus olseni from feces and marine sediment using immunological and molecular techniques. *J. Invertebr. Pathol.*, *105*, 261–69.

[121] Porte, C., Janer, G., Lorusso, L. C., Ortiz-Zarragoitia, M., Cajaraville, M. P., Fossi, M. C. & Canesi, L. (2006). Endocrine disruptors in marine organisms: Approaches and perspectives. *Comp. Biochem. Physiol. Part C*, *143*, 303–315.

[122] Rigonato, J., Mantovani, M. S. & Jordão, B. Q. (2005). Comet assay comparison of different Corbicula fluminea (Mollusca) tissues for the detection of genotoxicity. *Genetic. Mol. Biol.*, *28*, 464-468.

[123] Rittschof, D. & McClellan-Green, P. (2005). Molluscs as multidisciplinary models in environment toxicology. *Mar. Pollut. Bull.*, *50*, 369-73.

[124] Riva, C., Bintelli, A., Cogni, D. & Provini, A. (2007). Evaluation of DNA damage induced by decabromodiphenyl ether (BDE-209) in hemocytes of *Dreissena*

polymorpha using the comet and micronucleus assays. *Environ. Mol. Muta*, *48*, 735-743.

[125] Roberts, S. B., Sunila, I. & Wikfors, G. H. (2012). Immune response and mechanical stress susceptibility in diseased oysters, Crassostrea *virginica*. *J. Comp. Physiol. B, Biochem. Systemic. Environ. Physiol.*, *182*, 41–48.

[126] Rocha, T. L., Gomes, T., Sousa, V. S., Mestre, N. C. & Bebianno, M. J. (2015). Ecotoxicological impact of engineered nanomaterials in bivalve molluscs: An overview. *Mar. Environ. Res.*, *111*, 74-88.

[127] Rocher, B., Le Goff, J., Peluhet, L., Briand, M., Manduzio, H., Gallois, J., Devier, M. H., Geffard, O., Gricourt, L., Augagneur, S., Budzinski, H., Pottier, D., André, V., Lebailly, P. & Cachot, J. (2006). Genotoxicant accumulation and cellular defense activation in bivalves chronically exposed to waterborne contaminants from the Seine River. *Aquat. Toxicol.*, *79*, 65-77.

[128] Samain, J. M. & McCombie, H. (Eds.) (2008). Summer mortality of Pacific oyster *Crassostrea gigas*: The Morest Project. Editorial Quae. Versailles, France.

[129] Sarkar, A., Ray, D., Shrivastava, A. N. & Sarker, S. (2006). Molecular biomarkers: their significance and application in marine pollution monitoring. *Ecotoxicology*, *15*, 333-40.

[130] Schmid, W. (1975). The micronucleus test. *Muta Res./Environ. Muta Relat Subjec*, *31*, 9-15.

[131] Schmidt, P., Gueguen, Y., Desmarais, E., Bachère, E. & De Lorgeril, J. (2010). Molecular diversity of antimicrobial effectors in the oyster *Crassostrea gigas*. *BMC evolutionary biol*, *10*(1), 23.

[132] Schmidt, A., Wehrmann, A. & Dittmann, S. (2008). Population dynamics of the invasive Pacific oyster *Crassostrea gigas* during the early stages of an outbreak in the Wadden Sea (Germany). *Helgoland Mar Res*, *62*(4), 367-376.

[133] Schulte, P. M. (2001). Environmental adaptations as windows on molecular evolution. *Comp. Biochem. Physiol. Part B*, *128*, 597-611.

[134] Serrano, M. A. S., Gonzalez Rey, M., Mattos, J. J., Flores-Nunes, F., Mello, A. C. P., Zacchi, F. L., Piazza, C. E., Siebert, M. N., Piazza, R. S., Alvarez-Muñoz, D., Rodriguez-Mozaz, S., Barceló, D., Bebianno, M. J., Gomes, C. H. A. M., Melo, C. M. R. & Bainy, A. C. D. (2015). Differential gene transcription, biochemical responses, and cytotoxicity assessment in Pacific oyster *Crassostrea gigas* exposed to ibuprofen. *Environ. Sci. Pollut. Res.*, *22*(22), 17375-85.

[135] Shaw, J. P., Large, A. T., Donkin, P., Evans, S. V., Staff, F. J., Livingstone, D. R., Chipman, J. K. & Peters, L. D. (2004). Seasonal variation in cytochrome P450 immunopositive protein levels, lipid peroxidation and genetic toxicity in digestive gland of the mussel *Mytilus edulis*. *Aquat. Toxicol.*, *67*, 325-336.

[136] Siu, W. H. L., Cao, J., Jack, R. W., Wu, R. S. S., Richardson, B. J., Xu, L. & Lam, P. K. S. (2004). Application of the comet and micronucleus assays to the detection of B[a] P genotoxicity in haemocytes of the green-lipped mussel (*Perna viridis*). *Aquat. Toxicol.*, *66*, 381-392.

[137] Smolders, R., Bervoets, L., DeCoen, W. & Blust, R. (2004). Cellular energy allocation in zebra mussels exposed along a pollution gradient: linking cellular effects to higher levels of biological organization. *Environ. Pollut.*, *129*, 99-112.

[138] Snyder, M. J. (2000). Cytochrome P450 enzymes in aquatic invertebrates: recent advances and future directions. *Aquat. Toxicol.*, *48*, 529–54.

[139] Sole, M., Porte, C. & Baiges, J. (1994). Mixed function oxygenase system components and antioxidant enzymes in different marine bivalves: Its relation contaminant body burdens. *Aquat. Toxicol.*, *30*, 271-283.

[140] Soto, M., Cajaraville, M. P. & Marigomez, I. (1996). Tissue and cell distribution of cooper, zinc and cadmium in the mussel *Mytilus galloprovincialis* determined by autometallography. *Tissue Cell*, *28*, 557-568.

[141] Sunila, I. & LaBanca, J. (2003). Apoptosis in the pathogenesis of infectious diseases of the eastern oyster *Crassostrea Virginica. Dis. Aquat. Org.*, *56*, 163–70.

[142] Tanabe, S. & Subramanian, A. (2003). Bioindicators suitable for monitoring POPs in developing countries. STAP Workshop on the use of bioindicators, biomarkers and analytical methods for the analysis of POPs in developing countries. Tsukuba, Japan.

[143] Tanguy, A., Boutet, I., Boudry, P., Degremont, L., Laroche, J. & Moraga, D. (2006). Molecular identification and expression of the phosphoglucomutase gene from the pacific oyster *Crassostrea gigas*. *Gene*, *382*, 20-27.

[144] Tanguy, A., Boutet, I., Conhomme, F., Boudry, P. & Moraga, D. (2002). Polymorphism of metallothionein genes in the pacific oyster *Crassostrea gigas* as a biomarker of response to metal exposure. *Biomarkers*, *7*, 439-450.

[145] Tanguy, A., Boutet, I., Laroche, J. & Moraga, D. (2005a). Molecular identification and expression study of differentially regulated genes in the Pacific oyster *Crassostrea gigas* in response to pesticide exposure. *FEBS J.*, *272*, 390-403.

[146] Tanguy, A., Boutet, I. & Moraga, D. (2005b). Molecular characterization of the glutamine synthetase gene in the Pacific oyster *Crassostrea gigas*: expression study in response to xenobiotic exposure and developmental stage. *BBA Gen Structure Expression*, *1681*, 116-125.

[147] Tanguy, A., Bierne, N., Saavedra, C., Pina, B., Bachère, E., Kube, M., Bazin, E., Bonhomme, F., Boudry, P., Boulo, V., Boutet, I., Cancela, L., Dossat, C., Favrel, P., Huvet, A., Jarque, S., Jollivet, D., Klages, S., Lapègue, S., Leite, R., Moal, J., Moraga, D., Reinhardt, R., Samain, J. F., Zouros, E. & Canario, A. (2008). Increasing genomic information in bivalves through new EST collections in four species: development of new genetic markers for environmental studies and genome evolution. *Gene*, *408*, 27-36.

[148] Tasumi, S. & Vasta, G. R. (2007). A galectin of unique domain organization from hemocytes of the eastern oyster (*Crassostrea Virginica*) is a receptor for the protistan parasite *Perkinsus marinus*. *J. Immunol.*, *179*, 3086.

[149] Theodorakis, C. W., Blaylock, B. G. & Shugart, L. R. (1997). Genetic ecotoxicology I: DNA integrity and reproduction in mosquito fish exposed to radionuclides. *Ecotoxicology*, *6*, 205-218.

[150] Vazquez Boucard, C., Álvarez Ruiz, P., Fregoso Escobedo, C., Anguiano, G., Duran Avelar, M. J., Serrano Pinto, V. & Escobedo Bonilla, C. M. (2010). Detection of white spot syndrome virus in the Pacific oyster *Crassostrea gigas*. *J. Invertebr. Pathol.*, *104*, 245-247.

[151] Vazquez Boucard, C., Anguiano, G., Mercier, L. & Rojas Del Castillo, E. (2014). Pesticides residues, heavy metals and DNA damage in sentinels oysters *Crassostrea gigas* from Sinaloa and Sonora, Mexico. *J. Toxicol. Environ. Part A*, *77*, 169-176.

[152] Vazquez Boucard, C., Escobedo Fregoso, C., Duran Avelar, M. J., Mercier, L., Llera Herrera, R., Escobedo Bonilla, C. & Vibanco Pérez, N. (2012). *Crassostrea* gigas oyster as a shrimp farm bioindicator of White Spot Syndrome Virus. *Dis. Aquat Org.*, *98*, 201-207.

[153] Verlecar, X. N., Jena, K. B., Chainy. G. B. N. (2008). Seasonal variation of oxidative biomarkers in gills and digestive gland of green-lipped mussel *Perna viridis* from Arabian Sea. *Estuar. Coast. Shelf Sci*, *76*, 745-752.

[154] Vidal-Linan, L., Bellas, J. & Beiras, R. (2010). Battery of biomarkers (GST, GPx, CAT and AChE): A standardization of sampling procedures in mussels, *Mytilus galloprovincialis*. *Comp. Biochem. Physiol. Part A*, *157*, S27–S28.

[155] Villela, I. V., De Oliveira, I. M., Da Silva, J. & Henríquez, J. A. P. (2006). DNA damage and repair in haemolymph cells of golden mussel (*Limnoperna fortunei*) exposed to environmental contaminants. *Muta Res./Gen. Toxicol. Environ. Muta*, *605*, 78-86.

[156] Vlahogianni, T., Dassenakis, M., Scoullos, M. J. & Valavanidis, A. (2007). Integrated use of biomarkers (superoxide dismutase, catalase and lipid peroxidation) in mussels *Mytilus galloprovincialis* for assessing heavy metals pollution in coastal areas from the Saronikos, Gulf of Greece. *Mar. Pollut. Bullet.*, *54*, 1361–1371.

[157] Wirgin, I. I. & Theodorakis, C. W. (2002). Molecular biomarkers in aquatic organisms: DNA damage and RNA expression. In Adams, S. M. (Ed.) Biological indicators of aquatic ecosystem stress. pp. 43-110. American Fisheries Society. Bethesda, Maryland.

[158] Xue, Q. C., Waldrop, G. L., Schey, K. L., Ithon, N., Ogawa, M., Cooper, R. K., Losso, J. N. & La Peyre, J. F. (2006). A novel slow-tight binding serine protease inhibitor from eastern oyster (*Crassostrea Virginica)* plasma inhibits perkinsin, the major extracellular protease of the oyster protozoan parasite *Perkinsus marinus*. *Comp. Biochem. Physiol. B*, *145*, 16–26.

[159] Yu, H., He, Y., Wang, X., Zhang, Q., Bao, Z. & Guo, X. (2011). Polymorphism in a serine protease inhibitor gene and its association with disease resistance in the eastern oyster (*Crassostrea Virginica*). *Fish Shellfish Immunol.*, *30*, 757–62.

[160] Zapata, M., Tanguy, A., David, E., Moraga, D. & Riquelme, C. (2009). Transcriptomic response *of Argopecten purpuratus* post/larvae to copper exposure under experimental conditions. *Gene*, *442*, 37-46.

[161] Zhao, X., Yu, H., Kong, L. & Li, Q. (2012). Transcriptomic responses to salinity stress in the Pacific oyster *Crassostrea gigas*. *PLoS One, 7*, e46244. DOI: 10.1371/ journal.pone.0046244.

In: Oysters and Clams
Editor: Jesús L. Romalde
ISBN: 978-1-63485-074-2

Chapter 9

HEAVY METALS AND PLATINUM GROUP METALS (PGMS) DETERMINATION IN OYSTERS, MUSSELS AND CLAMS AS BIO-MONITORS OF THE AQUATIC SYSTEM POLLUTION

Clinio Locatelli* and Dora Melucci
Department of Chemistry "G. Ciamician," University of Bologna, Bologna, Italy

ABSTRACT

A rapid diffusion of heavy metals as environmental contaminants has called the attention to their determination. Indeed, heavy metals, together with pesticides, are very dangerous pollutants owing to their bioaccumulation and toxicity. It is therefore necessary to determine these metals at trace and ultra-trace level especially in aquatic ecosystems to establish reasonable water quality criteria. Moreover, toxic metals accumulate in certain marine species and thus enter the aquatic food chain. In particular mussels and clams, but also oysters, shrimps, fish and algae, sequestrate and concentrate several metals from their aqueous environment, possibly becoming dangerous to human health in consequence of their consumption. In particular mussels, clams and oysters, being filtering organisms - an adult organism is able to filter also up to 4-5 L h^{-1}, depending on its weight - they require particular attention and inspections before being sold on the market.

In addition to this important and fundamental aspect of public health, the determination of heavy metals in mussels and clams, that are not only filtering organisms but also sessile species, can be usefully employed for bio-monitoring campaigns, that evaluate the long-term trend of the pollution load of an aquatic ecosystem, information that evidently cannot be provided by punctual determinations.

For completely mapping environmental pollution, the sampling duration and cadence are very important. In our opinion, the use of bio-monitors (mussels and clams in the present case), just proposed by several authors, but certainly not scientifically supported, is possible only in the case of a long sampling plan.

* Corresponding author: Clinio Locatelli. Department of Chemistry "G. Ciamician," University of Bologna, Via F. Selmi 2, I-40126, Bologna (Italy). E-mail: clinio.locatelli@unibo.it.

In any case, the metal determination in mussels and clams evidently must be accurate, reproducible and especially it must show very low limits of detection.

The present work reports and discusses the different analytical methodologies for the voltammetric and spectroscopic determination of heavy metals (copper, lead, cadmium, zinc, mercury, arsenic, selenium, tin and antimony) in oysters, mussels and clams.

A separate section will be dedicated to the analytical procedure for the determination in such organisms of new emerging anthropic polluting traffic-related pollutants, i.e., the platinum group metals (platinum, palladium, rhodium, osmium, ruthenium and iridium).

Keywords: oyster, mussels, clams, heavy metals, platinum group metals (PGMs), bio-monitors, voltammetry, spectroscopy

INTRODUCTION

Metals are not only integral part of the food chain, but also of everything is taken and ingested by humans. Even if of different origins (natural or artificial, the latter including criminal acts), metals become in several cases very dangerous, also with irreversible effect for human life. In this sense, it is very important to determine these species at trace and ultra-trace concentration levels, especially considering the fact that, even at these levels of concentration, several metals can present high risks of toxicity to humans [1-3].

Moreover toxic metals accumulate in certain marine species, thus entering the food chain. In particular oysters, mussels and clams, but also fishes, shrimps and algae, were found to sequestrate and concentrate several metals from their aqueous environment. Evidently these species, covering a substantial part of the food chain, may be hazardous to human health. In particular oysters, mussels and clams, being filtering organisms – an adult organism is able to filter up to 5 L h^{-1}, depending on weight, size and water temperature – require special attention and surveys before they are placed on the market for sale.

In addition to this very important and fundamental aspect of public health, it should be emphasized a second point equally important: oysters, mussels and clams, which besides being filtering organisms are also sessile species, may be usefully employed as bio-monitors of metal pollution in a limited ecosystems [4-7]. This is true even if up today the literature has not described a rigorous and codified correlation between content of toxic metals in bio-monitor species and pollution of the aquatic ecosystem. In any case it is true, or at least very close to truth, the statement "*a high concentration of toxic metals in species considered as bio-monitors most likely implies a strong pollution of the same aquatic ecosystem.*" Consequently, the use of bio-monitors to evaluate the pollutant load of an aquatic ecosystem is considered with great suspicion by most of the scientific community, even if, in our opinion, this procedure is the only possibility to obtain a long period environmental information and to evaluate in the long-term the pollution load trend of an aquatic ecosystem; this information evidently cannot be provided by punctual determinations. In fact, the determination of pollutants level in organisms living for a long time in a given ecosystem certainly permits more complete information about the previous history of the same ecosystem. Several, but single samplings and determinations of metal content in water at different times may give only punctual information, without any information on the element concentration trend in the considered ecosystem. However, it should be highlighted that, for drawing complete environmental pollution information from bio-monitoring campaigns, the sampling duration and cadence are very important.

In our opinion the use of bio-monitors (oysters, mussels and clams in the present case), just proposed by many authors, is possible only in the case of a long sampling plan. Obviously this may be a disadvantage, but perhaps it is also the only possibility to obtain a long period environmental information.

Toxic metal determinations in oysters, mussels and clams, thus addressed either to establish a high food quality or to be employed in bio-monitoring campaigns, evidently must be accurate, and, especially, it must show very low limits of detection.

The work reports the most significant toxic metals that are present in oysters, mussels and clams, mainly due to environmental and anthropic causes, discussing analytical methods for their voltammetric and spectroscopic determination. Moreover a section has been inserted regarding a class of pollutants of new generation linked to vehicular traffic, i.e., the platinum group metals (PGMs), that, in the last decade, have aroused much interest from many authors.

METHODS

Experimental

Voltammetry

All voltammetric curves were recorded by an Amel Model 433 multipolarograph, employing a conventional three-electrode cell. As for the working electrode, a hanging mercury drop electrode (HMDE) was employed for the determination of Cu(II), Pb(II), Cd (II), Zn(II), As(III), Se(IV), Sn(II), Sb(III), Pt(II), Pd(II), Rh(III), Os(VIII), Ru(III)], a gold electrode (GE) (surface area: 0.785 mm^2, AMEL, Milan) was employed for the determination of Hg(II), and a glassy carbon electrode (GCE) (surface area: 7.065 mm^2, AMEL, Milan) was employed for the determination of Ir(III). An $Ag | AgCl | Cl^-_{satd.}$ electrode and platinum wire were chosen as reference and auxiliary electrode, respectively.

The Teflon voltammetric cell was rinsed every day with supra-pure concentrated nitric acid, diluted 1:1, in order to prevent any contamination. Standard additions were made by disposable plastic tips.

Keeping the temperature at 20.0 ± 0.5°C, the solutions were deaerated by water-saturated pure nitrogen for 5 min prior to measurements, while a nitrogen blanket was maintained above the solution during the analysis. The solutions were deaerated after each standard addition for 1 min. In the electrolysis step the solutions were stirred using a magnetic stirrer.

Spectroscopy

Atomic absorption spectrometric measurements were performed using a Perkin-Elmer Mod. A-Analyst 400 Atomic Absorption Spectrometer, equipped with a deuterium background corrector, Autosampler AS-72 and with HGA 800 graphite furnace. Single-element Lumina (Perkin-Elmer) hollow-cathode lamps were used. All measurements were carried out after studying the relevant ashing and atomization curves for each element [8], at the instrumental conditions reported in Table 1.

Table 1. Element instrument settings for the atomic absorption spectrometry (AAS) determination

Element	*Wave-length (nm)*	*Slit (nm)*	*Drying Temperature (°C)*	*Charring Temperature (°C)*	*Atomization Temperature (°C)*	*Matrix Modifiers*
Cu(II)	324.8	0.7	105	1350	2450	0.015 mg Pd + 0.03 mg $Mg(NO_3)_2$
Pb(II)	283.3	0.7	105	950	1900	0.3 mg $NH_4H_2PO_4$ + 0.01 mg $Mg(NO_3)_2$
Cd(II)	228.8	0.5	105	1050	1800	0.3 mg $NH_4H_2PO_4$ + 0.03 mg $Mg(NO_3)_2$
Zn(II)	213.9	0.5	105	900	1900	0.07 mg $Mg(NO_3)_2$
Sn(II)	224.6	0.7	105	1450	2250	0.015 mg Pd + 0.01 mg $Mg(NO_3)_2$
Pt(II)	265.9	0.5	105	1050	2700	
Pd(II)	247.6	0.5	105	950	2650	
Rh(III)	343.5	0.3	105	1100	2550	
Os(VIII)	290.9	0.2	105	750	2700	
Ru(III)	349.9	0.3	105	1350	2450	
Hg(II)	253.7	0.7	Determined by CV-AAS			
As(III)	193.7	0.6	Determined by CV-AAS			
Se(IV)	196.0	1.5	Determined by CV-AAS			
Sb(III)	217.6	0.5	Determined by CV-AAS			

Reagents, Reference Solutions and Standard Reference Materials

All acids and chemicals were suprapure grade (Merck, Germany). Acidic stock solutions of all the elements (1000 mg L^{-1}, Sigma-Aldrich, Germany) and Osmium tetroxide solution (4 wt% in water, Sigma-Aldrich, Germany) were employed in the preparation of reference solutions at varying concentrations, using for diluting water demineralised through a Milli-Q system.

Oyster Tissue NIST-SRM 1566a (from National Institute of Standards and Technology, Gaithersburg, MD, US) and Mussel Tissue BCR-CRM 278 (from Institute for Reference Materials and Measurements, European Commission, Joint Research Centre, Geel, Belgium) were employed as Standard Reference Materials (SRMs).

Sampling Area and Sampling Sites

Oysters, mussels and clams were sampled during summer 2014 in the Goro Bay (Province of Ferrara, Italy; salinity: 2.7%; average temperature from September to October: 23-25°C). This is a very important area devoted to the fishing and breeding of mussels and clams for food, but with considerable pollution problems due to its location in proximity of the Po River mouth, which is the carrier towards the sea of a large amount of industrial and domestic waste water.

Moreover, the waterways are also strongly influenced by atmospheric pollution load and then they result to be also carriers of air pollutants, especially including those related to vehicular traffic as in particular the platinum group metals (PGMs).

In the Goro Bay flow three branches of the Po river delta mouth, i.e., Po of Volano (site A), Canal Bianco (site B) and Po of Goro (site C).

Samplings of mussels and clams were carried out in front of the points where such branches flow into the bay and in the central area (site D) of the bay itself, while an additional sampling (site E), for eventual comparisons, was chosen at open sea.

All samplings were performed using a rubber boat without motor to avoid any kind of pollution.

Sample Preparation before the Instrumental Determination

About 3-4 kg of *Crassostrea gigas* and 7-8 kg of *Mytilus galloprovincialis* and of *Tapes philippinarum* were collected in five sites (see section "Sampling Area and Sampling Sites"), four within the mouth of the Po river (Goro Bay, Italy) and the fifth at open Adriatic Sea, taken to the laboratory and prepared for analyses.

They were opened with a plastic appliance and the organisms were carefully extracted and placed in polyethylene containers, previously treated with suprapure HNO_3 diluted in 1:1 proportion with water and followed by repeated rinsing for 48 h with Milli-Q water in order to avoid any contamination. The samples were frozen and then lyophilised for 30 h. After that treatment, the samples were homogenized thoroughly in an agate mortar.

For all three kinds of real samples (oysters, mussels and clams), the same sample mineralization procedure was carried out.

The sample preparation of Oyster Tissue NIST-SRM 1566a, Mussel Tissue BCR-CRM 278, and of real samples of oysters, mussels and clams was the following: approximately 0.5-1.0 g, accurately weighed, was placed in a platinum crucible and dissolved in 4 mL 69%w/w HNO3 + 4 mL 37%w/w HCl + 5 mL 98%w/w H2SO4 at 130-150°C. The mixture was evaporated to dryness and, after cooling, soluble salts were dissolved in 25 mL of the electrolyte employed in the determination of the different metals.

Limits of Detection

In the aqueous reference solution, in the solutions obtained by digestion of the standard reference material and of oysters, mussels and clams sampled in the Goro Bay, the limits of detection (LOD_s) for both techniques (voltammetry and spectroscopy) were obtained by the equation $LOD = K\ s_{y/x}/b$ [9], where $s_{y/x}$ and b are the estimated standard deviation and the slope of the analytical calibration function of each element, respectively, with a 99.7% (K = 3) confidence level [10].

In the case of voltammetric technique, since the analytical calibration functions were determined by standard addition method, it was possible to obtain the LODs directly also in the real matrix.

Metals of Interest

The present work will concern and discuss the most significant elements that are present in oysters, mussels and clams mainly for environmental and anthropic causes, and the different analytical methodologies for their voltammetric and spectroscopic determination.

Table 2 reports the metals investigated together with the supporting electrolytes employed for the voltammetric determination. The same solutions were also used for the spectroscopic determinations.

Moreover, for each metal or group of metals, a brief introduction is reported. This to emphasize the most recent, but also the most significant and interesting works relevant to the various analytical procedures and techniques for determining the same metals in oysters, mussels and clams.

Table 2. Metals investigated and relevant supporting electrolytes employed in the voltammetricdetermination

Metals	*Supporting electrolyte*
Cu(II) Pb(II) Cd(II) Zn(II)	0.15 mol L^{-1} KCl
Hg(II)	0.5 mol L^{-1} $HClO_4$ + 0.15 mol L^{-1} NaCl + 7.5x10^{-3} mol L^{-1} EDTA-Na_2
As(III) Se(IV)	0.5 mol L^{-1} ammonia-ammonium chloride buffer pH 9.1 + 9.6x10^{-4} mol L^{-1} EDTA-Na_2
Sn(II) Sb(III)	0.3 mol L^{-1} ammonium citrate buffer pH 6.3 + 7.9x10^{-4} mol L^{-1} EDTA-Na_2
Pt(II) Pd(II) Rh(III) Sequential determination	0.25 mol L^{-1} HCl + 0.7x10^{-3} mol L^{-1} formaldehyde + 1.32x10^{-3} mol L^{-1} hydrazine (formazone complex) [determination of Pt(II) and Rh(III)]. 0.50 mol L^{-1} acetate buffer pH 3.9 + 1.89x10^{-4} mol L^{-1} dimethylglyoxime (DMG) [determination of Pd(II) and Rh(III)].
Os(VIII) Ru(III)	0.45 mol L^{-1} acetate buffer pH 4.9 + 9.6x10^{-2} mol L^{-1} $NaBrO_3$ + 1.5 x10^{-4} mol L^{-1} EDTA-Na_2
Ir(III)	0.4 mol L^{-1} acetate buffer pH 4.9 + 8.7x10^{-2} mol L^{-1} $NaBrO_3$ + 0.15 mol L^{-1} KCl+ 6.9x10^{-5} mol L^{-1} cetyltrimethylammonium bromide (CTAB)

Determination of Cu(II), Pb(II), Cd(II) and Zn(II)

Certainly Cu(II), Pb(II), Cd(II) and Zn(II) are the most studied, and many work are present in literature regarding both analytical procedures for the analytical determination of such metals in oysters, mussels and clams and the employment of these filtering organisms for bio-monitoring studies.

Without doubt the most widely used technique is the spectroscopy in all its versions [11-29]. Other techniques as Neutron Activation Analysis (NAA) [30-32] or Voltammetry [16, 33-37] are seldom employed.

Moreover, it is important to highlight two papers proposing the use of Energy Dispersive X-Ray Fluorescence (EDXRF) [38] and of Laser Ablation Inductively Coupled Plasma Mass Spectrometry (LA-ICP-MS) [39] for the determination of the four elements in oysters, mussels and clams.

Finally, very interesting is the work by Wu and coworkers [40] that proposes an "artificial mussel" for monitoring heavy metals in marine environments.

Analytical Procedure for the Voltammetric Determination of Cu(II), Pb(II), Cd(II) and Zn(II)

0.15 mol L^{-1} KCl was employed as supporting electrolyte, which was the aqueous reference solution.

10 mL sample aliquots of supporting electrolyte or of solutions obtained in the mineralization step of the standard reference material and of real samples were pipetted into the voltammetric cell and deaerated for 5 min by bubbling water-saturated pure nitrogen. The determination of Cu(II), Pb(II), Cd(II) and Zn(II) was carried out by Square Wave Anodic Stripping Voltammetry (SWASV) using HMDE as working electrode.

The voltammetric experimental conditions are reported in Table 3.

Table 3. Instrumental parameters for the determination of Cu(II), Pb(II), Cd(II) and Zn(II) by SWASV[a]. Supporting electrolyte: 0.15 mol L^{-1} KCl

E_d	-1.150
E_i	-1.150
E_f	+0.100
t_d	300
t_r	10
dE/dt	100
ΔE	50
τ	0.010
ν	0.100
η	10
u	600

[a] E_d: deposition potential (V/Ag, AgCl, $Cl^-_{satd.}$); E_i: initial potential (V/Ag, AgCl, $Cl^-_{satd.}$); E_f: final potential (V/Ag, AgCl, $Cl^-_{satd.}$); t_d: electrodeposition time (s); t_r: delay time before the potential sweep (s); dE/dt: potential scan rate (mV/s); ΔE: superposed potential amplitude (mV); τ: sampling time (s); υ: wave period (s); η: wave increment (mV); u: stirring rate (r.p.m.).

The experimental peak potentials E_p (V/Ag|AgCl|$Cl^-_{satd.}$) of Cu(II), Pb(II), Cd(II) and Zn(II) are reported in Table 4.

The spectroscopic experimental conditions are reported in Table 1.

Results and Discussion

In this section, in a synthetic way, the fundamental parameters that characterize a correct analytical procedure — limits of detection, linear range, trueness and precision — are reported.

Limits of Detection and Linear Range

In the aqueous reference solution, in the solutions obtained by digestion of the standard reference materials and of real samples, the LOD_s (Table 5) for both techniques were calculated as described in the section "Limits of Detection."

At the experimental conditions employed, the linear range for all the elements in the aqueous reference solutions is <LOD-100.0 μg L^{-1} [Cu(II) and Zn(II)]; and <LOD-50.0 μg L^{-1} [Pb (II) and Cd(II)].

Table 4. Peak potentials (E_p, V/Ag | AgCl | $KCl_{sat.}$) in reference solutions and in solutions obtained by digestion of oysters, mussels and clams sampled in the Goro Bay

	Sampling sites[a]	*Cu(II)*	*Pb(II)*	*Cd(II)*	*Zn(II)*
Supporting electrolyte		-0.069 ± 0.010	-0.405 ± 0.015	-0.625 ± 0.010	-1.023 ± 0.015
Solution from digestion of Mussel Tissue BCR-CRM 278 SRM		-0.077 ± 0.015	-0.423 ± 0.010	-0.643 ± 0.015	-1.051 ± 0.015
Solution from digestion of Oyster Tissue NIST-SRM 1566a		-0.073 ± 0.010	-0.419 ± 0.015	-0.637 ± 0.015	-1.043 ± 0.015
Solution from digestion of Oysters	A	-0.061 ± 0.010	-0.415 ± 0.015	-0.628 ± 0.010	-1.037 ± 0.015
	B	-0.075 ± 0.015	-0.417 ± 0.010	-0.635 ± 0.015	-1.040 ± 0.010
	C	-0.063 ± 0.015	-0.409 ± 0.015	-0.639 ± 0.015	-1.047 ± 0.015
	D	-0.070 ± 0.010	-0.403 ± 0.010	-0.630 ± 0.015	-1.035 ± 0.015
	E	/	/	/	/
Solution from digestion of Mussels	A	-0.079 ± 0.015	-0.407 ± 0.015	-0.626 ± 0.010	-1.053 ± 0.015
	B	-0.068 ± 0.015	-0.421 ± 0.015	-0.645 ± 0.015	-1.038 ± 0.010
	C	-0.060 ± 0.015	-0.417 ± 0.010	-0.638 ± 0.015	-1.029 ± 0.010
	D	-0.071 ± 0.010	-0.413 ± 0.010	-0.629 ± 0.015	-1.045 ± 0.015
	E	-0.067 ± 0.015	-0.408 ± 0.015	-0.636 ± 0.010	-1.036 ± 0.015
Solution from digestion of Clams	A	-0.072 ± 0.010	-0.428 ± 0.015	-0.649 ± 0.015	-1.041 ± 0.015
	B	-0.076 ± 0.010	-0.411 ± 0.010	-0.634 ± 0.015	-1.031 ± 0.010
	C	-0.062 ± 0.015	-0.425 ± 0.015	-0.626 ± 0.010	-1.050 ± 0.015
	D	-0.078 ± 0.015	-0.413 ± 0.015	-0.647 ± 0.015	-1.029 ± 0.015
	E	-0.065 ± 0.015	-0.417 ± 0.010	-0.631 ± 0.015	-1.049 ± 0.015

[a] In the case of Oyster samplings in site E, the absence of element concentration data is due to lack of sample present in the same sampling site.

The determined values are the mean of 5 independent determinations. Confidence level: 95%.

Table 5. LOD$_s$[a] of Cu(II), Pb(II), Cd(II) and Zn(II) determined in the aqueous reference solution (μg L^{-1}), in the solutions obtained by digestion of Mussel Tissue BCR-CRM 278 and Oyster Tissue NIST-SRM 1566a SRM, and in the solutions obtained by digestion of real samples (calculated in μg L^{-1} and expressed in μg kg^{-1})

	Sampling sites[b]	*Cu(II)*	*Pb(II)*	*Cd(II)*	*Zn(II)*
Supporting electrolyte		0.69	0.23	0.13	0.77
Solution obtained by digestion of Mussel Tissue BCR-CRM 278 SRM		29.3	8.7	5.5	41.3
Solution obtained by digestion of Oyster Tissue NIST-SRM 1566a		35.7	9.6	6.3	49.6
Solution obtained by digestion of Oysters	A	25.3	10.9	7.7	53.1
	B	27.5	12.8	5.9	51.3
	C	32.7	15.3	7.2	57.4
	D	31.0	13.6	6.9	54.8
	E	/	/	/	/
Solution obtained by digestion of Mussels	A	71.3	35.5	25.0	75.3
	B	69.0	31.2	23.8	72.5
	C	76.6	36.7	29.9	77.9
	D	73.7	34.0	26.7	74.9
	E	63.5	28.0	20.1	70.1
Solution obtained by digestion of Clams	A	50.3	19.3	17.5	65.3
	B	45.4	18.7	13.7	61.1
	C	49.5	21.5	19.3	69.6
	D	47.3	20.1	17.1	67.0
	E	40.7	17.0	10.9	59.7

[a] In the case of spectroscopic measurements, the LODs (μg L^{-1}) in the aqueous reference solution were: 2.83 [Cu(II)], 0.96 [Pb(II)], 0.69 [Cd(II)] and 3.05 [Zn(II)]. Considering a sample weight exactly equal to 0.5 g (see section "Sample Preparation before the Instrumental Determination"), the same limits of detection (in μg kg^{-1}) were 123.7 [Cu(II)], 37.5 [Pb(II)], 29.6 [Cd(II)] and 169.0 [Zn(II)].

[b] In the case of Oyster samplings in site E, the absence of element concentration data is due to lack of sample present in the same sampling site.

The determined values are the mean of 5 independent determinations; confidence level: 95%.

Table 6. Accuracy of the analytical procedure. The determined values are the mean of 5 independent determinations

Oyster Tissue NIST-SRM 1566a

		Voltammetry			*Spectroscopy*		
Elements	*Certified Values*	*Determined Values*	*e (%)*	*s_r (%)*	*Determined Values*	*e (%)*	*s_r (%)*
Cu(II)	66.3	69.9 ± 4.1	+5.4	5.3	62.5 ± 4.7	-5.7	5.8
Pb(II)	0.371	0.350 ± 0.027	-5.7	5.7	0.396 ± 0.035	+6.7	5.4
Cd(II)	4.15	4.39 ± 0.31	+5.8	5.5	4.42 ± 0.40	+6.5	5.7
Zn(II)	830	877 ± 55	+5.7	5.6	775 ± 69	-6.6	5.5

Mussel Tissue BCR-CRM 278

		Voltammetry			*Spectroscopy*		
Elements	*Certified Values*	*Determined Values*	*e (%)*	s_r *(%)*	*Determined Values*	*e (%)*	s_r *(%)*
Cu(II)	9.60 ± 0.16	9.07 ± 0.61	-5.5	5.6	8.96 ± 0.77	-6.7	5.9
Pb(II)	1.91 ± 0.04	1.80 ± 0.17	-5.8	5.5	2.05 ± 0.19	+7.3	5.5
Cd(II)	0.34 ± 0.02	0.36 ± 0.03	+5.9	5.3	0.37 ± 0.04	+8.8	5.7
Zn(II)	76 ± 2	80.0 ± 4.9	+5.3	5.7	71.0 ± 5.5	-6.6	5.8

Confidence level: 95%. Concentrations in $\mu g\ g^{-1}$.

Table 7. Mean values of Cu(II), Pb(II), Cd(II) and Zn(II) ($\mu g\ g^{-1}$) for oysters, mussels and clams sampled in the Goro Bay

Voltammetry	*Sampling sites*[a]	*Cu(II)*	*Pb(II)*	*Cd(II)*	*Zn(II)*
Oysters	A	45.7 ± 2.5	15.5 ± 1.2	10.0 ± 0.6	187.1 ± 13.1
	B	40.1 ± 2.7	13.1 ± 1.0	8.6 ± 0.6	177.5 ± 12.3
	C	49.3 ± 2.9	19.0 ± 1.1	10.5 ± 0.7	199.6 ± 10.4
	D	44.9 ± 2.3	16.0. ± 1.0	9.7 ± 0.6	186.5 ± 14.0
	E	/	/	/	/
Mussels	A	50.0 ± 2.6	19.7 ± 1.3	9.6 ± 0.8	196.2 ± 13.5
	B	49.6 ± 2.8	17.3 ± 1.2	9.0 ± 0.6	190.0 ± 11.2
	C	55.1 ± 2.7	12.9 ± 1.0	12.9 ± 0.7	208.7 ± 13.7
	D	52.6 ± 2.9	19.1 ± 1.2	11.0 ± 0.7	196.9 ± 11.9
	E	39.7 ± 2.0	5.3 ± 0.4	2.1 ± 0.3	171.2 ± 10.3
Clams	A	37.5 ± 2.1	10.8 ± 0.7	6.7 ± 0.3	177.1 ± 13.8
	B	35.3 ± 2.3	9.6 ± 0.8	6.3 ± 0.3	169.8 ± 11.6
	C	42.1 ± 2.2	13.0 ± 0.9	7.9 ± 0.4	196.3 ± 10.9
	D	38.0 ± 1.9	11.7 ± 0.8	7.1 ± 0.3	181.7 ± 12.7
	E	30.1 ± 2.0	4.9 ± 0.5	1.3 ± 0.2	155.2 ± 9.6
Spectroscopy	*Sampling sites*[a]	*Cu(II)*	*Pb(II)*	*Cd(II)*	*Zn(II)*
Oysters	A	47.9 ± 2.6	14.6 ± 1.3	9.5 ± 0.7	196.9 ± 12.9
	B	38.0 ± 2.5	12.3 ± 1.1	9.1 ± 0.7	186.7 ± 13.0
	C	47.1 ± 3.0	20.0 ± 1.2	10.0 ± 0.9	205.3 ± 11.3
	D	46.0 ± 2.7	16.9 ± 1.2	9.3 ± 0.5	198.0 ± 13.1
	E	/	/	/	/
Mussels	A	47.0 ± 2.8	18.8 ± 1.2	10.3 ± 0.8	207.3 ± 13.5
	B	46.9 ± 2.7	16.5 ± 1.3	9.5 ± 0.7	197.1 ± 12.0
	C	52.8 ± 3.0	13.6 ± 1.1	13.5 ± 0.9	217.1 ± 14.1
	D	50.0 ± 3.1	20.2 ± 1.3	11.6 ± 0.7	206.5 ± 12.6
	E	38.1 ± 1.9	5.6 ± 0.5	2.3 ± 0.3	163.3 ± 10.7
Clams	A	39.3 ± 1.9	11.4 ± 0.8	6.9 ± 0.4	189.1 ± 13.8
	B	37.0 ± 2.1	10.1 ± 0.7	6.5 ± 0.3	179.0 ± 12.5
	C	41.9 ± 2.0	12.3 ± 0.8	7.7 ± 0.3	205.1 ± 11.4
	D	38.9 ± 1.8	11.0 ± 0.9	7.3 ± 0.4	193.2 ± 13.0
	E	31.8 ± 1.9	5.3 ± 0.5	1.4 ± 0.2	149.6 ± 8.9

[a] In the case of Oyster samplings in site E, the absence of element concentration data is due to lack of sample present in the same sampling site.

The determined values are the mean of 5 independent determinations. Confidence level: 95%.

Quality Control and Quality Assessment

The method set up in aqueous reference solutions was applied to Oyster Tissue NIST-SRM 1566a and Mussel Tissue BCR-CRM 278 standard reference materials in order to confirm and verify the applicability of the analytical procedure. Trueness and precision were determined (Table 6).

At the experimental conditions employed, precision as repeatability [9] expressed as relative standard deviation (s_r%) on five independent determinations was satisfactory, being in all cases lower than 6%, while accuracy expressed as relative error (*e*%) was generally lower than 7% (Table 6).

Trueness and precision results for the spectroscopic measurements are also reported in Table 6.

Practical Application

Once the procedure for the Cu(II), Pb(II), Cd(II) and Zn(II) determination was set up, the method was applied to oysters, mussels and clams sampled in the Goro Bay (see section "Sampling Area and Sampling Sites").

The experimental results of both techniques are reported in Table 7.

Determination of Hg(II)

A large amount of mercury is released every year into the environmental ecosystems as a consequence of the human activities. The danger posed to human health by this form of contamination comes mainly from food, i.e., from the ability of this element to enter the natural alimentary chains, to accumulate in progressively larger quantities, and to reach high toxic concentrations in the tissues of organisms that play a role in the human diet, *a fortiori* for oysters, mussels and clams, marine filtering organisms.

For the mercury determination spectroscopic techniques [16, 19, 26, 28, 29, 35-37, 41-50] are prevalently employed, rarely voltammetric techniques [16, 33, 51].

Although not recent, interesting are the works that quantify the inorganic mercury in mussel samples by atomic fluorescence spectrometry (AFS) [52] and by NAA [53].

Analytical Procedure for the Voltammetric Determination of Hg(II)

0.5 mol L^{-1} $HClO_4$ + 0.15 mol L^{-1} NaCl + 7.5×10^{-3} mol L^{-1} EDTA-Na_2 was employed as supporting electrolyte (aqueous reference solution).

10 mL sample aliquots of supporting electrolyte or of solutions obtained in the mineralization step of the standard reference material and of real samples were pipetted into the voltammetric cell and deaerated for 5 min by bubbling water-saturated pure nitrogen. The determination of Hg(II) was carried out by SWASV using GE as working electrode.

The voltammetric experimental conditions are reported in Table 8.

The experimental peak potentials E_p (V/Ag|AgCl|$Cl^-_{satd.}$) of Hg(II) are reported in Table 9.

The spectroscopic experimental conditions are reported in Table 1.

Table 8. Instrumental parameters for the determination of Hg(II) by SWASV[a]

E_d	-0.050
E_i	+0.250
E_{des}	-0.800
E_f	+0.850
t_d	270
t_{ded}	30
t_r	10
dE/dt	100
ΔE	50
τ	0.010
ν	0.100
η	10
u	600

[a] E_d: deposition potential (V/Ag, AgCl, $Cl^-_{satd.}$); E_i: initial potential (V/Ag, AgCl, $Cl^-_{satd.}$); E_{des}: desorption potential (V/Ag, AgCl, $Cl^-_{satd.}$); E_f: final potential (V/Ag, AgCl, $Cl^-_{satd.}$); t_d: electrodeposition time (s); t_{des}: desorption time (s); t_r: delay time before the potential sweep (s); dE/dt: potential scan rate (mV/s); ΔE: superposed potential amplitude (mV); τ: sampling time (s); υ: wave period (s); η: wave increment (mV); u: stirring rate (r.p.m.).

Supporting electrolyte: 0.5 mol L^{-1} $HClO_4$ + 0.15 mol L^{-1} NaCl + 7.5×10^{-3} mol L^{-1} EDTA-Na_2.

Table 9. Peak potentials (E_p, V/Ag|AgCl|$KCl_{sat.}$) in the reference solutions and in solutions obtained by digestion of oysters, mussels and clams sampled in the Goro Bay

	Sampling sites[a]	*Hg(II)*
Supporting electrolyte		+0.527 ± 0.010
Solution obtained by digestion of Mussel Tissue BCR-CRM SRM		+0.569 ± 0.015
Solution obtained by digestion of Oyster Tissue NIST-SRM 1566a		+0.549 ± 0.015
Solution obtained by digestion of Oysters	A	+0.553 ± 0.015
	B	+0.547 ± 0.015
	C	+0.560 ± 0.010
	D	+0.565 ± 0.015
	E	/
Solution obtained by digestion of Mussels	A	+0.537 ± 0.010
	B	+0.543 ± 0.015
	C	+0.551 ± 0.010
	D	+0.540 ± 0.010
	E	+0.545 ± 0.015
Solution obtained by digestion of Clams	A	+0.539 ± 0.015
	B	+0.546 ± 0.010
	C	+0.554 ± 0.015
	D	+0.548 ± 0.010
	E	+0.544 ± 0.015

[a] In the case of Oyster samplings in site E, the absence of element concentration data is due to lack of sample present in the same sampling site.

The determined values are the mean of 5 independent determinations. Confidence level: 95%.

Results and Discussion

In this section, in a synthetic way, the fundamental parameters that characterize a correct analytical procedure — limits of detection, linear range, trueness and precision — are reported.

Limits of Detection and Linear Range

In the aqueous reference solution, in the solutions obtained by digestion of the standard reference materials and of real samples, the LOD_s (Table 10) for both techniques were calculated as described in section "Limits of Detection."

Quality Control and Quality Assessment

The method set up in aqueous reference solutions was applied to Oyster Tissue NIST-SRM 1566a and Mussel Tissue BCR-CRM 278 standard reference materials in order to confirm and verify the applicability of the analytical procedure.

Trueness and precision for the voltammetric and spectroscopic measurements were determined and reported in Table 11.

Table 10. LOD_s[a] of Hg(II) determined in the aqueous reference solution (μg L^{-1}), in the solutions obtained by digestion of Mussel Tissue BCR-CRM 278 and Oyster Tissue NIST-SRM 1566a standard reference materials, and in the solutions obtained by digestion of real samples (calculated in μg L^{-1} and expressed in μg kg^{-1})

	Sampling sites[b]	*Hg(II)*
Supporting electrolyte		0.087
Solution obtained by digestion of Mussel Tissue BCR-CRM 278 SRM		3.5
Solution obtained by digestion of Oyster Tissue NIST-SRM 1566a		3.9
Solution obtained by digestion of Oysters	A	4.3
	B	4.5
	C	4.7
	D	4.2
	E	/
Solution obtained by digestion of Mussels	A	5.3
	B	5.9
	C	6.0
	D	5.7
	E	5.5
Solution obtained by digestion of Clams	A	5.6
	B	5.0
	C	5.5
	D	5.3
	E	4.9

[a] In the case of spectroscopic measurements, the Hg(II) LOD (μg L^{-1}) in the aqueous reference solution is 0.83. Considering a sample weight exactly equal to 0.5 g (see section "Sample Preparation before the Instrumental Determination"), the same limit of detection (μg kg^{-1}) is 41.3.

[b] In the case of Oyster samplings in site E, the absence of element concentration data is due to lack of sample present in the same sampling site.

The determined values are the mean of 5 independent determinations; confidence level: 95%.

Table 11. Accuracy of the analytical procedure.
The determined values are the mean of 5 independent determinations

Oyster Tissue NIST-SRM 1566a

	Voltammetry			*Spectroscopy*		
Certified Values	*Determined Values*	*e (%)*	*s_r (%)*	*Determined Values*	*e (%)*	*s_r (%)*
64.2	67.7 ± 3.7	+5.5	5.4	60.3 ± 4.5	-6.1	5.7

Mussel Tissue BCR-CRM 278

	Voltammetry			*Spectroscopy*		
Certified Values	*Determined Values*	*e (%)*	*s_r (%)*	*Determined Values*	*e (%)*	*s_r (%)*
188 ± 7	177.1 ± 13.2	-5.9	5.5	175.0 ± 15.3	-6.9	5.6

Confidence level: 95%. Concentrations in $\mu g\ g^{-1}$.

Practical Application

Once the procedure for the Hg(II) determination was set up, the method was applied to oysters, mussels and clams sampled in the Goro Bay (see section "Sampling Area and Sampling Sites").

The experimental results relevant to both techniques are reported in Table 12.

Determination of As(III) and Se(IV)

The arsenic and selenium are common pollutants of coal. Consequently, the major industrial sources of pollution from arsenic and selenium are obviously foundries and coal power plants. Furthermore, since the coal is widely used in cement plants, at least in the recent past, it is likely to find these two elements in the air and in areas of fallout of these implants with a consequent pollution of all environmental matrices. It should be added that, unlike arsenic is highly toxic, selenium is a semi-essential element, since it is essential to the development of animal and vegetal life, but very toxic beyond certain levels.

For the determination of As(III) and Se(IV) in oysters, mussels and clams, the literature reports prevalently spectroscopic works [14, 26, 31, 54-62]. Also for the determination of these elements, rarely NAA [30, 32, 63] and voltammetry [33] are employed, while interesting are the works that propose EDXRF [64, 65] and AFS [66, 67] as instrumental techniques.

Analytical Procedure for the Voltammetric Determination of As(III) and Se(IV)

0.5 mol L^{-1} ammonia-ammonium chloride buffer pH 9.1 + 9.6×10^{-4} mol L^{-1} EDTA-Na_2 was employed as supporting electrolyte, which was the aqueous reference solution.

As regards the composition of the supporting electrolyte, it should be emphasized that the presence of EDTA-Na_2 is necessary to obtain better resolution for the As(III) and Se(IV) peaks.

Table 12. Mean values of Hg(II) ($\mu g\ g^{-1}$) relevant to oysters, mussels and clams sampled in the Goro Bay. The determined values are the mean of 5 independent determinations

	Sampling sites[a]	*Voltammetry*	*Spectroscopy*
Oysters	A	5.4 ± 0.4	5.6 ± 0.3
	B	5.1 ± 0.3	4.9 ± 0.3
	C	5.5 ± 0.4	5.8 ± 0.4
	D	5.3 ± 0.4	5.0 ± 0.4
	E	/	/
Mussels	A	6.9 ± 0.5	7.3 ± 0.4
	B	6.5 ± 0.5	6.2 ± 0.4
	C	7.0 ± 0.4	6.7 ± 0.4
	D	6.7 ± 0.4	6.5 ± 0.3
	E	2.1 ± 0.1	2.0 ± 0.2
Clams	A	4.8 ± 0.3	4.6 ± 0.2
	B	4.2 ± 0.3	4.0 ± 0.3
	C	4.9 ± 0.2	5.1 ± 0.3
	D	4.7 ± 0.3	4.5 ± 0.2
	E	1.8 ± 0.2	1.7 ± 0.2

[a] In the case of Oyster samplings in site E, the absence of element concentration data is due to lack of sample present in the same sampling site.

Confidence level: 95%.

In the supporting electrolyte without the presence of EDTA-Na_2 the peak potentials E_p (V/Ag|AgCl|$Cl^-_{satd.}$) are: $E_{p\ As(III)} = -1.296 \pm 0.010$ and $E_{p\ Se(IV)} = -1.377 \pm 0.015$.

The problem is decidedly important, considering that the same problem is also present in the standard reference material and in the real samples collected in the sampling sites. Our methodological procedure proposes the possibility to shift the peaks towards more cathodic potential values by adding EDTA-Na_2. The presence of EDTA-Na_2 shifts towards more cathodic potential values either As(III) or Se(IV) [As(III): -1.309 ± 0.015; Se(IV): -1.487 ± 0.015 V *vs.* Ag|AgCl|$Cl^-_{satd.}$]. Evidently the new peak position of the elements allows their better resolution and then also their quantitative determination.

10 mL sample aliquots of supporting electrolyte or of solutions obtained in the mineralization step of the standard reference material and of real samples were pipetted into the voltammetric cell and deaerated for 5 min by bubbling water-saturated pure nitrogen. The determination of As(III) and Se(IV) was carried out by Differential Pulse Anodic Stripping Voltammetry (DPASV) using HMDE as working electrode.

The voltammetric experimental conditions are reported in Table 13.

The experimental peak potentials E_p (V/Ag|AgCl|$Cl^-_{satd.}$) of As(III) and Se(IV) are reported in Table 14.

The spectroscopic experimental conditions are reported in Table 1.

Table 13. Instrumental parameters for the determination of As(III) and Se(IV) by DPASV[a]. Supporting electrolyte: 0.5 mol L^{-1} ammonia-ammonium chloride buffer pH 9.1 + 9.6x10^{-4} mol L^{-1} EDTA-Na_2

E_d	-1.650
E_i	-1.650
E_f	-1.050
t_d	430
t_r	10
dE/dt	20
ΔE	50
τ	0.065
ν	0.250
r	600

[a] E_d: deposition potential (V/Ag, AgCl, $Cl^-_{satd.}$); E_i: initial potential (V/Ag, AgCl, $Cl^-_{satd.}$); E_f: final potential (V/Ag, AgCl, $Cl^-_{satd.}$); t_d: electrodeposition time (s); t_r: delay time before the potential sweep (s); dE/dt: potential scan rate (mV/s); ΔE: superposed potential amplitude (mV); τ: superposed pulse duration (s); υ: superposed pulse repetition (s); r: stirring rate (r.p.m.).

Table 14. Peak potentials (E_p, V/Ag|AgCl|$KCl_{sat.}$) in the reference solutions and in solutions obtained by digestion of oysters, mussels and clams sampled in the Goro Bay

	Sampling sites[a]	*As(III)*	*Se(IV)*
Supporting electrolyte		-1.309 ± 0.010	-1.487 ± 0.015
Solution obtained by digestion of Mussel Tissue BCR-CRM 278 SRM		-1.301 ± 0.015	-1.489 ± 0.015
Solution obtained by digestion of Oyster Tissue NIST-SRM 1566a		-1.316 ± 0.010	-1.496 ± 0.015
Solution obtained by digestion of Oysters	A	-1.307 ± 0.015	-1.488 ± 0.010
	B	-1.299 ± 0.010	-1.481 ± 0.015
	C	-1.310 ± 0.015	-1.490 ± 0.015
	D	-1.315 ± 0.015	-1.495 ± 0.010
	E	/	/
Solution obtained by digestion of Mussels	A	-1.312 ± 0.015	-1.493 ± 0.015
	B	-1.303 ± 0.015	-1.497 ± 0.010
	C	-1.297 ± 0.015	-1.485 ± 0.015
	D	-1.308 ± 0.010	-1.491 ± 0.015
	E	-1.314 ± 0.015	-1.496 ± 0.015
Solution obtained by digestion of Clams	A	-1.295 ± 0.010	-1.484 ± 0.015
	B	-1.311 ± 0.015	-1.485 ± 0.015
	C	-1.319 ± 0.015	-1.499 ± 0.010
	D	-1.305 ± 0.015	-1.491 ± 0.015
	E	-1.302 ± 0.015	-1.482 ± 0.015

[a] In the case of Oyster samplings in site E, the absence of element concentration data is due to lack of sample present in the same sampling site.

The determined values are the mean of 5 independent determinations. Confidence level: 95%.

Results and Discussion

In this section, in a synthetic way, the fundamental parameters that characterize a correct analytical procedure — limits of detection, linear range, trueness and precision — are reported.

Limits of Detection and Linear Range

In the aqueous reference solution, in the solutions obtained by digestion of the standard reference materials and of real samples, the LOD_s (Table 15) for both techniques were calculated as described in section "Limits of Detection."

At the experimental conditions employed, the linear range for all the elements in the aqueous reference solutions is <LOD-45.0 $\mu g\ L^{-1}$.

Table 15. LOD_s[a] of As(III) and Se(IV) determined in the aqueous reference solution ($\mu g\ L^{-1}$), in the solutions obtained by digestion of Mussel Tissue BCR-CRM 278 and Oyster Tissue NIST-SRM 1566a standard reference materials, and in the solutions obtained by digestion of real samples (calculated in $\mu g\ L^{-1}$ and expressed in $\mu g\ kg^{-1}$)

	Sampling sites[b]	*As(III)*	*Se(IV)*
Supporting electrolyte		0.27	0.19
Solution obtained by digestion of Mussel Tissue BCR-CRM 278 SRM		10.1	9.6
Solution obtained by digestion of Oyster Tissue NIST-SRM 1566a		11.3	9.1
Solution obtained by digestion of Oysters	A	12.1	10-1
	B	11.7	9.7
	C	12.5	10.5
	D	10.6	10.3
	E	/	/
Solution obtained by digestion of Mussels	A	11.4	9.4
	B	12.3	9.9
	C	10.9	10.6
	D	11.5	9.7
	E	11.9	10.0
Solution obtained by digestion of Clams	A	12.7	10.2
	B	11.8	9.5
	C	12.4	9.8
	D	11.2	10.4
	E	12.6	9.3

[a] In the case of spectroscopic measurements, the As(III) and Se(IV) LOD_s ($\mu g\ L^{-1}$) in the aqueous reference solution are 1.43 and 1.15, respectively. Considering a sample weight exactly equal to 0.5 g (see section "Sample Preparation before the Instrumental Determination"), the same limits of detection ($\mu g\ kg^{-1}$) are 55.9 and 49.6 for As(III) and Se(IV), respectively.

[b] In the case of Oyster samplings in site E, the absence of element concentration data is due to lack of sample present in the same sampling site.

The determined values are the mean of 5 independent determinations; confidence level: 95%.

Quality Control and Quality Assessment

The method set up in aqueous reference solutions was applied to Oyster Tissue NIST-SRM 1566a and Mussel Tissue BCR-CRM 278 standard reference materials in order to confirm and verify the applicability of the analytical procedure.

Trueness and precision for the voltammetric and spectroscopic measurements were determined and reported in Table 16.

Table 16. Accuracy of the analytical procedure. The determined values are the mean of 5 independent determinations

Oyster Tissue NIST-SRM 1566a

		Voltammetry			*Spectroscopy*		
Element	*Certified Values*	*Determined Values*	*e (%)*	*s_r (%)*	*Determined Values*	*e (%)*	*s_r (%)*
As(III)	14.0	14.8 ± 0.9	+5.7	5.3	14.9 ± 1.0	+6.4	5.6
Se(IV)	2.21	2.35 ± 0.15	+6.3	5.7	2.05 ± 0.18	-7.2	5.8

Mussel Tissue BCR-CRM 278

		Voltammetry			*Spectroscopy*		
Element	*Certified Values*	*Determined Values*	*e (%)*	*s_r (%)*	*Determined Values*	*e (%)*	*s_r (%)*
As(III)	5.9 ± 0.2	6.2 ± 0.4	+5.1	5.4	5.5 ± 0.5	-6.8	5.9
Se(IV)	1.66 ± 0.04	1.57 ± 0.10	-5.4	5.7	1.55 ± 0.13	-6.6	5.7

Confidence level: 95%. Concentrations in $\mu g\ g^{-1}$.

Practical Application

Once the procedure for the As(III) and Se(IV) determination was set up, the method was applied to oysters, mussels and clams sampled in the Goro Bay (see section "Sampling Area and Sampling Sites").

The experimental results relevant to both techniques are reported in Table 17.

Determination of Sn(II) and Sb(III)

Tin is mainly employed in the production of canned goods for the preservation of food, while antimony is used in the paint and glazes, rubbers, ceramics and in the military industries. These types of industries are therefore the primary pollution source for all the environmental matrices. With regard to the toxicity, tin is a trace element semi-essential element: essential for growth but, in excessive amounts, it can be very toxic because it interferes with the absorption of calcium and zinc. On the contrary, antimony is extremely toxic for human health.

Moreover, these two elements, as for example mercury, tend to bio-accumulate strongly in marine organisms, and for this reason their determination, especially at trace and ultra-trace level concentration, becomes extremely important to control the content of the two metals in the matrices that enter the food chain, just like oysters, mussels and clams.

Works on the Sn (II) and Sb (III) determination in oysters, mussels and clams are not numerous, and very often the investigation concerns the organic tin compounds. They propose spectroscopic [68-70], NAA [71], voltammetric [72] and AFS [73] techniques, while very interesting is the paper by Ferrarello and coworkers [74] that employs size-exclusion chromatography and double focusing ICP-MS detection as instrumental technique that allows the resolution of several spectral interferences that can not be resolved by quadrupole ICP-MS.

Table 17. Mean values of As(III) and Se(IV) (concentrations expressed in μg g^{-1}) relevant to oysters, mussels and clams sampled in the Goro Bay

		Voltammetry		*Spectroscopy*	
	Sampling sites[a]	*As(III)*	*Se(IV)*	*As(III)*	*Se(IV)*
Oysters	A	0.69 ± 0.4	0.57 ± 0.3	0.72 ± 0.4	0.60 ± 0.3
	B	0.61 ± 0.3	0.55 ± 0.3	0.64 ± 0.3	0.59 ± 0.3
	C	0.73 ± 0.4	0.63 ± 0.4	0.70 ± 0.4	0.67 ± 0.4
	D	0.67 ± 0.4	0.59 ± 0.4	0.68 ± 0.4	0.63 ± 0.4
	E	/	/	/	/
Mussels	A	0.91 ± 0.5	0.71 ± 0.4	0.96 ± 0.5	0.69 ± 0.4
	B	0.83 ± 0.5	0.67 ± 0.4	0.87 ± 0.5	0.64 ± 0.4
	C	0.96 ± 0.4	0.77 ± 0.4	0.91 ± 0.4	0.73 ± 0.4
	D	0.87 ± 0.4	0.72 ± 0.3	0.92 ± 0.4	0.68 ± 0.3
	E	0.49 ± 0.5	0.40 ± 0.4	0.46 ± 0.5	0.43 ± 0.4
Clams	A	0.55 ± 0.3	0.31 ± 0.2	0.58 ± 0.3	0.32 ± 0.2
	B	0.49 ± 0.3	0.29 ± 0.3	0.53 ± 0.3	0.31 ± 0.3
	C	0.59 ± 0.2	0.35 ± 0.3	0.56 ± 0.2	0.37 ± 0.3
	D	0.54 ± 0.3	0.32 ± 0.2	0.57 ± 0.3	0.34 ± 0.2
	E	0.23 ± 0.2	0.13 ± 0.3	0.21 ± 0.2	0.15 ± 0.3

[a] In the case of Oyster samplings in site E, the absence of element concentration data is due to lack of sample present in the same sampling site.

The determined values are the mean of 5 independent determinations. Confidence level: 95%.

Analytical Procedure for the Voltammetric Determination of Sn(II) and Sb(III)

0.3 mol L^{-1} ammonium citrate buffer pH 6.3 + 7.9x10^{-4} mol L^{-1} EDTA-Na_2 was employed as supporting electrolyte, which was the aqueous reference solution.

Also in this case, as regards the composition of the supporting electrolyte, it should be emphasized that the presence of EDTA-Na_2 is necessary to obtain better resolution for the Sn(II) and Sb(III) peaks in presence of Pb(II).

Indeed, in the marine ecosystems Pb(II) is always and inevitably present. Unfortunately it shows to have a peak potential very close to those of Sn(II) and Sb(III).

In the supporting electrolyte without the presence of EDTA-Na_2 the peak potentials E_p (V/Ag|AgCl|$Cl^-_{satd.}$) are: $E_{p\ Pb(II)}$ = -0.542 ± 0.015, $E_{p\ Sn(II)}$ = -0.649 ± 0.010 and $E_{p\ Sb(III)}$ = -0.755 ± 0.010.

The problem is decidedly important, considering that the same problem is also present in the standard reference material and in the real samples collected in the sampling sites. Also in this case, our methodological procedure proposes the possibility to shift the interferent peaks towards more cathodic potential values by adding EDTA-Na_2. Indeed, the presence of EDTA-Na_2 shifts towards more cathodic potential the values of all the three elements: Pb(II): -0.569 ± 0.010; Sn(II): -0.723 ± 0.015 and Sb(III): -0.877 ± 0.015 V *vs.* Ag|AgCl|$Cl^-_{satd.}$. Evidently the new peak position of the elements allows their resolution and then also their quantitative determination.

Evidently the proposed analytical procedure, with the addition of EDTA-Na_2, allows to determine also Pb(II); it is sufficient to set instrumentally the final potential E_i (see Table 18) at -0.300 V *vs.* Ag|AgCl|Cl^-_{satd}, so as to extend the potential scan.

10 mL sample aliquots of supporting electrolyte or of solutions obtained in the mineralization step of the standard reference material and of real samples were pipetted into the voltammetric cell and deaerated for 5 min by bubbling water-saturated pure nitrogen. The determination of Sn(II) and Sb(III) was carried out by SWASV using HMDE as working electrode.

The voltammetric experimental conditions are reported in Table 18.

The experimental peak potentials E_p (V/Ag|AgCl|$Cl^-_{satd.}$) of Sn(II) and Sb(III) are reported in Table 19.

The spectroscopic experimental conditions are reported in Table 1.

Results and Discussion

In this section, in a synthetic way, the fundamental parameters that characterize a correct analytical procedure — limits of detection, linear range, trueness and precision — are reported.

Table 18. Instrumental parameters for the determination of Sn(II) and Sb(III) by SWASV[a]

E_d	-1.100
E_i	-1.100
E_f	-0.450
t_d	360
t_r	10
dE/dt	100
ΔE	50
τ	0.010
ν	0.100
η	10
u	600

[a] E_d: deposition potential (V/Ag, AgCl, $Cl^-_{satd.}$); E_i: initial potential (V/Ag, AgCl, $Cl^-_{satd.}$); E_f: final potential (V/Ag, AgCl, $Cl^-_{satd.}$); t_d: electrodeposition time (s); t_r: delay time before the potential sweep (s); dE/dt: potential scan rate (mV/s); ΔE: superposed potential amplitude (mV); τ: sampling time (s); υ: wave period (s); η: wave increment (mV); u: stirring rate (r.p.m.).

Supporting electrolyte: 0.3 mol L^{-1} ammonium citrate buffer pH 6.3 + 7.9×10^{-4} mol L^{-1} EDTA-Na_2.

Limits of Detection and Linear Range

In the aqueous reference solution, in the solutions obtained by digestion of the standard reference materials and of real samples, the LOD_s (Table 20) for both techniques were calculated as described in section "Limits of Detection."

At the experimental conditions employed, the linear range for all the elements in the aqueous reference solutions is <LOD-50.0 μg L^{-1}.

Table 19. Peak potentials (E_p, V/Ag | AgCl | $KCl_{sat.}$) in the reference solutions and in solutions obtained by digestion of oysters, mussels and clams sampled in the Goro Bay

	Sampling sites[a]	*Sn(II)*	*Sb(III)*
Supporting electrolyte		-0.723 ± 0.015	-0.877 ± 0.015
Solution obtained by digestion of Mussel Tissue BCR-CRM SRM		-0.731 ± 0.010	-0.883 ± 0.015
Solution obtained by digestion of Oyster Tissue NIST-SRM 1566a		-0.715 ± 0.015	-0.880 ± 0.010
Solution obtained by digestion of Oysters	A	-0.719 ± 0.015	-0.891 ± 0.010
	B	-0.729 ± 0.015	-0.886 ± 0.015
	C	-0.725 ± 0.015	-0.879 ± 0.015
	D	-0.721 ± 0.010	-0.885 ± 0.015
	E	/	/
Solution obtained by digestion of Mussels	A	-0.716 ± 0.010	-0.878 ± 0.010
	B	-0.728 ± 0.015	-0.890 ± 0.015
	C	-0.720 ± 0.015	-0.883 ± 0.015
	D	-0.727 ± 0.015	-0.889 ± 0.010
	E	-0.730 ± 0.015	-0.881 ± 0.015
Solution obtained by digestion of Clams	A	-0.724 ± 0.010	-0.882 ± 0.015
	B	-0.718 ± 0.010	-0.893 ± 0.010
	C	-0.724 ± 0.015	-0.885 ± 0.010
	D	-0.717 ± 0.015	-0.892 ± 0.015
	E	-0.722 ± 0.015	-0.887 ± 0.015

[a] In the case of Oyster samplings in site E, the absence of element concentration data is due to lack of sample present in the same sampling site.

The determined values are the mean of 5 independent determinations. Confidence level: 95%.

Quality Control and Quality Assessment

The method set up in aqueous reference solutions was applied to Oyster Tissue NIST-SRM 1566a and Mussel Tissue BCR-CRM 278 standard reference materials in order to confirm and verify the applicability of the analytical procedure.

Trueness and precision for the voltammetric and spectroscopic measurements were determined and reported in Table 21.

Practical Application

Once the procedure for the Sn(II) and Sb(III) determination was set up, the method was applied to oysters, mussels and clams sampled in the Goro Bay (see section "Sampling Area and Sampling Sites").

The experimental results relevant to both techniques are reported in Table 22.

Table 20. LOD$_s$[a] of Sn(II) and Sb(III) determined in the aqueous reference solution (μg L^{-1}), in the solutions obtained by digestion of Mussel Tissue BCR-CRM 278 and Oyster Tissue NIST-SRM 1566a standard reference materials, and in the solutions obtained by digestion of real samples (calculated in μg L^{-1} and expressed in μg kg^{-1})

	Sampling sites[b]	*Sn(II)*	*Sb(III)*
Supporting electrolyte		0.49	0.31
Solution obtained by digestion of Mussel Tissue BCR-CRM 278 SRM		19.8	14.3
Solution obtained by digestion of Oyster Tissue NIST-SRM 1566a		21.3	15.1
Solution obtained by digestion of Oysters	A	23.2	15.9
	B	20.5	14.8
	C	22.3	15.5
	D	21.7	16.3
	E	/	/
Solution obtained by digestion of Mussels	A	20.8	14.0
	B	21.5	15.7
	C	22.7	14.5
	D	22.0	16.0
	E	17.3	10.9
Solution obtained by digestion of Clams	A	20.9	16.1
	B	21.3	15.3
	C	22.9	14.9
	D	22.1	15.6
	E	15.7	9.1

[a] In the case of spectroscopic measurements, the Sn(II) and Sb(III) LOD$_s$ (μg L^{-1}) in the aqueous reference solution are 2.49 and 1.77, respectively. Considering a sample weight exactly equal to 0.5 g (see section "Sample Preparation before the Instrumental Determination"), the same limits of detection (μg kg^{-1}) are 109.6 and 157.7 for Sn(II) and Sb(III), respectively.

[b] In the case of Oyster samplings in site E, the absence of element concentration data is due to lack of sample present in the same sampling site.

The determined values are the mean of 5 independent determinations; confidence level: 95%.

Determination of Platinum Group Metals (PGMs), Part I: Pt(II), Pd(II)), Rh(III)

Anthropic activities are certainly the cause of the continuous and constant increase of heavy metals in the environment. Among these, the vehicle emissions are undoubtedly mainly responsible, especially with regard to the strong increase of the concentrations of platinum group metals (PGMs) in all the environmental matrices [75, 76] with great risk for human health [77, 78]. Indeed, in the latest years PGMs have received great attention, because emissions are the main - if not only - source responsible for a significant and constant increase in the environment of such elements. In fact, the compelling and always increasing use of autocatalytic converters, containing PGMs, has resulted in a decrease of pollutants like carbon monoxide, nitrogen oxides, unburned hydrocarbons and especially of lead in exhaust gases from motor vehicles, but, contemporaneously, it is the cause of a widespread distribution in the environment of fine particulate matter and dust at high PGMs concentration originating from deterioration or abrasion of the bulk catalysts due to the incorrect "stop and go" use of the vehicles.

Table 21. Accuracy of the analytical procedure. The determined values are the mean of 5 independent determinations

Oyster Tissue NIST-SRM 1566a

		Voltammetry			*Spectroscopy*		
Element	*Certified Values*	*Determined Values*	*e (%)*	*s_r (%)*	*Determined Values*	*e (%)*	*s_r (%)*
Sn(II)	3.0*	3.2 ± 0.3	+6.7	5.7	2.7 ± 0.3	-10.0	5.9
Sb(III)	0.010*	<LOD	-	-	<LOD	-	-

Mussel Tissue BCR-CRM 278

		Voltammetry			*Spectroscopy*		
Element	*Certified Values*	*Determined Values*	*e (%)*	*s_r (%)*	*Determined Values*	*e (%)*	*s_r (%)*
Sn(II)	6.9**	6.5 ± 0.5	-5.8	5.5	6.4 ± 0.5	-7.2	6.1
Sb(III)	7.7**	7.3 ± 0.5	-5.2	5.6	8.2 ± 0.6	+6.5	5.9

* Values not certified, given for information only.

** Spiked-sample concentration.

Confidence level: 95%. Concentrations in $\mu g\ g^{-1}$.

In this context, after the first massive use of platinum, palladium and rhodium in the construction of autocatalytic converters, there has been a gradual reduction of these PGMs and a growing use of osmium, together with ruthenium and iridium.

In recent years, and more and more frequently, these elements have been employed as alloying with platinum, palladium and rhodium in the manufacture of autocatalytic converters, in order to withstand high temperatures and high wear, so increasing the life time of the same autocatalytic converters.

In the case of Pt(II), Pd(II) and Rh(III), spectroscopy is the technique most widely employed [79-83], while very interesting is the paper by Essumang and coworkers [84] that proposes NAA as instrumental technique.

As regards the voltammetric techniques, the literature reports prevalently the works of two research team: the former (Zimmermann and coworkers [85-90]) that develops the research on the use of mussels as biomonitors for Pt(II), Pd(II) and Rh(III), the latter (Locatelli and coworkers [91-104]) that proposes analytic voltammetric methods for the determination either of Pt(II), Pd(II), Rh(III) or of Os(VIII), Ru(III), Ir(III).

Indeed, in our laboratories since some years it is underway an important line of research on the development of analytical methods for the voltammetric determination of PGMs in all the environmental matrices, starting with those most investigated (platinum, palladium, rhodium) [91-97] to finish those generally considered, at least until a few years ago, less important from the environmental pollution point of view (osmium, ruthenium and iridium) [98-104].

Table 22. Mean values of Sn(II) and Sb(III) (μg g^{-1}) relevant to oysters, mussels and clams sampled in the Goro Bay

		Voltammetry		*Spectroscopy*	
	Sampling sites[a]	*Sn(II)*	*Sb(III)*	*Sn(II)*	*Sb(III)*
Oysters	A	2.7 ± 0.3	1.5 ± 0.2	2.5 ± 0.2	1.4 ± 0.2
	B	2.4 ± 0.2	1.2 ± 0.2	2.3 ± 0.2	1.3 ± 0.2
	C	2.6 ± 0.2	1.8 ± 0.3	2.7 ± 0.3	1.7 ± 0.3
	D	2.9 ± 0.3	1.4 ± 0.2	2.6 ± 0.3	1.5 ± 0.2
	E	/	/	/	/
Mussels	A	3.5 ± 0.3	2.2 ± 0.2	3.7 ± 0.3	2.3 ± 0.3
	B	3.1 ± 0.3	2.1 ± 0.3	2.9 ± 0.3	2.0 ± 0.2
	C	3.8 ± 0.2	2.5 ± 0.2	3.6 ± 0.3	2.4 ± 0.3
	D	3.4 ± 0.2	2.3 ± 0.2	3.5 ± 0.2	2.2 ± 0.2
	E	1.3 ± 0.2	0.5 ± 0.1	1.2 ± 0.2	0.6 ± 0.2
Clams	A	1.5 ± 0.2	0.8 ± 0.2	1.6 ± 0.2	0.9 ± 0.2
	B	1.3 ± 0.2	0.6 ± 0.1	1.4 ± 0.3	0.6 ± 0.2
	C	1.7 ± 0.2	1.0 ± 0.2	1.8 ± 0.2	0.9 ± 0.2
	D	1.4 ± 0.2	0.7 ± 0.2	1.5 ± 0.2	0.8 ± 0.2
	E	0.6 ± 0.1	0.2 ± 0.1	0.5 ± 0.1	0.3 ± 0.1

[a] In the case of Oyster samplings in site E, the absence of element concentration data is due to lack of sample present in the same sampling site.

The determined values are the mean of 5 independent determinations. Confidence level: 95%.

Analytical Procedure for the Voltammetric Determination of Pt(II), Pd(II)) and Rh(III)

The voltammetric determination of Pt(II), Pd(II)) and Rh(III) here proposed is performed sequentially and requires the use of two supporting electrolytes:

a. 0.25 mol L^{-1} HCl + 0.7x10^{-3} mol L^{-1} formaldehyde + 1.32x10^{-3} mol L^{-1} hydrazine (formazone complex) [determination of Pt(II) and Rh(III)].
b. 0.50 mol L^{-1} acetate buffer pH 3.9 + 1.89x10^{-4} mol L^{-1} dimethylglyoxime (DMG) [determination of Pd(II) and Rh(III)].

The two supporting electrolytes may be employed separately for the determination of Pt(II) and Rh(III) (supporting electrolyte *a*) or Pd(II) and Rh(III) (supporting electrolyte *b*), but they may be used sequentially without problems for the determination of all three elements: just set the adsorption $E_{ads.}$ and initial E_i potentials at -0.500 V *vs.* Ag, AgCl, $Cl^-_{satd.}$, and the final potential E_f at -1.350 V *vs.* Ag, AgCl, Cl^-_{satd}. In this last case, the analytical procedure is the following:

- 10 mL sample aliquots of 0.25 mol L^{-1} HCl + $0.7x10^{-3}$ mol L^{-1} formaldehyde + $1.32x10^{-3}$ mol L^{-1} hydrazine supporting electrolyte or of solutions obtained in the mineralization step of the standard reference material and of real samples were pipetted into the voltammetric cell and deaerated for 5 min by bubbling water-saturated pure nitrogen. The determination of Pt(II) and Rh(III) was carried out by Square Wave Adsorptive Stripping Voltammetry (SWAdSV) using HMDE as working electrode.
- Subsequently, 10 mL of 1.0 mol L^{-1} acetate buffer pH 4.1 + $3.77x10^{-4}$ mol L^{-1} DMG were added (final pH 3.9, final DMG concentration $1.89x10^{-4}$ mol/L). The determination of Pd(II) and Rh(III) was carried out by SWAdSV.
- The voltammetric experimental conditions are reported in Table 23.

The experimental peak potentials E_p (V/Ag | AgCl | $Cl^-_{satd.}$) of Pt(II), Pd(II) and Rh(III) in the two supporting electrolytes are reported in Table 24.

The spectroscopic experimental conditions are reported in Table 1.

Results and Discussion

In this section, in a synthetic way, the fundamental parameters that characterize a correct analytical procedure — limits of detection, linear range, trueness and precision — are reported.

Table 23. Instrumental parameters for the determination of Pt(II), Pd(II) and Rh(III) by SWAdSV[a]

	Pt(II)-Rh(III)	*Pd(II)-Rh(III)*
	0.25 mol L^{-1} HCl + $0.7x10^{-3}$ mol L^{-1} formaldehyde + $1.32x10^{-3}$ mol L^{-1} hydrazine	0.50 mol L^{-1} acetate buffer pH 3.9 + $1.89x10^{-4}$ mol L^{-1} dimethylglyoxime
E_{ads}	-0.700	-0.500
E_i	-0.700	-0.500
E_f	-1.350	-1.350
t_{ads}	390	390
t_r	10	10
dE/dt	100	100
ΔE	50	50
τ	0.010	0.010
ν	0.100	0.100
η	10	10
u	600	600

[a] E_{ads}: adsorption potential (V/Ag, AgCl, $Cl^-_{satd.}$); E_i: initial potential (V/Ag, AgCl, $Cl^-_{satd.}$); E_f: final potential (V/Ag, AgCl, $Cl^-_{satd.}$); t_{ads}: electroadsorption time (s); t_r: delay time before the potential sweep (s); dE/dt: potential scan rate (mV/s); ΔE: superposed potential amplitude (mV); τ: sampling time (s); υ: wave period (s); η: wave increment (mV); u: stirring rate (r.p.m.).

Limits of Detection and Linear Range

In the aqueous reference solution, in the solutions obtained by digestion of the standard reference materials and of real samples, the LOD_s (Table 25) for both techniques were calculated as described in section "Limits of Detection."

At the experimental conditions employed, the linear range for all the elements in the aqueous reference solutions is <LOD-10.0 μg L-1.

Table 24. Peak potentials (E_p, V/Ag|AgCl|$KCl_{sat.}$) in the reference solutions and in solutions obtained by digestion of oysters, mussels and clams sampled in the Goro Bay

		Pt(II)	*Rh(III)*	*Pd(II)*	*Rh(III)*
	Sampling sites[a]	0.25 mol L^{-1} HCl + 0.7x10^{-3} mol L^{-1} formaldehyde + 1.32x10^{-3} mol L^{-1} hydrazine		0.50 mol L^{-1} acetate buffer pH 3.9 + 1.89x10^{-4} mol L^{-1} dimethylglyoxime	
Supporting electrolyte		-0.963 ± 0.010	-1.129 ± 0.010	-0.719 ± 0.015	-1.149 ± 0.015
Solution obtained by digestion of Mussel Tissue BCR-CRM 278 SRM		-0.977 ± 0.015	-1.153 ± 0.015	-0.743 ± 0.015	-1.161 ± 0.010
Solution obtained by digestion of Oyster Tissue NIST-SRM 1566a		-0.969 ± 0.015	-1.161 ± 0.015	-0.737 ± 0.010	-1.158 ± 0.015
Solution obtained by digestion of Oysters	A	-0.975 ± 0.010	-1.149 ± 0.015	-0.751 ± 0.015	-1.153 ± 0.010
	B	-0.965 ± 0.015	-1.135 ± 0.015	-0.729 ± 0.015	-1.169 ± 0.015
	C	-0.968 ± 0.015	-1.159 ± 0.010	-0.735 ± 0.010	-1.171 ± 0.015
	D	-0.981 ± 0.015	-1.147 ± 0.015	-0.748 ± 0.015	-1.159 ± 0.015
	E	-	-	-	-
Solution obtained by digestion of Mussels	A	-0.959 ± 0.010	-1.148 ± 0.015	-0.738 ± 0.015	-1.177 ± 0.015
	B	-0.973 ± 0.015	-1.152 ± 0.015	-0.741 ± 0.015	-1.165 ± 0.010
	C	-0.980 ± 0.010	-1.135 ± 0.010	-0.727 ± 0.015	-1.180 ± 0.015
	D	-0.963 ± 0.015	-1.157 ± 0.015	-0.745 ± 0.010	-1.163 ± 0.015
	E	No signal	No signal	No signal	No signal
Solution obtained by digestion of Clams	A	-0.971 ± 0.015	-1.137 ± 0.010	-0.725 ± 0.015	-1.155 ± 0.010
	B	-0.974 ± 0.010	-1.146 ± 0.015	-0.740 ± 0.010	-1.167 ± 0.015
	C	-0.961 ± 0.015	-1.160 ± 0.015	-0.747 ± 0.015	-1.173 ± 0.015
	D	-0.970 ± 0.015	-1.145 ± 0.015	-0.734 ± 0.015	-1.181 ± 0.015
	E	No signal	No signal	No signal	No signal

[a] In the case of Oyster samplings in site E, the absence of element concentration data is due to lack of sample present in the same sampling site.

Spiked-sample concentration in reference solutions: 69.6 μg kg^{-1}.

The determined values are the mean of 5 independent determinations. Confidence level: 95%.

Quality Control and Quality Assessment

The method set up in aqueous reference solutions was applied to Oyster Tissue NIST-SRM 1566a and Mussel Tissue BCR-CRM 278 standard reference materials in order to confirm and verify the applicability of the analytical procedure.

Table 25. LOD$_s$[a] of Pt(II), Pd(II) and Rh(III) determined in the aqueous reference solution (μg L^{-1}), in the solutions obtained by digestion of Mussel Tissue BCR-CRM 278 and Oyster Tissue NIST-SRM 1566a standard reference materials, and in the solutions obtained by digestion of real samples (calculated in μg L^{-1} and expressed in μg kg^{-1})

		Pt(II)	*Rh(III)*	*Pd(II)*	*Rh(III)*
	Sampling sites[b]	0.25 mol L^{-1} HCl + 0.7x10^{-3} mol L^{-1} formaldehyde + 1.32x10^{-3} mol L^{-1} hydrazine		0.50 mol L^{-1} ac. buffer pH 3.9 + 1.89x10^{-4} mol L^{-1} dimethylglyoxime	
Supporting electrolyte		0.035	0.043	0.026	0.039
Solution obtained by digestion of Mussel Tissue BCR-CRM 278 SRM		1.2	1.4	1.0	1.5
Solution obtained by digestion of Oyster Tissue NIST-SRM 1566a		1.5	1.7	1.1	1.3
Solution obtained by digestion of Oysters	A	1.6	1.8	1.2	1.4
	B	1.4	1.5	1.4	1.3
	C	1.6	1.6	1.0	1.1
	D	1.5	1.7	1.1	1.2
	E	-	-	-	-
Solution obtained by digestion of Mussels	A	1.7	1.8	1.0	1.1
	B	1.3	1.5	1.2	1.3
	C	1.5	1.5	0.9	0.9
	D	1.4	1.6	1.3	1.5
	E	No signal	No signal	No signal	No signal
Solution obtained by digestion of Clams	A	1.4	1.6	0.9	1.0
	B	1.5	1.7	1.0	1.1
	C	1.7	1.8	1.1	1.4
	D	1.6	1.5	0.9	1.3
	E	No signal	No signal	No signal	No signal

[a] In the case of spectroscopic measurements, the LODs (μg L^{-1}) were: 0.119 [Pt(II)], 0.121 [Rh(III)]. in 0.25 mol L^{-1} HCl + 0.7x10^{-3} mol L^{-1} formaldehyde + 1.32x10^{-3} mol L^{-1} hydrazine supporting electrolyte. and 0.101 [Pd(II)], 0.123 [Rh(III)] in 0.50 mol L^{-1} acetate buffer pH 3.9 + 1.89x10^{-4} mol L^{-1} dimethylglyoxime supporting electrolyte. Considering a sample weight exactly equal to 0.5 g (see section "Sample Preparation before the Instrumental Determination"), the same limits of detection (μg kg^{-1}) were 4.0 [Pt(II)]; 4.2 [Rh(III)] and 3.8 [Pd(II)]; 4.3 [Rh(III)] for the former and the latter supporting electrolyte, respectively. [b] In the case of Oyster samplings in site E, the absence of element concentration data is due to lack of sample present in the same sampling site.
The determined values are the mean of 5 independent determinations; confidence level: 95%.

Trueness and precision for the voltammetric and spectroscopic measurements were determined and reported in Table 26.

Table 26. Accuracy of the analytical procedure. The determined values are the mean of 5 independent determinations

Oyster Tissue NIST-SRM 1566a

	Voltammetry			*Spectroscopy*		
Elements	*Determined Values*	*e (%)*	s_r *(%)*	*Determined Values*	*e (%)*	s_r *(%)*
Pt(II)*	73.3 ± 4.2	+5.3	5.3	73.9 ± 4.6	+6.2	5.9
Rh(II)*	73.5 ± 4.3	+5.6	5.4	65.5 ± 4.7	-6.3	6.0
Pd(II)**	73.6 ± 4.7	+5.7	5.9	74.1 ± 5.3	+6.5	6.1
Rh(II)**	65.9 ± 4.0	-5.3	5.7	64.9 ± 5.0	-6.8	5.7

Mussel Tissue BCR-CRM 278

	Voltammetry			*Spectroscopy*		
Elements	*Determined Values*	*e (%)*	s_r *(%)*	*Determined Values*	*e (%)*	s_r *(%)*
Pt(II)*	65.8 ± 4.7	-5.5	5.7	65.4 ± 4.9	-6.0	5.9
Rh(II)*	73.7 ± 4.6	+5.9	5.6	74.3 ± 5.3	+6.8	6.3
Pd(II)**	73.4 ± 4.9	+5.5	5.8	64.8 ± 5.4	-6.9	6.0
Rh(II)**	65.7 ± 4.5	-5.6	5.9	74.2 ± 5.5	+6.6	6.2

* 0.25 mol L^{-1} HCl + 0.7x10^{-3} mol L^{-1} formaldehyde + 1.32x10^{-3} mol L^{-1} hydrazine as supporting electrolyte.

** 0.50 mol L^{-1} acetate buffer pH 3.9 + 1.89x10^{-4} mol L^{-1} dimethylglyoxime as supporting electrolyte.

Confidence level: 95%. Concentrations in μg g^{-1}. In all cases the spiked-sample concentration was 69.6 μg kg^{-1}.

Practical Application

Once the procedure for the Pt(II), Pd(II) and Rh(III) determination was set up, the method was applied to oysters, mussels and clams sampled in the Goro Bay (see section "Sampling Area and Sampling Sites").

The experimental results relevant to both techniques are reported in Table 27.

Determination of Platinum Group Metals (PGMs), Part II: Os(VIII) and Ru(III)

As regards the determination of Os(VIII) and Ru(III) in filtering marine organisms, at least to our knowledge, the literature does not report works. This work is therefore the first that discusses a critical comparison between voltammetric and spectroscopic methods for the determination of such PGMs in oysters, mussels and clams.

Analytical Procedure for the Voltammetric Determination of Os(VIII) and Ru(III)

0.45 mol L^{-1} acetate buffer pH 4.9 + 9.6x10^{-2} mol L^{-1} $NaBrO_3$ + 1.5 x10^{-4} mol L^{-1} EDTA-Na_2 was employed as supporting electrolyte and aqueous reference solution.

Table 27. Mean values of Pt(II), Pd(II) and Rh(III) (μg g^{-1}) relevant to oysters, mussels and clams sampled in the Goro Bay

Voltammetry	*Sampling sites*[a]	*Pt(II)*[*]	*Rh(III)*[*]	*Pd(II)*[**]	*Rh(III)*[**]
Oysters	A	7.3 ± 0.3	7.8 ± 0.3	8.0 ± 0.3	7.9 ± 0.3
	B	6.9 ± 0.3	7.6 ± 0.2	7.6 ± 0.2	7.7 ± 0.3
	C	7.4 ± 0.2	8.1 ± 0.3	7.9 ± 0.2	7.9 ± 0.2
	D	7.0 ± 0.2	7.9. ± 0.2	7.6 ± 0.2	8.0 ± 0.2
	E	/	/	/	/
Mussels	A	9.1 ± 0.2	8.3 ± 0.3	9.1 ± 0.3	8.5 ± 0.3
	B	8.7 ± 03	8.0 ± 0.2	8.4 ± 0.2	8.1 ± 0.2
	C	9.3 ± 0.3	8.2 ± 0.3	8.7 ± 0.3	8.0 ± 0.3
	D	9.0 ± 0.3	8.1 ± 0.3	8.6 ± 0.2	8.4 ± 0.3
	E	5.3 ± 0.2	4.9 ± 0.2	4.0 ± 0.2	4.7 ± 0.2
Clams	A	6.1 ± 0.2	6.0 ± 0.3	6.8 ± 0.3	6.3 ± 0.3
	B	5.8 ± 0.3	5.7 ± 0.3	6.0 ± 0.2	5.5 ± 0.3
	C	6.3 ± 0.3	5.9 ± 0.2	6.4 ± 0.3	6.1 ± 0.2
	D	6.0 ± 0.2	5.7 ± 0.2	6.2 ± 0.2	5.9 ± 0.2
	E	4.1 ± 0.2	3.9 ± 0.2	4.3 ± 0.2	3.7 ± 0.2
Voltammetry	*Sampling sites*[a]	*Pt(II)*[*]	*Rh(III)*[*]	*Pd(II)*[**]	*Rh(III)*[**]
Oysters	A	7.6 ± 0.3	8.1 ± 0.3	8.3 ± 0.3	8.2 ± 0.3
	B	6.6 ± 03	7.4 ± 0.2	7.7 ± 0.2	7.5 ± 0.2
	C	7.2 ± 0.3	8.0 ± 0.2	8.1 ± 0.2	8.1 ± 0.3
	D	7.2 ± 0.2	7.7 ± 0.2	7.8 ± 0.3	7.8 ± 0.2
	E	/	/	/	/
Mussels	A	9.3 ± 0.2	8.6 ± 0.3	8.9 ± 0.3	8.8 ± 0.3
	B	9.0 ± 0.3	8.2 ± 0.3	8.3 ± 0.3	8.3 ± 0.2
	C	9.6 ± 0.3	7.9 ± 0.3	9.0 ± 0.3	8.2 ± 0.3
	D	9.2 ± 0.2	8.3 ± 0.2	8.5 ± 0.2	8.1 ± 0.3
	E	5.1 ± 0.2	4.7 ± 0.2	4.1 ± 0.2	4.8 ± 0.2
Clams	A	5.9 ± 0.2	6.3 ± 0.3	6.7 ± 0.2	6.1 ± 0.2
	B	5.6 ± 0.2	6.0 ± 0.3	5.8 ± 0.3	5.7 ± 0.3
	C	6.6 ± 0.3	5.7 ± 0.3	6.1 ± 0.3	5.8 ± 0.3
	D	6.1 ± 0.2	5.9 ± 0.3	6.0 ± 0.2	5.7 ± 0.2
	E	4.0 ± 0.2	3.7 ± 0.2	4.5 ± 0.2	3.8 ± 0.2

[a] In the case of Oyster samplings in site E, the absence of element concentration data is due to lack of sample present in the same sampling site.

[*] 0.25 mol L^{-1} HCl + 0.7x10^{-3} mol L^{-1} formaldehyde + 1.32x10^{-3} mol L^{-1} hydrazine as supporting electrolyte.

[**] 0.50 mol L^{-1} acetate buffer pH 3.9 + 1.89x10^{-4} mol L^{-1} dimethylglyoxime as supporting electrolyte.

The determined values are the mean of 5 independent determinations. Confidence level: 95%.

Also in this case, as regards the composition of the supporting electrolyte, it should be emphasized that the presence of EDTA-Na_2 is necessary to obtain better resolution for the Os(VIII) and Ru(III) peaks in presence of Cu(II) and Pb(II).

Indeed, in the marine ecosystems Cu(II) and Pb(II) are always and inevitably present. Unfortunately, under the experimental conditions employed in the Os(VIII) and Ru(III) determination, Cu(II) and Pb(II) also had well defined reversible voltammetric peaks, peak potentials being -0.149 ± 0.015 and -0.296 ± 0.010 V *vs.* Ag|AgCl|KCl $_{satd}$, respectively. Inevitably, this means that, especially for Ru(III), the problem of interference is strong, considering that Os(VIII) and Ru(III) show to have -0.023 ± 0.015 and -0.227 ± 0.015 V *vs.* Ag|AgCl|KCl $_{satd}$ peak potential values, respectively.

The problem is decidedly important, considering that the same problem is also present in the standard reference material and in the real samples collected in the sampling sites. Our methodological procedure proposal, already reported in previous works [102, 104], proposes the possibility to shift the interferent peaks towards more cathodic potential values by adding EDTA-Na_2. Indeed, the peak potentials of the four elements, in presence of EDTA-Na_2 have the following values: Os(VIII): -0.099 ± 0.015; Ru(III): -0.305 ± 0.015; Cu(II): -0.577 ± 0.015 and Pb(II): -0.796 ± 0.015 V *vs.* Ag|AgCl|$Cl^-_{satd.}$. Evidently the new peak position of the elements allows their resolution and then also their quantitative determination.

Moreover, the proposed analytical procedure with the addition of EDTA-Na_2 allows to determine also Cu(II) and Pb(II): it is sufficient to set instrumentally the final potential E_f (see Table 23) at -1.050 V *vs.* Ag|AgCl|Cl^-_{satd}, so as to extend the potential scan.

10 mL sample aliquots of supporting electrolyte or of solutions obtained in the mineralization step of the standard reference material and of real samples were pipetted into the voltammetric cell and deaerated for 5 min by bubbling water-saturated pure nitrogen. The determination of Os(VIII) and Ru(III) was carried out by Differential Pulse Catalytic Voltammetry (DPCV) using HMDE as working electrode.

The voltammetric experimental conditions are reported in Table 28.

The experimental peak potentials E_p (V/Ag|AgCl|$Cl^-_{satd.}$) of Os(VIII) and Ru(III) are reported in Table 29.

The spectroscopic experimental conditions are reported in Table 1.

Results and Discussion

In this section, in a synthetic way, the fundamental parameters that characterize a correct analytical procedure — limits of detection, linear range, trueness and precision — are reported.

Limits of Detection and Linear Range

In the aqueous reference solution, in the solutions obtained by digestion of the standard reference materials and of real samples, the LOD_s (Table 30) for both techniques were calculated as described in section "Limits of Detection."

At the experimental conditions employed, the linear range for all the elements in the aqueous reference solutions is <LOD-5.0 μg L^{-1}.

Quality Control and Quality Assessment

The method set up in aqueous reference solutions was applied to Oyster Tissue NIST-SRM 1566a and Mussel Tissue BCR-CRM 278 standard reference materials in order to confirm and verify the applicability of the analytical procedure.

Table 28. Instrumental parameters for the determination of Os(VIII) and Ru(III) by DPCV[a]. Supporting electrolyte: 0.45 mol L^{-1} acetate buffer pH 4.9 + 9.6x10^{-2} mol L^{-1} $NaBrO_3$ + 1.5 x10^{-4} mol L^{-1} EDTA-Na_2

E_d	-0.015
E_i	+0.150
E_f	-0.550
t_d	420
t_r	10
dE/dt	20
ΔE	50
τ	0.065
ν	0.250
r	600

[a] E_d: deposition potential (V/Ag, AgCl, $Cl^-_{satd.}$); E_i: initial potential (V/Ag, AgCl, $Cl^-_{satd.}$); E_f: final potential (V/Ag, AgCl, $Cl^-_{satd.}$); t_d: electrodeposition time (s); t_r: delay time before the potential sweep (s); dE/dt: potential scan rate (mV/s); ΔE: superposed potential amplitude (mV); τ: superposed pulse duration (s); υ: superposed pulse repetition (s); r: stirring rate (r.p.m.).

Table 29. Peak potentials (E_p, V/Ag|AgCl|$KCl_{sat.}$) in the reference solutions and in solutions obtained by digestion of oysters, mussels and clams sampled in the Goro Bay

	Sampling sites[a]	*Os(VIII)*	*Ru(III)*
Supporting electrolyte		-0.099 ± 0.015	-0.305 ± 0.015
Solution obtained by digestion of Mussel Tissue BCR-CRM 278 SRM		-0.107 ± 0.010	-0.313 ± 0.015
Solution obtained by digestion of Oyster Tissue NIST-SRM 1566a		-0.103 ± 0.015	-0.309 ± 0.010
Solution obtained by digestion of Oysters	A	-0.106 ± 0.015	-0.301 ± 0.010
	B	-0.109 ± 0.015	-0.307 ± 0.015
	C	-0.112 ± 0.015	-0.304 ± 0.015
	D	-0.100 ± 0.010	-0.293 ± 0.015
	E	/	/
Solution obtained by digestion of Mussels	A	-0.110 ± 0.010	-0.307 ± 0.010
	B	-0.097 ± 0.015	-0.296 ± 0.015
	C	-0.104 ± 0.015	-0.297 ± 0.015
	D	-0.115 ± 0.015	-0.301 ± 0.010
	E	No signal	No signal
Solution obtained by digestion of Clams	A	-0.101 ± 0.010	-0.299 ± 0.015
	B	-0.112 ± 0.010	-0.306 ± 0.010
	C	-0.095 ± 0.015	-0.302 ± 0.010
	D	-0.105 ± 0.015	-0.311 ± 0.015
	E	No signal	No signal

[a] In the case of Oyster samplings in site E, the absence of element concentration data is due to lack of sample present in the same sampling site.

The determined values are the mean of 5 independent determinations. Confidence level: 95%.

Spiked-sample concentration in reference solutions: 53.7 μg kg^{-1}.

Trueness and precision for the voltammetric and spectroscopic measurements were determined and reported in Table 31.

Practical Application

Once the procedure for the Os(VIII) and Ru(III) determination was set up, the method was applied to oysters, mussels and clams sampled in the Goro Bay (see section “Sampling Area and Sampling Sites”).

The experimental results relevant to both techniques are reported in Table 32.

Table 30. LOD$_s$[a] of Os(VIII) and Ru(III) determined in the aqueous reference solution (μg L^{-1}), in the solutions obtained by digestion of Mussel Tissue BCR-CRM 278 and Oyster Tissue NIST-SRM 1566a standard reference materials, and in the solutions obtained by digestion of real samples (calculated in μg L^{-1} and expressed in μg kg^{-1})

	Sampling sites[b]	*Os(VIII)*	*Ru(III)*
Supporting electrolyte		0.023	0.019
Solution obtained by digestion of Mussel Tissue BCR-CRM 278 SRM		0.81	0.53
Solution obtained by digestion of Oyster Tissue NIST-SRM 1566a		0.73	0.64
Solution obtained by digestion of Oysters	A	0.77	0.55
	B	0.65	070
	C	0.53	0.62
	D	0.72	0.68
	E	/	/
Solution obtained by digestion of Mussels	A	0.83	0.53
	B	0.71	0.47
	C	0.62	0.70
	D	0.75	0.65
	E	No signal	No signal
Solution obtained by digestion of Clams	A	0.77	0.58
	B	0.67	0.67
	C	0.92	0.90
	D	0.63	0.75
	E	No signal	No signal

[a] In the case of spectroscopic measurements, the Os(VIII) and Ru(III) LOD$_s$ (μg L^{-1}) in the aqueous reference solution are 0.123 and 0.077, respectively. Considering a sample weight exactly equal to 0.5 g (see section “Sample Preparation before the Instrumental Determination”), the same limits of detection (μg kg^{-1}) are 4.1 and 3.0 for Os(VIII) and Ru(III), respectively.

[b] In the case of Oyster samplings in site E, the absence of element concentration data is due to lack of sample present in the same sampling site.

The determined values are the mean of 5 independent determinations; confidence level: 95%.

Table 31. Accuracy of the analytical procedure. The determined values are the mean of 5 independent determinations

Oyster Tissue NIST-SRM 1566a

	Voltammetry			*Spectroscopy*		
Element	*Determined Values*	*e (%)*	s_r *(%)*	*Determined Values*	*e (%)*	s_r *(%)*
Os(VIII)	56.9 ± 3.5	+6.0	5.3	57.0 ± 3.7	+6.1	5.7
Ru(III)	56.7 ± 3.4	+5.6	5.6	57.3 ± 3.9	+6.7	5.9

Mussel Tissue BCR-CRM 278

	Voltammetry			*Spectroscopy*		
Element	*Determined Values*	*e (%)*	s_r *(%)*	*Determined Values*	*e (%)*	s_r *(%)*
Os(VIII)	56.8 ± 3.6	+5.8	5.5	50.1 ± 4.0	-6.7	6.0
Ru(III)	56.7 ± 3.2	+5.6	5.4	57.1 ± 3.8	+6.3	6.1

Confidence level: 95%. In all cases the spiked-sample concentration was 53.7 $\mu g\ kg^{-1}$.

Table 32. Mean values of Os(VIII) and Ru(III) ($\mu g\ g^{-1}$) relevant to oysters, mussels and clams sampled in the Goro Bay

		Voltammetry		*Spectroscopy*	
	Sampling sites[a]	*Os(VIII)*	*Ru(III)*	*Os(VIII)*	*Ru(III)*
Oysters	A	6.1 ± 0.3	4.9 ± 0.2	6.4 ± 0.3	5.1 ± 0.3
	B	6.3 ± 0.2	4.1 ± 0.2	6.5 ± 0.3	4.3 ± 0.3
	C	5.8 ± 0.3	4.7 ± 0.2	6.1 ± 0.3	4.6 ± 0.2
	D	5.5 ± 0.2	4.5 ± 0.3	5.3 ± 0.3	4.8 ± 0.4
	E	/	/	/	/
Mussels	A	6.5 ± 0.3	5.3 ± 0.3	6.8 ± 0.4	5.5 ± 0.2
	B	5.9 ± 0.3	4.6 ± 0.2	5.6 ± 0.3	4.5 ± 0.2
	C	6.9 ± 0.4	5.5 ± 0.3	7.3 ± 0.4	5.7 ± 0.3
	D	6.3 ± 0.3	5.0 ± 0.3	6.6 ± 0.3	5.3 ± 0.4
	E	<LOD	<LOD	<LOD	<LOD
Clams	A	4.7 ± 0.2	3.8 ± 0.2	4.9 ± 0.3	3.6 ± 0.3
	B	3.9 ± 0.3	3.0 ± 0.2	4.1 ± 0.2	2.9 ± 0.2
	C	4.6 ± 0.3	3.5 ± 0.3	4.3 ± 0.3	3.7 ± 0.2
	D	4.5 ± 0.2	3.7 ± 0.2	4.7 ± 0.3	3.5 ± 0.3
	E	<LOD	<LOD	<LOD	<LOD

[a] In the case of Oyster samplings in site E, the absence of element concentration data is due to lack of sample present in the same sampling site.

The determined values are the mean of 5 independent determinations. Confidence level: 95%.

Determination of Platinum Group Metals (PGMs), Part III: Ir(III)

As in the case of Os(VIII) and Ru(III), at least to our knowledge, the literature does not report works regarding the determination of Ir(III) in filtering marine organisms. This work is therefore the first that discusses a critical comparison between voltammetric and spectroscopic methods for the determination of such metal in oysters, mussels and clams.

Analytical Procedure for the Voltammetric Determination of Ir(III)

0.4 mol L^{-1} acetate buffer pH 4.9 + 8.7x10^{-2} mol L^{-1} $NaBrO_3$ + 0.15 mol L^{-1} KCl + 6.9x10^{-5} mol L^{-1} cetyltrimethylammonium bromide (CTAB) was employed as supporting electrolyte.

10 mL sample aliquots of supporting electrolyte or of solutions obtained in the mineralization step of the standard reference material and of real samples were pipetted into the voltammetric cell and deaerated for 5 min by bubbling water-saturated pure nitrogen. The determination of Ir(III) was carried out by Square Wave Catalytic Adsorptive Stripping Voltammetry (SWCAdSV) using GCE as working electrode.

The voltammetric experimental conditions are reported in Table 33.

The experimental peak potentials E_p (V/Ag | AgCl | $Cl^-_{satd.}$) of Ir(III) are reported in Table 34. The spectroscopic experimental conditions are reported in Table 1.

Results and Discussion

In this section, in a synthetic way, the fundamental parameters that characterize a correct analytical procedure — limits of detection, linear range, trueness and precision — are reported.

Table 33. Instrumental parameters for the determination of Ir(III) by SWCAdSV[a]

E_{ads}	+0.850
E_i	+0.850
E_f	+0.400
t_{ads}	490
t_r	10
dE/dt	100
ΔE	50
τ	0.010
ν	0.100
η	10
u	600

[a] E_{ads}: adsorption potential (V/Ag, AgCl, $Cl^-_{satd.}$); E_i: initial potential (V/Ag, AgCl, $Cl^-_{satd.}$); E_f: final potential (V/Ag, AgCl, $Cl^-_{satd.}$); t_{ads}: electroadsorption time (s); t_r: delay time before the potential sweep (s); dE/dt: potential scan rate (mV/s); ΔE: superposed potential amplitude (mV); τ: sampling time (s); υ: wave period (s); η: wave increment (mV); u: stirring rate (r.p.m.).

Supporting electrolyte: 0.4 mol L^{-1} acetate buffer pH 4.9 + 8.7x10^{-2} mol L^{-1} $NaBrO_3$ + 0.15 mol L^{-1} KCl + 6.9x10^{-5} mol L^{-1} cetyltrimethylammonium bromide (CTAB).

Limits of Detection and Linear Range

In the aqueous reference solution, in the solutions obtained by digestion of the standard reference materials and of real samples, the LOD_s (Table 35) for both techniques were calculated as described in section "Limits of Detection."

At the experimental conditions employed, the linear range for all the elements in the aqueous reference solutions is <LOD-5.0 μg L^{-1}.

Table 34. Peak potentials (E_p, V/Ag|AgCl|$KCl_{sat.}$) in the reference solutions and in solutions obtained by digestion of oysters, mussels and clams sampled in the Goro Bay

	Sampling sites[a]	*Ir(III)*
Supporting electrolyte		+0.625 ± 0.010
Solution obtained by digestion of Mussel Tissue BCR-CRM 278 SRM		+0.651 ± 0.015
Solution obtained by digestion of Oyster Tissue NIST-SRM 1566a		+0.647 ± 0.015
Solution obtained by digestion of Oysters	A	+0.639 ± 0.010
	B	+0.647 ± 0.015
	C	+0.643 ± 0.015
	D	+0.652 ± 0.015
	E	/
Solution obtained by digestion of Mussels	A	+0.649 ± 0.015
	B	+0.637 ± 0.015
	C	+0.655 ± 0.010
	D	+0.645 ± 0.015
	E	No signal
Solution obtained by digestion of Clams	A	+0.658 ± 0.015
	B	+0.663 ± 0.010
	C	+0.641 ± 0.015
	D	+0.648 ± 0.015
	E	No signal

[a] In the case of Oyster samplings in site E, the absence of element concentration data is due to lack of sample present in the same sampling site.

The determined values are the mean of 5 independent determinations. Confidence level: 95%.

Spiked-sample concentration in reference solutions: 77.5 μg kg^{-1}.

Quality Control and Quality Assessment

The method set up in aqueous reference solutions was applied to Oyster Tissue NIST-SRM 1566a and Mussel Tissue BCR-CRM 278 standard reference materials in order to confirm and verify the applicability of the analytical procedure.

Trueness and precision for the voltammetric and spectroscopic measurements were determined and reported in Table 36.

Practical Application

Once the procedure for the Ir(III) determination was set up, the method was applied to oysters, mussels and clams sampled in the Goro Bay (see section "Sampling Area and Sampling Sites").

The experimental results relevant to both techniques are reported in Table 37.

Table 35. LOD$_s$[a] of Ir(III) determined in the aqueous reference solution (μg L^{-1}), in the solutions obtained by digestion of Mussel Tissue BCR-CRM 278 and Oyster Tissue NIST-SRM 1566a standard reference materials, and in the solutions obtained by digestion of real samples (calculated in μg L^{-1} and expressed in μg kg^{-1})

	Sampling sites[b]	*Ir(III)*
Supporting electrolyte		0.027
Solution obtained by digestion of Mussel Tissue BCR-CRM 278 SRM		1.4
Solution obtained by digestion of Oyster Tissue NIST-SRM 1566a SRM		1.6
Solution obtained by digestion of Oysters	A	1.3
	B	1.5
	C	1.7
	D	1.5
	E	/
Solution obtained by digestion of Mussels	A	1.4
	B	1.2
	C	1.5
	D	1.5
	E	1.6
Solution obtained by digestion of Clams	A	1.5
	B	1.3
	C	1.6
	D	1.4
	E	1.7

[a] In the case of spectroscopic measurements, the Ir(III) LOD (μg L^{-1}) in the aqueous reference solution is 0.107. Considering a sample weight exactly equal to 0.5 g (see section "Sample Preparation before the Instrumental Determination"), the same limit of detection (μg kg^{-1}) is 4.2.

[b] In the case of Oyster samplings in site E, the absence of element concentration data is due to lack of sample present in the same sampling site.

The determined values are the mean of 5 independent determinations; confidence level: 95%.

Table 36. Accuracy of the analytical procedure. The determined values are the mean of 5 independent determinations

Oyster Tissue NIST-SRM 1566a

Voltammetry			*Spectroscopy*		
Determined Values	*e (%)*	*s_r (%)*	*Determined Values*	*e (%)*	*s_r (%)*
81.9 ± 5.6	+5.7	5.3	82.4 ± 5.9	+6.3	5.7

Mussel Tissue BCR-CRM 278

Voltammetry			*Spectroscopy*		
Determined Values	*e (%)*	*s_r (%)*	*Determined Values*	*e (%)*	*s_r (%)*
72.9 ± 4.9	-5.9	5.5	82.5 ± 6.0	+6.5	5.9

Confidence level: 95%. Concentrations in μg g^{-1}. Spiked-sample concentration: 77.5 μg kg^{-1}.

Table 37. Mean values of Ir(III) ($\mu g\ g^{-1}$) relevant to oysters, mussels and clams sampled in the Goro Bay

	Sampling sites[a]	*Voltammetry*	*Spectroscopy*
Oysters	A	6.1 ± 0.3	6.4 ± 0.4
	B	5.9 ± 0.3	5.6 ± 0.4
	C	6.3 ± 0.4	6.0 ± 0.3
	D	6.2 ± 0.3	6.1 ± 0.3
	E	/	/
Mussels	A	7.7 ± 0.4	8.0 ± 0.5
	B	6.9 ± 0.3	7.2 ± 0.4
	C	7.1 ± 0.4	6.8 ± 0.3
	D	7.5 ± 0.3	7.3 ± 0.3
	E	<LOD	<LOD
Clams	A	5.0 ± 0.2	4.8 ± 0.3
	B	4.1 ± 0.2	<LOD
	C	5.8 ± 0.3	6.1 ± 0.4
	D	5.5 ± 0.3	5.3 ± 0.3
	E	<LOD	<LOD

[a] In the case of Oyster samplings in site E, the absence of element concentration data is due to lack of sample present in the same sampling site.

The determined values are the mean of 5 independent determinations. Confidence level: 95%.

REFERENCES

[1] Locatelli, C. 2008. Metals. In: Yolanda Picò, Y. (Ed.) *Comprehensive Analytical Chemistry*, Vol. 51, Ch. 16, pp. 571-598, Elsevier Publ., Amsterdam.

[2] Jovic, M., Stankovic, S. 2014. Human exposure to trace metals and possible public health risks via consumption of mussels *Mytilus galloprovincialis* from the Adriatic coastal area. *Food Chem. Toxicol.* 70, 241-251.

[3] Olmedo, P., Pla, A., Hernandez, A.F., Barbier, F.,Ayouni, L., Gil, F. 2013. Determination of toxic elements (mercury, cadmium, lead, tin and arsenic) in fish and shellfish samples. Risk assessment for the consumers. *Environ. Int.* 59, 63-72.

[4] Lynch, J.M., Wiseman, A. (Eds.) 2011. *Environmental Biomonitoring – The Biotechnology Ecotoxicology Interface*, Cambridge University Press, Cambridge, UK.

[5] Gupta, S.K., Singh, J. 2011. Evaluation of mollusc as sensitive indicator of heavy metal pollution in aquatic system. A review. *IIOAB J.* 2(1), 49-57.

[6] Phillips, D.J.H., Rainbow, P.S. 1994. *Biomonitoring of Trace Aquatic Contaminants*, Chapman and Hall Publ., London, UK.

[7] Yanez-Sedeno, P., Agui, L., Villalonga, R., Pingarron, J.M. 2014. Biosensors in forensic analysis. A review. *Anal. Chim. Acta* 823, 1-19.

[8] Welz, B., Sperling, M. 1999. *Atomic Absorption Spectrometry*, 3rd Edition, Wiley VCH, Weinheim.

[9] Miller, J.C., Miller, J.N. 2010. *Statistics and Chemometrics for Analytical Chemistry*, 6th Edition, Pearson Education Ltd. Publ., Ashford Colour Press Ltd., Gosport, UK.

[10] International Union of Pure and Applied Chemistry - Analytical Chemistry Division, 1978. Nomenclature, symbols, units and their usage in spectrochemical analysis—II. Data interpretation. *Spectrochim. Acta* 33B, 241-245.

[11] Berhoumi, B., Clerandeau, C., Gourves, P-Y., Le Menach, K., El Megdiche, Y., Peluhert, L., Budzinski, H., Baudrimont, M., Driss, M.R., Cachot, J. 2014. Pollution biomonitoring in the Bizerte lagoon (Tunisia), using combined chemical and biomarker analyses in grass goby, *Zosterisessor ophiocefalus* (Teleostei, Gobiidae). *Mar. Environ. Res.* 101, 184-195.

[12] George, R., Martin, G.D., Nair, S.M., Chandramohanakumar, N. 2013. Biomonitoring of trace metal pollution using the bivalve molluscs, *Villorita cyprinoids*, from the Cochin backwaters. *Environ. Monit. Assess.* 185, 10317-10331.

[13] Alfonso, J.A., Handt,H., Mora, A., Vasquez., Y., Azocar, J., Marcano, E. 2013. Temporal distribution of heavy metal concentrations in oysters *Crassostrea rhizophorae* from the centralVenezuelan coast. *Mar. Pollut. Bull.* 73, 394-398.

[14] Kanduca, T., Medakovic, D., Hamer, B. 2011. *Mytilus galloprovincialis* as a bioindicator of environmental conditions: the case of the eastern coast of the Adriatic Sea. *Isot. Environ. Health S.* 47, 42-61.

[15] Roncevic, S., Svedruzic, L.P., Smetisko, J., Medakovic, D. 2010. ICP-AES analysis of metal content in shell of mussel *Mytilus galloprovincialis* from Croatian coastal waters. *Int. J. Environ. Anal. Chem.* 90, 620-632.

[16] Locatelli, C., Melucci, D. 2010. Sequential voltammetric determination of mercuri(II) and toxic metals in environmental bio-monitors: application to mussels and clams. *Int. J. Environ. Anal. Chem.* 90, 49-63.

[17] Sarkar, S.K. Cabral, H., Chatterjee, M., Cardoso, I., Bhattacharya, A.K., Satpathy, K.K., Alam, M.A. 2008. Biomonitoring of heavy metals using the bivalve molluscs in Sunderban mangrove wetland, northeast coast of Bay of Bengal (India): possible risks to human health. *Clean - Soil Air Water* 36, 187-194.

[18] Kucuksezgin, F., Kayatekin, R.M., Uluturhan, E., Uysal, N., Acikgoz, O., Gonenc, S. 2008. Preliminary investigation of sensitive biomarkers of trace metal pollution in mussel (*Mytilus galloprovincialis*) from Izmir Bay (Turkey). *Environ. Monit. Assess.* 141, 339-345.

[19] Protasowicki, M., Dural, M. 2008. Trace metals in the shells of blue mussels (*Mytilus edulis*) from the Poland coast of Baltic sea. *Environ. Monit. Assess.* 141, 329-337.

[20] Turkmen, M., Ciminli, C. 2007. Determination of metals in fish and mussel species by inductively coupled plasma-atomic emission spectrometry. *Food Chem.* 103, 670-675.

[21] Conti, M.E., Iacobucci, M., Cecchetti, G. 2007. A biomonitoring study: trace metals in seagrass, algae and mollusco in a marine reference ecosystem (Southern Tyrrhenian Sea). *Int. J. Environ. Pollut.* 29, 308-332.

[22] Alla, A.A., Mouneyrac, C., Durou, C., Moukrim, A., Pellerin, J. 2006. Tolerance and biomarkers as a useful tools for assessing environmental quality in the Oued Souss estuary (Bay of Agadir, Marocco). *Comp. Biochem. Phys. C* 143, 23-29.

[23] Daby, D. 2006 Coastal pollution and potential biomonitors of metals in Mauritius. *Water Air Soil Pollut.* 174, 63-91.

[24] Hamed, M.A., Emara, A.M. 2006. Marine molluscs as biomonitors for heavy metal levels in the Gulf of Suez, Red Sea. *J. Marine Syst.* 60, 220-234.

[25] Conti, M.E., Iacobucci, M., Cecchetti, G. 2005. A statistical approach applied to trace metal data from biomonitoring studies. *Int. J. Environ. Pollut.* 23, 29-41.

[26] Airas, S., Duinker, A., Julshamn, K. 2004. Copper, zinc, arsenic, cadmium, mercury and lead in blue mussels (*Mytilus edulis*) in the Bergen Harbor area, Western Norway. *Bull. Environ. Contam. Toxicol.* 73, 276-284.

[27] Conti, M.E., Cecchetti, G. 2003. A biomonitoring study: trace metals in algae and molluscs from Tyrrhenian coastal areas. *Environ. Res.* 93, 99-112.

[28] Locatelli, C. 2003. Heavy metal determinations in algae, mussels and clams. Their possible employment for assessingthe sea water quality criteria. *J. Phys. JP* 107, 785-788.

[29] Locatelli, C., Fagioli, F. 1997 Determination at trace level of heavy metals in mussels and clams. *Ann. Chim. (Rome)* 87, 535-538.

[30] Catharino, M.G.M., Vasconcellos, M.B.A., Kirschbaum, A.A., Gasparro, M.R., Minei, C.C., de Sousa, E.C.P.M., Seo, D., Moreira, E.G., 2012. Biomonitoring of coastal regions of Sao Paulo State, Brazil, using mussels *Perna perna*. *J. Radioanal. Nucl. Chem.* 291, 113-117.

[31] Yusof, A.M., Yanta, N.F., Wood, A.K.H. 2004. The use of bivalves as bio-indicators in the assessment of marine pollution along a coastal area. *J. Radioanal. Nucl. Chem.* 259, 119-127.

[32] Fukushima, M., Tamate, H., Nakano, Y. 2003. Trace element determination in soft tissues of marine bivalves by activation analysis. *J. Radioanal. Nucl. Chem.* 255, 231-234.

[33] Anzano, J., Lasheras, R-J., Bonilla, B., Bonilla, A., Lanaja, J., Peribanez, M.A., Gracia-Salinas, M-J., Anwar, J., Shafique, U. 2011. Determination of trace metals by voltamperometry in zebra mussel (*Dreissena polymorpha*) employed as environmental bio-indicator. *Green Chem. Lett. Rev.* 4, 261-267.

[34] Anastasiadou, Z.D., Jannakoudakis, P.D., Girousi, S.T. 2010. Square wave anodic stripping voltammetry determination of eco-toxic metals in samples of biological and environmental importance. *Cent. Eur. J. Chem.* 8, 999-1008.

[35] Locatelli, C., Fabbri, D., Torsi, G. 2001. Heavy metal determinations in algae and clams and their possibile employment for assessing the sea water qualità criteria. *Ann. Chim. (Rome)* 91, 425-434.

[36] Locatelli, C. 1994. Proposal of new analytical procedures for heavy metal determinations in mussels, clams and fishes. *Food Addit. Contam. A* 17, 769-774.

[37] Fagioli, F., Locatelli, C., Landi, S. 1994. Heavy metals in the Goro Bay: sea water, sediments and mussels. *Ann. Chim. (Rome)* 84, 129-140.

[38] Juncharoenwongsa, N., Siriprom, W., Kaewkhao, J., Choeysuppaket, A., Limsuwan, P., Phachana, K. 2011. A biomarkers study: trace metal elements in *Paphia undulate* shell for assessing pollution of coastal area. *Proc. Eng.* 8, 80-84.

[39] Santos, M.C., Wagner, M., Wu, B., Scheider, J., Oehlmann, J., Cadore, S., Becker, J.S. 2009. Biomonitoring of metal contamination in a marine prosobranch snail (*Nassarius reticulates*) by imaging laser ablation inductively coupled plasma mass spectrometry (LA-ICP-MS). *Talanta* 80, 428-433.

[40] Wu, R.S.S., Lau, T.C., Fung, W.K.M., Ko, P.H., Leung, K.M.Y. 2007. An "artificial mussel" for monitoring heavy metals in marine environments. *Environ. Poll.* 145, 104-110.

[41] Lemos, V.A., dos Santos, L.O. 2014. A new method for preconcentration and determination of mercury in fish, shellfish and saliva by cold vapour atomic absorption spectrometry. *Food Chem.* 149, 203-207.

[42] Aguilar, C.A., Montalvo, C., Rodriguez, L., Ceròn, J.G., Ceròn, R.M. 2012. American oyster (Crassostrea virginica) and sediments as a coastal zone pollution monitor by heavy metals. *Int. J. Environ. Sci. Technol.* 9, 579-586.

[43] Liu, X., Zhang, L., You, L., Yu, J., Cong, M., Wang, Q., Li, F., Li, L., Zhao, J., Li, C., Wu, H. 2011. Assessment of clam *Ruditapes philippinarum* as heavy metal bioindicators using NMR-based metabolomics. *Clean - Soil Air Water* 39, 759-766.

[44] Besada, V., Andrade, J.M., Schultze, F., Gonzalez, J.J. 2011. Monitoring of heavy metals in wild metals (*Mytilus galloprovincialis*) from the Spanish North-Atlantic coast. *Cont. Shelf Res.* 31, 457-465.

[45] Ipolyi, I., Massanisso, P., Sposato, S., Fodor, P., Morabito, R. 2004. Concentration levels of total and methylmercury in mussel samples collected along the coast of Sardinia Island (Italy). *Anal. Chim. Acta* 505, 145-151.

[46] Beiras, R., Fernandez, N., Gonzalez, J.J., Besada, V., Schultze, F. 2002. Mercury concentration in seawater, sediments and wild mussels from the coast of Galicia (NW Spain). *Mar. Pollut. Bull.* 44, 345-349.

[47] Zachariadis, G.A., Anthemidis, A.N., Caniou, I., Stratis, J.A. 2001. Determination of lead, cadmium and mercury in surface marine sediments and mussels. *Int. J. Environ. Anal. Chem.* 80, 153-166.

[48] Kawaguchi, T., Porter, D., Bushek, D., Jones, B. 1999. Mercury in the American oyster Crassostrea virginia in South Carolina, US, and public health concerns. *Mar. Poll. Bull.* 38, 324-327.

[49] Trombini, C., Fabbri, D., Lombardo, M., Bassura, I., Zavoli, E., Horvat, M. 2003. Mercuri and methylmercury contamination in superficial sediments and clams of a coastal lagoon (Pialassa Baiona, Ravenna, Italy). *Cont. Shelf Res.* 23, 1821-1831.

[50] Luis, E.S., Balagot, A.H. Villaflor, A.C., Sanchez, F.C., Develles, E.N. 1989. Pb, Cd and Hg contents of bivalves collected during the different seasons of the year. *Food Chem.* 32, 239-255.

[51] Silva, N.A.F., Lopes, M.I., Leitao, R.A.E., Silva, H.F.A., Matos, M.J. 2005. Simultaneous monitoring of toxic metals on White Poplar (*Populus*) by SWASV. *J. Brazil. Chem. Soc.* 16(6A), 1275-1282.

[52] Liang, N-L., Shi, J-B., He, B., Jiang, G-B., Yuan, C-G. 2003. Investigation of methylmercury and total mercury contamination in mollusc samples collected from coastal sites along the Chinese Bohai Sea. *J. Agr. Food Chem.* 51, 7373-7378.

[53] Yusof, A.M., Rahman, N.A., Wood, A.K.H. 1994. The accumulation and distribution of trace metals in some localized marine species. *Biol. Trace Elem. Res.* 43-45, 239-249.

[54] Maria-Cervantes, A., Jimenez-Carceles, F.J., Alvarez-Rogel, J. 2009. As, Cd, Cu, Mn, Pb and Zn contents in sediments and mollusks (*Hexaplex trunculus* and *Tapes decussatus*) from coastal zones of a Mediterranean lagoon (Mar Menor, SE Spain) affected by mining wastes. *Water Air Soil Poll.* 200, 289-304.

[55] Lavilla, I., Vilas, P., Bendicho, C. 2008. Fast determination of arsenic, selenium, nickel and vanadium in fish and shellfish by electrothermal atomic absorption spectrometry following ultrasound-assisted extraction. *Food Chem.* 106, 403-409.

[56] Peltier, G.L., Meyer, J.L., Jagoe, C.H., Hopkins, W.A. 2008 Using trace element concentrations in *Corbicula fluminea* to identify potential sources of contamination in an urban river. *Environ. Poll.* 154, 283-290.

[57] Sloth, J.J, Julshamn, K. 2006. Survey of total and inorganic arsenic content in blue mussels (*Mytilus edulis L.*) from Norwegian fiords: revelation of unusual high levels of inorganic arsenic. *J. Agr. Food Chem.* 56, 1269-1273.

[58] Hirata, S., Toshimitsu, H., Aihara, M. 2006. Determination of arsenic species in marine samples by HPLC-ICP-MS. *Anal. Sci.* 22, 39-43.

[59] Amodio-Cocchieri, R., Amoroso, S., Arnese, A., Cirillo, T., Montuori, P., Triassi, M. 2003. Pollution of mercury, arsenic, lead, chromium, cadmium and polycyclic aromatic Hydrocarbons of fish and mussels from the Gulf of Naples, Italy. *Bull. Environ. Contam. Toxicol.* 71, 551-560.

[60] Soros, C., Bodò, E.T., Fodor, P., Morabito, R. 2003. The potential of arsenic speciation in molluscs for environmental monitorino. *Anal. Bioanal. Chem.* 377, 25-31.

[61] Plessi, M., Bertelli, D., Monzani, A. 2001. Mercury and selenium content in selected seafood. *J. Food Compos. Anal.* 14, 461-467.

[62] Brunori, C., De La Calle, M.B., Angelone, M., Morabito, R. 1999. Determination of total selenium in mussel and wheat samples after tetramethylammonium hydroxide digestion. *Ann. Chim. (Rome)* 89, 873-880.

[63] Ibrahim, N. 1995. Trace element content in Malaysian cockles (*Anadara granosa*). *Food Chem.* 54, 133-135.

[64] Santos, I., Diniz, M.S., Carvalho, M.L., Santos, J.P. 2014. Assessment of essential elements and heavy metals content on *Mytilus galloprovincialis* from river Tagus estuary. *Biol. Trace Elem. Res.* 159, 233-240.

[65] Orescanin, V., Lovrencic, I., Mikelic, L., Barisic, D., Matasin, Z., Lulic, S., Pezelj, D. 2006. Biomonitoring of heavy metals and arsenic on the east coast of the Middle Adriatic Sea using *Mytilus galloprovincialis*. *Nucl. Instrum. Meth. B* 245, 495-500.

[66] Leermakers, M., Baeyens, W., De Gieter, M., Smedts, B., Meert, C., De Bisschop, H.C., Morabito, R., Quevauviller, Ph. 2006. Toxic arsenic compounds in environmental samples: speciation and validation. *Trends Anal. Chem.* 25, 1-10.

[67] Schaeffer, R., Soeroes, C., Ipolyi, I., Fodor, P., Thomaidis, N.S. 2005. Determination of arsenic species in sea food samples from the Aegean Sea by liquid chromatography—(photo-oxidation)—hydride generation—atomic fluorescence spectrometry. *Anal. Chim. Acta* 547, 109-118.

[68] Fernandez, N., Fernandez-Boan, M., Verisimo, P., Freire, J. 2013 Assessing the spatial variability, level and source of organic chemical contaminants in bivalve fishing grounds on the Galician coast (NW Spain). *Mar. Poll. Bull.* 74, 291-301.

[69] Bujan-Villar, M.D., Barciela-Alonso, M.C., Aboal-Somoza, M., Bermejo-Barrera, B. 2011. Total antimony determination in Galician mussel samples (*Mytilus galloprovincialis*) by slurry sampling — graphite furnace atomic absorption spectrometry. Preliminary results. *Atom. Spectrosc.* 32, 34-40.

[70] Mckie, J.C., 1987. Determination of total tin and tributyltin in marine biological materials by electrothermal atomic absorption spectrometry. *Anal. Chim. Acta* 197(C), 303-308.

[71] Amer, H.A., Shawky, S. 2002. Radiochemical neutron activation analysis of total trace tin and determination of its differentiated compounds in aquatic biota. *Radiochim. Acta* 90, 351-355.
[72] Abollino, O., Aceto, M., La Gioia, C., Sarzanini, C., Mentasti, E. 2002 Voltammetric determination and speciation of inorganic and organometallic tin. *Electoanalysis* 14, 1090-1097.
[73] D. Sanchez-Rodas, D., W.T. Corns, W.T., B. Chen, B., P.B. Stockwell, P.B. 2010. Atomic fluorescence spectrometry: a suitable detection technique in speciation studies for arsenic, selenium, antimony and mercury. *J. Anal. Atom. Spectrom.* 25, 933-946.
[74] Ferrarello, C.N., Fernandez de la Campa, M.D.R., Muniz, C.S., Sanz-Medel, A. 2000. Metal distribution patterns in the mussel *Mytilus edulis cytosols* using size-exclusion chromatography and double focusing ICP-MS detection. *Analyst* 125, 2223-2229.
[75] Pawlak, J., Lodyga-Chruscinska, E., Chrustowicz, J. 2014. Fate of platinum metals in the environment. *J. Trace Elem. Med. Biol.* 28, 247-254.
[76] Dubiella-Jackowska, A., Kudlak, B., Polkowska, Z., Namiesnik, J. 2009. Environmental fate of traffic-derived platinum group metals. *Crit. Rev. Anal. Chem.* 39, 251-271.
[77] Wang, Y., Li, X. 2012. Health risk of platinum group elements from automobile catalysts – 2012 International Symposium on Safety Science and Technology, *Proc. Eng.* 45, 1004-1009.
[78] Ravindra, K., Bencs, L., Van Grieken, R. 2004. Platinum group elements in the environment and their health risk. *Sci. Total Environ.* 318, 1-43.
[79] Ruchter, N., Sures, B. 2015. Distribution of platinum and other traffic related metals in sediments and clams (*Corbicula sp.*). *Water Res.* 70, 313-324.
[80] Somerset, V., Van Der Horst, C., Silwana, B., Walters, C., Iwuoha, E. 2015. Biomonitoring and evaluation of metal concentrations in sediment and crab samples from the North-West Province of South Africa. *Water Air Soil Poll.* 226, 2329, DOI 10.1007/s11270-015-2329-2.
[81] Essumang, D.K. 2010. First determination of the levels of platinum group metals in *Manta birostris* (*Manta ray*) caught along the Ghanian coastline. *Bull. Environ. Contam. Toxicol.* 84, 720-725.
[82] Essumang, D.K., Dodoo, D.K., Adokoh, C.K., Sam, A., Doe, N.G. 2008. Bioaccumulation of platinum group metals on some fish species (*Oreochromis niloticus*, *Penaeus laspisulcates*, *Scylla serrate*, *Galaxias brevipinnis* and mollusc) in the Pra estuary of Ghana. *Toxicol. Environ. Chem.* 90, 625-638.
[83] El-Sikaily, A., Khaled, A, El Nemr, A. 2004. Heavy metals monitoring using bivalves from Mediterranean Sea and Red Sea. *Environ. Monit. Assess.* 98, 41-58.
[84] Essumang, D.K., Adokoh, C.K., Boamponsem, L. 2010. Levels of platinum group metals in selected species (*Sarotherodon melanotheron*, *Chonophorus lateristriga*, *Macrobrachium vollenhovenii* and *Crassostrea tulipa*) in some estuaries and lagoons along the coast of Ghana. *The Scientific World Journal* 10, 1971-1987.
[85] Thielen, F., Zimmermann, S., Sures, B. 2006. Biomonitoring of palladium in the environment using different accumulation indicators. In: Zereini, F., Alt, F (Editors) *Palladium Emissions in the Environment – Analytical Methods, Environmental Assessment and Health Effects*, Springer-Verlag Publ., Berlin-Heidelberg, Germany.

[86] Zimmermann, S., Messerschmidt, J., Von Bohlen, A., Sures, B. 2005. Uptake and bioaccumulation of platinum group metals (Pt, Pd, Rh) from automobile catalytic converter materials by the zebra mussel (*Dreissena polymorpha*). *Environ. Res.* 98, 203-209.

[87] Zimmermann, S., Messerschmidt, J., Von Bohlen, A., Sures, B. 2003. Determination of Pt, Pd and Rh in biological samples by electrothermal atomic absorption spectrometry as compared with adsorptive cathodic stripping voltammetry and total-reflection X-ray fluorescence analysis. *Anal. Chim. Acta* 498, 93-104.

[88] Zimmermann, S., Alt, F., Messerschmidt, J., Von Bohlen, A., Taraschewski, H., Sures, B. 2002. Biological availability of traffic-related platinum-group elements (palladium, platinum and rhodium) and other metals to the zebra mussel (*Dreissena polymorpha*) in water containing road dust. *Environ. Toxicol. Chem.* 21, 2713-2718.

[89] Sures, B., Zimmermann, S., Messerschmidt, J., Von Bohlen, A. 2002. Relevance and analysis of traffic related platinum group metals (Pt, Pd, Rh) in the aquatic biosphere, with emphasis on palladium. *Ecotoxicology* 11, 385-392.

[90] Zimmermann, S., Menzel, C.M., Berner, Z., Eckhardt, J-D., Stuben, D., Alt, F., Messerschmidt, J., Taraschewski, H., Sures, B. 2001. Trace analysis of platinum in biological samples: a comparison between sector field ICP-MS and adsorptive cathodic stripping voltammetry following different digestion procedures. *Anal. Chim. Acta* 439, 203-209.

[91] Locatelli, C. 2007. Voltammetric analysis of trace levels of platinum group metals – principles and applications. *Electroanalysis* 19, 2167-2175.

[92] Locatelli, C. 2007. Platinum(II), palladium(II), rhodium (III) and lead(II) voltammetric determination in sites differently influenced by vehicle traffic. *Ann. Chim. (Rome)* 97, 373-384.

[93] Locatelli, C. 2007. Voltammetric peak area as instrumental datum. A possibility to improve the determination at ultratrace level concentration of platinum group metals (PGMs) and lead. Application to particulate matter. *Electroanalysis* 19, 445-452.

[94] Locatelli, C. 2006. Possible interference in the sequential voltammetric determination at trace and ultratrace concentration level of platinum group metals (PGMs) and lead. Application to environmental matrices. *Electrochim. Acta* 52, 614-622.

[95] Locatelli, C. 2006. Simultaneous square wave stripping voltammetric determination of platinum group metals (PGMs) and lead at trace and ultratrace concentration level. Application to surface water. *Anal. Chim. Acta* 557, 70-77.

[96] Locatelli, C., Melucci, D., Torsi, G. 2005. Determination of platinum-group metals and lead in vegetable environmental bio-monitors by voltammetric and spectroscopic techniques: critical comparison. *Anal. Bioanal. Chem.* 382, 1567-1573.

[97] Locatelli, C. 2005. Platinum, rhodium, palladium and lead: elements linked to vehicle emissions. Their simultaneous voltammetric determination in superficial water. *Electroanalysis* 17, 140-147.

[98] Locatelli, C. 2014. Square wave catalytic adsorptive voltammetric determination of osmium, ruthenium and lead in vegetable environmental bio-monitors. *Int. J. Environ. Anal. Chem.* 94, 277-290.

[99] Locatelli, C. 2013. Use of peak area instrumental datum as possibility to improve the analytical sensitivity in the sequential voltammetric determination of ultra-trace iridium and lead in vehicle emission particulate matter. *Microchem. J.* 110, 99-106.

[100] Locatelli, C. 2013. Iridium and lead as vehicle emission pollutants: their sequential voltammetric determination in vegetable environmental bio-monitors. *Microchem. J.* 106, 282-288.

[101] Locatelli, C. 2012. Ultratrace osmium, ruthenium and lead in airborne particulate matter: peak area as instrumental datum to improve their simultaneous voltammetric determination. *Electroanalysis* 24, 2273-2282.

[102] Locatelli, C. 2012 Simultaneous determination of osmium, ruthenium, copper and lead by electrocatalytic voltammetry. Application to superficial waters. *Microchem. J.* 102, 54-60.

[103] Locatelli, C. 2011. Catalytic-adsorptive stripping voltammetric determination of ultratrace iridium(III). Application to fresh- and sea-water. *Talanta* 85, 546-550.

[104] Locatelli, C. 2011. Sequential voltammetric determination of ultratrace osmium, ruthenium and iridium Application to superficial water. *Electroanalysis* 23, 1329-1336.

In: Oysters and Clams
Editor: Jesús L. Romalde
ISBN: 978-1-63485-074-2

Chapter 10

OYSTER SUMMER MORTALITY: THE EXAMPLE OF A MULTIFACTORIAL SYNDROME

Jesús L. Romalde*
Departamento de Microbiología y Parasitología, CIBUS-Facultad de Biología, Universidade de Santiago de Compostela, Santiago de Compostela, Spain

ABSTRACT

The occurrence of massive mortality outbreaks in cultured bivalves throughout the world, with high economic and social impact, has been generally associated with a range of bacterial and viral pathogens. Since 2008, successive disease outbreaks associated with high temperatures (above 18ºC) have been reported in Pacific oyster (*Crassosstrea gigas*) populations from France and other countries, including Ireland, Spain, Mexico, New Zealand and Australia among others. The syndrome, known as "Summer Mortality" resulted in huge losses in production, threatening the oyster-farming industry in some areas. Several hypothesis about the etiology of the summer mortality have been proposed including pathogenic vibrios (*Vibrio aestuarianus* and *V. splendidus*) and viruses (oyster herpesvirus). In addition, the influence of environmental (dissolved oxygen, harmful algae, pollutants, etc.) and host (physiological condition, reproductive status, glycogen metabolism, etc.) factors on the appearance of the disease has also been studied. The more recent studies seem to indicate that the syndrome is the result of a complex relationship among all the mentioned parameters, where co-infection *Vibrio*-herpesvirus is needed to induce the full expression of the disease and the envionmental factors may have an impact on the host, the pathogen and the host/pathogen interactions.

Keywords: summer mortality, bacteria, *Vibrio aestuarianus*, virus, oyster herpervirus, environmental factors, host factors

* Corresponding author: Jesús L. Romalde. Departamento de Microbiología y Parasitología. CIBUS-Facultad de Biología. Universidade de Santiago de Compostela. Campus Vida s/n. 15782, Santiago de Compostela. Spain. E-mail: jesus.romalde@usc.es.

Introduction

Bivalve aquaculture industry is an expanding sector with high social and economic impact in different regions around the world. In 2013, the world aquaculture production of molluscs reached 15.5 million tons with a value of US$ 20.7 billion [FAO, http://www.fao.org/fishery/statistics/en]. China is by far the main mollusc producer with more than 12.7 million tons, followed by Japan and the Republic of Korea (0.33 and 0.29 million tons, respectively). Within Europe, Spain is the major producer followed by France and Italy, all of them among the top-10 producers of molluscs in the world [FAO, http://www.fao.org/fishery/statistics/en].

Disease in oysters has been steadily rising over the past decades, threatening the long-term survival of commercial and natural stocks. Our understanding and management of such diseases are of critical importance as aquaculture is an important aspect of dealing with the approaching worldwide food shortage.

One of the most threatening diseases affecting oysters is the syndrome known as "Summer Mortality." The first description of the summer mortality syndrome in the Pacific oyster (*Crassosstrea gigas*) was in Japan in the 1940s. The syndrome was, and continues to be, associated with high mortality rates in Pacific oysters and other bivalves around the world. The causes remain unknown, but a multifactorial etiology is suspected. These mass mortalities (50 to 60%, or more) all had certain common features: i) the affected animals are one year old or older, and were the largest and fastest growing; ii) the mortalities occur in eutrophic waters, mostly during the spawning season, but shortly thereafter in some regions; ii) deaths begin with the rise of temperatures to 21-22ºC or above and take place gradually over a spawning season; iv) there are no apparent association with hydrographical conditions other than temperature; and v) bacteria and/or viruses are present in the affected oysters.

The most accepted theory today is that the oyster summer mortality cannot be attributed to only one bacterial pathogen or to the oyster herpesvirus, but to a complex interaction between the physiological and/or genetic state of the host, environmental factors, the presence of various opportunistic infectious *Vibrio* species, and the concurrence of oyster herpesvirus [1, 2, 3].

In this chapter, all the actors playing a role in the appearance of summer mortality will be reviewed with the aim to give a complete picture of the complexity of the syndrome. In addition, the future trends for the study of the syndrome will be discussed.

Environmental Factors

Temperature

High temperature has been suggested as one of the main influential factors in summer mortality of oysters [4-6], probably due to its influence on the haemocyte behaviour [7]. In fact, several works have pointed out the decreases in phagocytosis and in percentage granulocytes at high temperatures [8-10].

One hypothesis that can explain the high mortalities observed during summer mortality episodes can be a negative sinergistic interaction between high temperatures and the energetic cost of spawning [11-15]. In fact, Li et al. [15] pointed out that the thermal tolerance of oysters is partially dependent on the glycogen storage. Therefore, energy expended during reproduction or other stressful conditions (the more stressed animals use more energy to counteract the stress) compromises the termotolerance. On the other hand, oyster under heat shock situation suffer impacts on cell-mediated immunity, which can lead to a high susceptibility to different pathogens [15].

In the current context of global warming and climate change, alterations in seasonality and severity of the summer mortality syndrome could be expected.

Salinity

Young Pacific oysters tolerate a wide salinity range varying from 15 to 44 psu [16] although optimal growth conditions are between 20 and 25 psu [17, 18]. The maintenance of osmotic balance under a stressful fluctuating salinity regime entail a significant metabolic cost [17, 18], which in extreme circumstances may be the cause of mass mortalities [19]. It has been described that pulses of low salinity water after high rainfall may induce haemocyte mortality, while low (15 psu) and high (45 psu) sanility enhanced phagocytic activity [10, 19]. In any case, it has been demonstrated that extreme salinities are detrimental for oysters [19].

Nutrients

Excesive anthropogenic inputs of nutrients and organic matter into the shellfish growing areas, may cause eutrophication and hypoxia [10, 20]. Although the primary cause may be the catchment disturbance with implications for the siting of oyster farms [21], *C. gigas* mortality has been directly linked to eutrophication and increased gonad development [22] and to enriched water combined with high temperatures [23]. In the first case, it has been suggested that eutrophic conditions provoke physiological disorders and metabolic disturbance, which are exacerbated by high temperatures [24].

On the other hand, deteriorated water quality, including altered nutrient balance of phosphates and nitrates or increased ammonium, leads in many cases to hypoxia that has also been related to *C. gigas* mortalities [6, 25].

Later, Malham et al. [10] corroborated that high nutrients, including those obtained from agricultural run off, in combination with high temperatures can be one cause of the summer mortality phenomenon. In their laboratory experiments, the addition of inorganic nutrients (phosphates, nitrates and nitrites) caused significant oyster mortality at 21ºC but not at 12ºC. The authors concluded that *C. gigas* mortality is multifactorial and that water quality and elevated temperatures contributte to the significant mortalities.

Pollutants

Pesticides are among the most commonly detected pollutants in coastal waters [26-28] and, besides important ecological impact, may endage the growth, reproduction or survival of marine organisms [29]. The effects on disease susceptibility in vertebrates and invertebrates have also been demonstrated [30-33]. In the case of bivalves, including oysters, their filter-feeding nature may lead to the accumulation and/or metabolization of contaminants within their tissues [33, 34]. Therefore, the presence of pesticides in the oyster growing areas constitute an environmental risk for the molluscs that was documented in a number of studies [32, 35, 36].

The impact of contaminants on the immune system of bivalves is an issue of ecological, but also economical, concern since it may result in an increased susceptibility of affected organisms to pathogens [26]. Recently, Moreau et al. [37] experimentally demonstrated a decrease in oyster survival to herpesvirus infections after a previous contact with a mixture of 14 pesticides belonging to 4 pesticide groups (herbicides, fongicides, insecticides and molluscicides), due to the induction of adverse effects on the immune system of the oyster.

Rearing Practices

Proximity to sediment appears to be a consistent detrimental factor [38], probably due to its high organic content. One of the early observations was that, despite inter-annual and inter-stock variability, mortality of oysters reared on sediment was significantly higher (20-30%), than mortality of animals on racks (50-70 cm above the sediment) [39].

Later, after the hugh mortility events initiated in 2008, and in an attemp to reduce the spread of the disease, producers and other stakeholders take some health management strategies based on risk reduction through the restriction of transfer oyster stocks between sites during the summer months [40].

At the same time, the development of accurate and cost-effective tools for the detection of putative etiological agents, mainly oyster herpesvirus, became a priority for many reasearch teams [41-45]. In addition, a method based on a thermal challenge to detect latent infections was also optimized [46].

On the basis of the thermal challenge method, Normand et al. [47] carried out a study to determine if early rearing practices can be considered as potential risk factors for summer mortality. Results indicated that the use of large-sized spat from hatcheries, nursery-cultured in onshore facilities, reduces the probability of disease outbreaks, and that infected asymptomatic oysters probably play the role of carriers and reservoirs of the herpesvirus. Related to the latter, the use of the thermal challenge that induces viral proliferation can be an excellent tool for monitoring of oyster groups.

Bacterial Hypothesis

Bivalve Bacterial Diseases

One of the main problems in aquaculture of molluscs is the repetitive episodes of mortality, which seriously reduce the production. These outbreaks of disease affect larval and post-larval stages in hatcheries, as well as juvenile and adults cultured in natural environment. In the case of hatcheries, the massive mortalities involve the complete loss of the stocks of production, with serious economic consequences. In most of the cases, the studies have demonstrated that the problems are caused by bacterial pathologies, being the main etiological agents members of genus *Vibrio* [1, 48, 49].

Guillard [50] was the first to report evidence of the involvement of a *Vibrio* sp. in the disruption of the velum and internal tissues of the clam larvae *Mercenaria mercenaria*, which produced a mortality of 70% of the population. Few years later, Tubiash et al. [51] introduced the term "Bacillary Necrosis" to describe a disease affecting numerous bivalves including oysters (*Crassostrea virginica*, *Ostrea edulis*), clams (*Mercenaria mercenaria*, *Teredo navalis*) and scallops (*Argopecten irradians*), characterized by the appearance of swarms of bacteria on the margins of the larvae, which became progressively denser and after 8 hours, mortality occurred as a result of granular necrosis. The species *Vibrio alginolyticus*, *V. tubiashii* and *V. anguillarum* were recognised as the main causal agents of the bacillary necrosis in later studies [52, 53].

The species *V. tubiashii* reported as one of the causative agents of the bacillary necrosis [52, 53] has recently been described as a re-emergent pathogen in North America causing a decline of approximately 59% in larvae oyster production [54]. In addition, Hasegawa et al. [55] demonstrated that one of the critical factors in the pathogenicity of *V. tubiashii* in oysters (*Crassostrea gigas*) is the presence of a quorum sensing-regulated metalloprotease (VtpA), which is widespread among *Vibrio* species.

The studies of Tubiash et al. [51] also associated for the first time moribund adult bivalves (*M. mercenaria, C. virginica, Mytilus edulis* and *Mya arenaria*) with the species *V. tubiashii* and *V. alginolyticus*. Further cases of vibriosis in oyster were described by Elston and Leibovitz [56], detecting anomalies in the shell and alterations in the function of the ligament and digestive processes. However, the classical *Vibrio* infections are summer mortality in juvenile oysters and brown ring disease in adult clams (reviewed in chapter 7 of this book).

Vibrios and Oyster Summer Mortality

Lipp et al. [57] were the first to observe that during episodes of summer mortality the oyster haemolymph had high levels of vibrios that were causing death. Later studies identified *V. splendidus* as the causal agent of summer mortalities [58-61]. Garnier et al. [62] analysed the bacterial populations in the haemolymph of moribund oysters by phenotypic and molecular methods and found that the prevalent species were *V. aestuarianus* (56%), describing the new subspecies *V. aestuarianus* subsps. *francensis*, and *V. splendidus* (25%). More recently, Allain et al. [63] suggested the possible role of *V. harveyi* as etiological agent of summer mortality, since it was detected in most samples of affected oysters during the 2008 warm season, and was able to produce mortality in experimental challenges.

Vibrio aestuarianus was described by Tison and Seidler [64] during the characterization of a group of isolates obtained from seawater as well as healthy oysters, clams and crabs in the Pacific coast of US. However, its presence in moribund oysters in France was systematically reported from 2001 and, its pathogenicity for this mollusc was demonstrated by Labreuche et al. [2], that described the immunosuppressive activity of the extracellular products on the haemocyte functions. Since 2012, observations indicate a reemergence of this pathogen associated to cultured oyster [65]. However, *V. aestuarianus* has also been detected in different sites in Europe, including Baltic sea, Spain and Italy, during both environmental studies and bivalve monitoring, but most of the isolates obtained were classified as non-virulent [66].

For most of the cited *Vibrio* species, although pathogenic in experimental challenges, a direct causality could not be established with oyster mortality in nature. A possible explanation is the limited knowledge on the dynamic changes of bacterial populations within the host during disease progression. Recently, Lemire et al. [67] studied the oyster disease ecology of microdiverse *Vibrio* genotypes, applying a novelty approach that combines experimental ecology, a high-throughput infection assay and genome sequencing. The authors demonstrated that the onset of disease in oysters is associated with a progressive replacement of diverse benign colonizers by members of a virulent population which is phylogenetically coherent. Comparative genomics between virulent and nonvirulent populations allowed the identification of candidate virulence factors, specifically a putative outer membrane protein that is necessary for infection of oysters. In addition, results during experimental infections indicated that the presence of nonvirulent strains can facilitate the disease onset, contributing to increase the mortality. Therefore, in the case of *V. aestuarianus*, it seems that are the populations, and not the clones, the true units of pathogenesis in oysters.

VIRAL HYPOTHESIS

Bivalve Viral Diseases

A variety of viral families has been described in bivalves, including *Picornaviridae*, *Retroviridae*, *Birnaviridae*, or *Reoviridae*, among others [68, 69]. Although such agents have been reported worldwide associated with massive mortality outbreaks in economically important species, few studies confirmed their pathogenicity of these agents or determined the viral characterization.

Gill necrosis virus, haemocyte infection virus and oyster velar virus are among the irido-like viruses detected in oysters, *C. angulata* and *C. gigas* [70-72]. Papova-like viruses have been reported from different bivalve species, including *C. virginica*, C. gigas or *Pinctada maxima* [73-75]. On the other hand, there have been many reports on small virus-like particles in a variety of bivalves, such as mussels (*Mytilus edulis*, *M. galloprovincialis*, *Perna canaliculus*), oysters (*Pinctada fucata martensii*, *P. margaritifera*), scallops (*Pecten novaezelandiae*), clams (*Paphies ventricosa, Ruditapes decussatus*), and cockles (*Cerastoderma edule*) [76-82].

A viral etiology for disseminated neoplasia (DN), a proliferative cell disorder of the circulatory system, was also proposed after finding retrovirus in neoplastic *M. arenaria*, together with the fact that non-neoplastic clams that were inoculated with virus isolated from neoplastic clams developed DN [83, 84]. Such relationship was also proposed in the cases of neoplasia in cockles (*C. edule*) and golden carpet shell clams (*Venerupis aurea*) among other bivalve species [85, 86].

Herpes Virus and Summer Mortality

Herpes-like viral infections were fisrt described in the eastern oyster *C. virginica* from the east coast of the US in 1972 [87]. Later, herpes-like viral infections causing significant mortalities were reported in Pacific oysters (*C. gigas*) larvae in New Zealand and France [88, 89]. Similar viruses have been reported in other oyster species, including flat oysters (*Ostrea edulis, O. angasi* and *O. chilensis*) in Australia and France, Portuguese and Sumione oysters (*C. angulata, C. rivularis*), clams (*Ruditapes phillipinarum, R. decussatus*) and scallops (*Pecten maximus*) [90-97].

Since 1993 high mortalities of spat and juveniles of Pacific oyster were regularly observed during summer periods in different French locations [98, 99], but in summer of 2008 mortalities were widespread to all production areas in the country killing billions of young oysters [100]. The recently described Ostreid herpesvirus-1 (OsHV-1), the only member of Family *Malacoherpesviridae* [101], was detected in most samples of moribund oysters together with some *Vibrio* species [100].

The molecular characterization of the virus began in the late 1990's when Le Deuff and Renault [102] purified and and characterized the genome of a viral strain from Japanese oyster. Soon after, different molecular detection tools were available [103-105] facilitating the rapid identification and confirmation of OsHV-1 in bivalves and allowing the description of different variants [106, 107]. Segarra et al. [100], studying the strains from the 2008 mass mortalities, described the variant OsHV-1 μvar with polymorphisms in regions C and IA, which was considered as an emerging genotype. This variant was detected in subsequent years in France [108] and also in other countries in Europe, including Ireland, Spain, England, Italy and the Netherlands [109-114], and outside Europe in Australia, New Zealand and Korea [115-117]. The OsHV-1 μvar differs from OsHV-1 in a deletion fo 12 base pairs in ORF4 of the genome as well as in a higher virulence potential [110]. Virulence was demonstrated in experimental challenges, and horizontal transmission corroborated in cohabitation experiments [118]. On the other hand, the presence of virus in populations not presenting mortalities or in healthy individuals has also been reported [42, 43, 112], which for some authors indicated the possibility of existence of healthy carriers.

Histopathological studies showed a variety of alterations in affected individuals including the presence of hypertrophied nuclei with peripheral condensed chromatin [110], or an apparent increase of focal or diffuse cellularity, resembling haemocytosis, in the mantle or interstitial connective tissue [116]. However, such alterations are not specific for herpesvirus presence. Molecular techniques, including quantitative PCR and *in situ* detection, allowed to determine that gills and mantle are the target organs for viral replication [42, 43]. Schikorski et al. [118] suggested that infective viral particles enter first through the digestive gland and haemolymphatic system, and then viral particles are transported by haemolymph to the different target organs where an intense viral replication occurs leading to develoment of the disease.

Host Factors

One of the first observations of the summer mortality syndrome evidencing the implications of host factors was the different susceptibility of oysters related to age. Thus, mortality affects spat and juveniles but not adult oysters [119]. Nevertheless, viral DNA and proteins can be detected in adult oysters in the absence of mortality [120]. Gonadal maturation is also an intrinsic factor of risk in summer mortality [22, 121, 122]. Thus, triploid oysters, with reduced maturation, suffered the lowest mortalities and presented higher potential defence capacities than did diploids. In some rare observations, triploids demonstrating an unusually high reproductive state were as susceptible to summer mortality as diploids.

On the other hand, genetic variability is thought to be a major determinant in sensitivity to summer mortality [123], suggesting that the resistance is hereditable. This variability allowed the selection of oyster families with increased tolerance to stress and disease through selective breading [24, 124]. The availability of the different oyster families together with the development in the last years of molecular approaches to study the the differential gene expression constituted the perfect basis to design studies aiming to disentangle the host factors and/or the host-pathogen interactions in the summer mortality episodes.

Quantitative PCR, microarray-based identification, suppression substractive hybridization, digital gene expression, and QTL analysis, among other approaches have been employed to study the genes differentially expressed in resistant oyster families or as response to heat shock stress and *Vibrio* and/or ostreid herpesvirus infections [5, 125-137].

The first works already pointed out the differences in the expression of genes related with energy generation and immune function between resistant and susceptible progenies to *Vibrio* infection [125]. Thus, haemocyte phagocytosis and adhesive capacities were affected during the bacterial infection, but also a deregulation of the haemocyte oxidative metabolism could be observed, providing an opportunity to the pathogen by impairing haemocyte functions and survival [126]. Samain et al. [5], in both field and laboratory trials, observed that susceptible oyster invest more energy in reproduction and stay longer without spawning, showing also a higher respiration rate and an earlier decrease in adsorption efficiency during gametogenesis. Differences were also detected before the mortality episodes or after a thermal increase in the adhesive capacity of haemocytes, reactive oxygen species production, hyalinocyte concentrations or heat shock protein levels. All these facts suggested possible links between higher reproductive effort, partial spawning, specific stress response to temperature and hypoxia, and the infection process. More recently, de Lorgeril et al. [131] studied the transcriptome profiling after a *Vibrio* infection by means of digital expression analysis. They could identify genes related to signal transduction of immune response, adhesion and communication, of defence mechanisms of phagocytosis, actin cytosqueleton reorganization, cell trafficking and autophagy, as well as antioxidant and anti-apoptotic reactions. The results obtained support those of previous studies but allowed to go a step foreword in the understanding the complex molecular processes implicated in the oyster response.

The more recent studies employed dual transcriptomics of pathogen-host interactions, specifically OsHV-1. This approach offers an excellent chance to study the interactions occurring during an infection at the molecular level. Segarra et al. [135] suggested apoptosis as one of the main mechanisms involve in disease reistance in adult individuals. On the other hand, Rosani et al. [137]) identified most induced genes in *C. gigas*, including cytochrome P450, lectins, big defensin, hsp70, and follistatin. The role of the cytokine and interferon pathways was also demonstrated. Finally, apoptosis and autophagy related genes were also identified as playing a role in resistance to OsHV-1 infections.

Other authors investigated the resistance of other bivalves to summer mortality. Thus, Pezzati et al. [138] comparatively studied the interactions of *V. aestuarianus* with haemolymph of *C. gigas* and the mussel *Mytilus galloprovincialis* in order to identify the components related with resistance to infection. They found that vibrios were sensitive to the bactericidal activity of haemolymph from mussel but not from oyster, and identified an opsonin, related with the extrapallial protein of the mussel, that mediates in the interactions *Vibrio*-mussel haemocytes. In the presence of such opsonin, oyster haemocytes killed the bacteria almos as efficiently as mussel phagocytes. The authors pointed out that species-specific serum components imply differential interactions with vibrios, resulting in different sensitivities to the antibacterial activity of haemolymph, and that deciphering the molecular mechanisms implicated in the bactericidal activity of haemolymph may be important to define new strategies to control oyster summer mortality.

CONCLUSION

Through the present chapter, I have tried to highlight the complexity of the summer mortality syndrome pointing out the different aspects that have been related with the disease from its first reports. Such complexity, involving environmental and host factors as well as bacterial and viral pathogens, has hindered its understanding due to the difficulty of inclusion of all these aspects in a single study. Today, it seems clear that future studies must be interdisciplinary involving ecologists, microbiologists, immunologists and producers in order to fully decode the appearance of the syndrome.

One example is the recent work by Petton and coworkers [139], using specific-pathogen-free oysters infected in an estuarine environment, which observed that rapid *Vibrio* colonization followed by viral replication precedes oyster death, and that a high viral load is not sufficient to induce the full expression of the disease in absence of bacteria. In addition, they also found that the oyster physiology is a key feature of the disease, and that the oyster susceptibility may be acquired (as a result of environmental factors) or innate (derived from genetic traits). On the other hand, the authors observed that the farmed animals present more risk of disease transmission than wild animals and that geographic origin of oysters influences the carriage of pathogens.

More interdisciplinary studies are needed in order to confirm all these hypothesis. The availability of a variety of appropriate tools may help to develop such research, essential to get a full understanding of the problem and, therefore, to design effective measures and strategies to control summer mortalities affecting bivalves.

REFERENCES

[1] Paillard, C., Le Roux, F., Borrego, J.J. 2004. Bacterial disease in marine bivalves, a review of recent studies: Trends and evolution. *Aquat. Living Resour.* 17, 477-498.

[2] Labreuche, Y., Soudant, P., Goncalves, M., Lambert, C., Nicolas, J.L. 2006. Effects of extracellular products from the pathogenic *Vibrio aestuarianus* strain 01/32 on lethality and cellular immune responses of the oyster *Crassostrea gigas*. *De.v Comp. Immunol.* 30, 367-379.

[3] Pruzzo, C., Gallo, G., Canesi, L. 2005. Persistence of vibrios in marine bivalves: the role of interactions with haemolymph components. *Environ. Microbiol.* 6, 761-772.

[4] Cheney, D.P., MacDonald, B.F., Elston, R.A. 2000. Summer mortality of Pacific oysters, *Crassostrea gigas* (Thunberg): initial findings on multiple environmental stressors in Puget Sound, Washington. *J. Shellfish Res.* 19, 353-359.

[5] Samain, J.-F., Dégremont, L., Soletchnik, P., Haure, J., Bédier, E., Ropert, M., Moal, J., Huvet, A., Bacca, H., Van Wormhoudt, A., Delaporte, M., Costil, K., Pouvreau, S., Lambert, C., Boulo, V., Soudant, V., Nicolas, J.L., Le Roux, F., Renault, T., Gagnaire, B., Geret, F., Boutet, I., Burgeot, T., Boudry, P. 2007. Genetically based resistance to summer mortality in the Pacific oyster (*Crassostrea gigas*) and its relationshipwith physiological, immunological characteristics and infection processes. *Aquaculture* 268, 227-243.

[6] Soletchnik, P., Ropert, M., Maurizé, J., Fleury, P.F., Le Coz, F. 2007. Relationship between oyster mortality patterns and environmental data from monitoring databases along the coasts of France. *Aquaculture* 271, 384-400.

[7] Fisher, S.W. 1988. Environmental influence on host response: environmental influence on bivalve hemocyte function. *Am. Fish. Soc. Spec. Publ.* 18, 225-237.

[8] Fisher S.W., Auffret, M., Balouet, G. 1987. Response of European flat oyster (*Ostrea edulis*) hemocytes to acute salinity and temperature changes. *Aquaculture* 67, 179-190.

[9] King, J.W., Malham, S.K., Skov, M.W., Cotter, E., Latchford, J.W., Culloty, S.C., Beaumont, A.R. 2006. Growth of *Crassostrea gigas* spat and juveniles under different enviornmental conditions at two sites in Wales. *Aquat. Living Resour.* 19, 289-297.

[10] Malham, S.K., Cotter, E., O'Keeffe, S., Lynch, S., Culloty, S.C., King, J.W., Latchford, J.W., Beaumont, A.R. 2009. Summer mortality of the Pacific oyster, *Crassostrea gigas*, in the Irish sea: the influence of temperature and nutrients on health and survival. *Aquaculture* 287, 128-138.

[11] Berthelin, C., Kellner, K., mathieu, M. 2000. Histological characterisation and glucose incorporation into glycogen of the Pacific oyster *Crassostrea gigas* cells. *Mar. Biotechnol.* 2, 136-145.

[12] Cho, S.M., Jeong, W.G. 2005. Spawning impact on lysosomal stability of the Pacific oyster, *Crassostrea gigas*. *Aquaculture* 244, 383-387.

[13] Mori, K., Tamate, H.K., Inai, T.K., Itikawa, O. 1965. Studies on the mass mortality of the oyster in Matsushima Bay. V. Changes in the metabolism of lipids and glycogen of the oyster during the stages of sexual maturation and spawning. *Bull. Tohoko Reg. Fish Res. Lab.* 25, 66-68.

[14] Perdue, J., Beattie, J.H., Chew, K.K. 1981. Some relationships between gametogenetic cycle and summer mortality phenomenon in the Pacific oyster *Crassostrea gigas* in the Washington state. *J. Shellfish Res.* 1, 9-16.

[15] Li, Y., Qin, J.G., Abbott, C.A., Li, X., Benkendorff, K. 2007. Synergistic impacts of heat shock and spawning on the physiology and immune health of *Crassostrea gigas*: an explanation for summer mortality in Pacific oysters. *Am. J. Physiol. Regul. Integr. Comp. Physiol.* 293, R2353-R2362.

[16] Nell, J.A., Holliday, J.E. 1988. Effects of salinity on the growth and survival of Sydney rock oyster (*Saccostrea commercialis*) and Pacific oyster (*Crassostrea gigas*) larvae and spat. *Aquaculture* 68, 39-44.

[17] Brown, J.R., Hartwick, E.B. 1988a. Influences of temperature, salinity and available food upon suspended culture of the Pacific oyster, *Crassostrea gigas*. I. Absolute and allometric growth. *Aquaculture* 70, 231-251.

[18] Brown, J.R., Hartwick, E.B. 1988b. Influences of temperature, salinity and available food upon suspended culture of the Pacific oyster, *Crassostrea gigas*. II. Condition index and survival. *Aquaculture* 70, 253-267.

[19] Gagnaire, B., Froulin, H., Moreau, K., Thomas-Guyon, H., Renault, T. 2006. The effect of temperature and salinity on haemocyte activities of Pacific oyster, *Crassostrea gigas* (Thunberg). *Fish Shellfish Immunol.* 20, 536-547.

[20] Gray, J.S., Wu, R.S.S., Or, Y.Y. 2002. Effect of hypoxia and organic enrichment on the coastal marine environment. *Mar. Ecol. Prog. Ser.* 238, 249-279.

[21] Paterson, K.J., Schreider, M., Zimmerman, K.D. 2003. Effect of catchment disturbance on seston quality and quantity in Brisbane water, NSW. *Estuar. Coast Shelf. Sci.* 56, 355-368.

[22] Mori, K. 1979. Effects of artifical eutrophication on the metabolism of the Japanese oyster, *Crassostrea gigas*. *Mar. Biol.* 53, 361-369.

[23] Lipovsky, V.P., Chew, K.K. 1972. Mortality of Pacific oysters (*Crassostrea gigas*): the influence of the temperature and enriched seawater on oyster survival. *Proc. Natl. Shellfish Ass.* 62, 72-82.

[24] Beattie, J.H., Chew, K.K., Hershberger, W.K. 1980. Differential survival of selected strains of Pacific oyster *Crassostrea gigas* during summer mortality. *Proc. Natl. Shellfish Ass.* 70, 184-189.

[25] Soletchnik, P., Faury, N., Razet, D., Goulletquer, P. 1998. Hydrobiology of the Marennes-Oléron bay. Seasonal indices and analysis of trends from 1978 to 1995. *Hydrobiologia* 386, 131-146.

[26] Renault, T. 2011. Effects of pesticides on marine bivalves: what do we know and what do we need to know?. In: Stoytcheva, M. (Ed.) *Pesticides in the Modern World-Risks and Benefits*. InTech, DOI: 10. 5772/19155.

[27] Masuda, M., Satsuma, K., Sato, K. 2012. An environmental fate study of methoxychlor using water-sediment model system. *Biosci. Biotechnol. Biochem.* 76, 73-77.

[28] Savage, C., Thrush, S.F., Lohrer, A.M., Hewitt, J.E. 2012. Ecosystem services trascended boundaries: estuaries provide resource subsidies and influence functional diversity in coastal benthic communities. *PloS One* 7, e42708.

[29] Banerjee, B.D., Koner, B.C., Ray, A. 1996. Immunotoxicity of pesticides: perspectives and trends. *Indian J. Exp. Biol.* 34, 723-733.

[30] Yin, L., Lu, L., Prasad, K., Richfield, E.K., Unger, E.L., Xu, J., Jones, B.C. 2011. Genetic-based, differential susceptibility to paraquat neurotoxicity in mice. *Neurotoxicol. Teratol.* 33, 415-421.

[31] Baker, N.J., Bancroft, B.A., Garcia, T.S. 2013. A meta-analysis of the effects of pesticides and fertilizers on survival and growth of amphibians. *Sci. Total Environ.* 449, 150-156.

[32] Gagnaire, B., Gay, M., Huvet, A., Daniel, J.-Y., Saulnier D., Renault, T. 2007. Combination of a pesticide exposure and a bacterial challenge: in vivo effects on immune response of Pacific oyster, *Crassostrea gigas* (Thunberg). *Aquat. Toxicol. Amst. Neth.* 84, 92-102.

[33] Ramu, K., Kajiwara, N., Isobe, T., Takahashi, S., Kim, E.-Y., Min, B.-Y., We, S.U., Tanabe, S. 2007. Spatial distribution and accumulation of brominated flame retardants, polychlorinated biphenyls and organochlorine pesticides in blue mussels (*Mytilus edulis*) from coastal waters of Korea. *Environ. Pollut.* 148, 562-569.

[34] Bernal-Hernández, Y.Y., Medina-Díaz, I.M., Robledo-Marenco, M.L., Velázquez-Fernández, J.B., Girón-Pérez, M.I., Ortega-Cervantes, L., Maldonado-Vázquez, W.A., Rojas-García, A.E. 2010. Acetylcholinesterase and metallothionein in oysters (*Crassostrea corteziensis*) from a subtropical Mexican Pacific estuary. *Ecotoxicology* 19, 819-825.

[35] Gagnaire, B., Renault, T., Bouilly, K., Lapegue, S., Thomas-Guyon, H. 2003. Study of atrazine effects on Pacific oyster, *Crassostrea gigas*, haemocytes. *Curr. Pharm. Des.* 9, 193-199.

[36] Tanguy, A., Boutet, I., Laroche, J., Moraga, D. 2005. Molecular identification and expression study of differentially regulated genes in the Pacific oyster *Crassostrea gigas* in response to pesticide exposure. *FEBS J.* 272, 390-403.

[37] Moreau, P., Faury, N., Burgeot, T., Renault, T. 2015. Pesticides and ostreid herpesvirus 1 infection in the Pacific oyster, *Crassostrea gigas*. *PLoS One* 10, e0130628.

[38] Soletchnik, P., Lambert, C., Costil, K. 2005. Summer mortality of *Crassostrea gigas* (Thunberg) in relation to rearing environmental conditions. *J. Shellfish Res.* 24, 197-207.

[39] Soletchnik, P., Le Moine, O., Faury, N., Razet, D., Geairon, P., Goulletquer, P. 1999. Mortalité de l'huître *Crassostrea gigas* dans le bassin de Marennes-Oléron: étude de la variabilité spatiale de son environnement et de sa biologie par un système d'informations géographiques (SIG) (Summer mortality of the oyster in the bay Marennes-Oléron: spatial variability of environment and biology using a geographical information system (GIS)). *Aquat. Living Resour.* 12, 131-143.

[40] Murray, A.G., Marcos-Lopez, M., Collet, B., Munro, L.A. 2012. A review of the risk posed to Scottish mollusc aquaculture from *Bonamia*, *Marteilia* and oyster herpes virus. *Aquaculture* 370-371, 7-13.

[41] Martenot, C., Oden, E., Travaille, E., Malas, J.P., Houssin, M. 2010. Comparison of two real-time PCR methods for detection of ostrid herpes virus 1 in the Pacific oyster *Crassostrea gigas*. *J. Virol. Methods* 170, 86-89.

[42] Arzul, I., Renault, T., Thébault, A., Gérard, A. 2002. Detection of oyster herpesvirus DNA and proteins in asymptomatic *Crassostrea gigas* adults. *Virus Res.* 84, 151-160.

[43] Pepin, J.F., Riou, A., Renault, T. 2008. Rapid and sensitive detection of ostreid herpesvirus 1 in oyster samples by real-time PCR. *J. Virol. Methods* 149, 269-276.

[44] Saulnier, D., De Decker, S., Haffner, P. 2009. Real-time PCR assay for rapid detection and quantification of *Vibrio aestuarianus* in oyster and seawater: a useful tool for epidemiological studies. *J. Microbiol. Methods* 77, 191-197.

[45] McCleary, S., Hensilwood, K. 2015. Novel quantitative TaqMan® MGB real-time PCR for sensitive detection of *Vibrio aestuarianus* in *Crassostrea gigas*. *Dis. Aquat. Org.* 114, 239-248.

[46] Petton, B. Pernet, F., Robert, R., Boudry, P. 2013. Temperature influence on pathogen transmission and subsequent mortalities in juvenile Pacific oysters *Crassostrea gigas*. *Aquacult. Environ. Interact.* 3, 257-273.

[47] Normand, J., Blin, J.-L., Jouaux, A. 2014. Rearing practices identified as risk factors for ostreid herpesvirus 1 (OsHV-1) infection in Pacific oyster *Crassostrea gigas* spat. *Dis. Aquat. Org.* 110, 201-211.

[48] Beaz-Hidalgo, R., Balboa, S., Romalde, J.L., Figueras, M.J. 2010. Diversity and pathogenicity of *Vibrio* species in cultured bivalve molluscs. *Environ. Microbiol. Reports* 2, 34-43.

[49] Romalde, J.L., Diéguez, A.L., Lasa, A., Balboa, S. 2014. New *Vibrio* species associated to molluscan microbiota: a review. *Front. Microbiol.* 4, 413.

[50] Guillard, R.R.L. 1959. Further evidence of the destruction of bivalve larvae by bacteria. *Biol. Bull.* 117, 258-266.

[51] Tubiash, H.S., Chanley, P.E., Leifson, E. 1965. Bacillary necrosis, a disease of larval and juvenile bivalve mollusks. I. Etiology and epizootiology. *J. Bacteriol.* 90, 1036-1044.

[52] Tubiash, H.S., Otto, S.V. 1986. Bacterial problems in oysters. A review. In: Vivarès CP, Bonami JR, Jasper E. (Eds.) *Pathology in Marine Aquaculture*. Bredene, Belgium, *Eur. Aquacult. Soc. Spec. Publ.* 9: 233-242.

[53] Tubiash, H.S., Colwell, R.R., Sakazaki, R. 1970. Marine vibrios associated with bacillary necrosis, a disease of larval and juvenile bivalve molluscs. *J. Bacteriol.* 103, 272-273.

[54] Elston, R.A., Hasegawa, H., Humphrey, K.L., Polyak, I.K., Häse, C.C. 2008. Re-emergence of *Vibrio tubiashii* in bivalve shellfish aquaculture: severity, environmental drivers, geographic extent and management. *Dis. Aquat. Org.* 82, 119-134.

[55] Hasegawa, H., Lind, E.J., Boin, M.A., Häse, C. 2008. The extracellular metalloprotease of *Vibrio tubiashii* is a major virulence factor for Pacific oyster (*Crassostrea gigas*) larvae. *Appl. Environ. Microbiol.* 74, 4101-4110.

[56] Elston, R., Leibovitz, L. (1980). Pathogenesis of experimental vibriosis in larval American oysters, *Crassostrea virginica*. *Can. J. Fish Aquat. Sci.* 37, 964-978.

[57] Lipp, P.R., Brown, B., Liston, J., Chew, K. 1976. Recent findings on the summer diseases of Pacific oysters. *Proc. Natl. Shellfish Ass.* 65, 9-10.

[58] Lacoste, A., Jalabert, F., Malham, S., Cueff, A., Gelebart, F., Cordevant, C., Lange, M., Poulet, S.A. 2001. A *Vibrio splendidus* strain is associated with summer mortality of juvenile oysters *Crassostrea gigas* in the Bay of Morlaix (North Brittany, France). *Dis. Aquat. Org.* 46, 139-145.

[59] Waechter, M., Le Roux, F., Nicolas, J.L., Marissal, E., Berthe, F. 2002. Characterization of pathogenic bacteria of the cupped oyster *Crassostrea gigas*. *C. R. Biol.* 325, 231.

[60] Gay, M., Berthe, F.C.J., Le Roux, F. 2004. Screening of *Vibrio* isolates to develop an experimental infection model in the Pacific oyster *Crassostrea gigas*. *Dis. Aquat. Org.* 59, 49-56.

[61] Gay, M., Renault, T., Pons, A.M., Le Roux, F. 2004. Two *Vibrio splendidus* related strains collaborate to kill *Crassostrea gigas*: taxonomy and host alterations. *Dis. Aquat. Org.* 62, 65-74.

[62] Garnier, M., Labreuche, Y., García, C., Robert, M., Nicolas, J.L. 2007. Evidence for the involvement of pathogenic bacteria in summer mortalities of the Pacific oyster *Crassostrea gigas*. *Microb. Ecol.* 53, 187-196.

[63] Allain, G., Arzul, I., Chollet, B., Cobret, L., de Decker, S., Faury, N., Ferrand, S., François, C., Garcia, C., Haffner, P., Joly, J.P., Michel, J., Miossec, L., Morga, B., Nicolas, J.L., Omnes, E., Pépin, J.F., Robert, M., Saulnier, D., Schikorski, D., Segarra, A., Tourbiez, D., Renault, T. 2009. Summer mortality outbreaks of French Pacific oysters, *Crassostrea gigas* in 2008: research and detection of pathogens. Proc. 14th EAFP International Conference. Diseases of Fish and Shellfish. Prague, Czech Republic. pp. 127.

[64] Tison, D.L., Seidler, R.J. 1983. *Vibrio aestuarianus* – a new species from estuarine waters and sehllfish. *Int. J. Syst. Bacteriol.* 33, 699-702.

[65] Solomieu, V.B., Renault, T., Travers, M.-A. 2015. Mass mortality in bivalves and the intricate case of the Pacific oyster, *Crassostrea gigas*. *J. Invertebr. Pathol.* 131, 2-10.

[66] Goudenège, D., Travers, M.A., Lemire, A., Haffner, P., Labreuche, Y., Tourbiez, D., Petton, B., Mangenot, S., Calteau, A., Mazel, D., Nicolas, J.L., Jacq, A., Le Roux, F. 2014. A single regulatory gene is sufficient to alter *Vibrio aestuarianus* pathogenicity. *Environ. Microbiol.* 17, 4189-4199.

[67] Lemire, A., Goudenège, D., Versigny, T., Petton, B., Calteau, A., Labreuche, Y., Le Roux, F. 2015. Populations, not clones, are the unit of vibrio pathogenesis in naturally infected oysters. *ISME J* 9, 1523-1531.

[68] Renault, T., Novoa, B. 2004. Viral infection among bivalves. *Aquat. Living Resour.* 17, 397-409.

[69] Ren, W., Chen, H., Renault, T., Cai, Y., Bai, C., Wang, C., Huang, J. 2013. Complete genome sequence of acute viral necrosis virus associated with massive mortalitiy outbreaks in the Chinese scallop, *Chlamys farreri*. *Virology J.* 10, 110.

[70] Comps, M. 1980. Les infections virales associés aux épizooties des huîtres du genre *Crassostrea* (Viral infections associated with epizootics of oysters of the genus *Crassostrea*). International Council for the Exploration of the Sea. Spec. Meeting on Diseases of Commercially Important Marine Fish and Shellfish. Copenhagen N°6.

[71] Comps, M., Bonami, J.R. 1977. Infection virale associé à des mortalitès chez l'huître *Crassostrea angulata* (Viral infection associated with mortalities in the oyster *Crassostrea angulata*). *Th. C. R. Acad. Sci. D.* 285, 1139-1140.

[72] Elston, R., Wilkinson, M.T. 1985. Pathology, management and diagnosis of oyster velar virus disease (OVVD). *Aquaculture* 48, 189-210.

[73] McGladdery, S.E., Stephenson, M.F. 1994. A viral infection of the gonads of eastern oyster (*Crassostrea virginica*) from Atlantic Canada. *Bull. Aquat. Ass. Can.* 94, 84-86.

[74] Norton, J.H., Shepherd, M.A., Prior, H.C. 1993. Papova-like virus infection of the golden-lipped pearl oyster, *Pinctada maxima*, from the Torres Strait, Australia. *J. Invertebr. Pathol.* 62, 198-200.

[75] Choi, D.L., Lee, N.-S., Choi, H.J., Park, M.-A., McGladdery, S.E., Park, M.S. 2004. Viral gametocytic hypertrophy caused by a papova-like virus infection in the Pacific oyster *Crassostrea gigas* in Korea. *Dis. Aquat. Org.* 59, 205-209.

[76] Rasmussen, L.P.D. 1986. Virus-associated granulocytomas in the marine mussel, *Mytilus edulis*, from three sites in Denmark. *J. Invertebr. Pathol.* 48, 117-123.

[77] Jones, J.B., Scotti, P.D., Dearing, S.C., Wesney, B. 1996. Virus-like particles associated with marine mussel mortalities in New Zealand. *Dis. Aquat. Org.* 25, 143-149.

[78] Hine, P.M., Wesney, B. 1997. Virus-like particles associated with cytopathology in the digestive gland epithelium of scallops *Pecten novaezelandiae* and toheroa *Paphies ventricosum*. *Dis. Aquat. Org.* 29, 197-204.

[79] Miyazaki, T., Goto, K., Kobayashi, T., Kageyama, T., Miyata, M. 1999. Mass mortalities associated with a virus disease in Japanese pearl oysters *Pinctada fucata martensii*. *Dis. Aquat. Org.* 37, 1-12.

[80] Comps, M., Herbaut, C., Fougerousse, A. 1999. Virus-like particles in pearl oyster *Pinctada maragaritifera*. *Bull. Eur. Ass. Fish Pathol.* 19, 85-88.

[81] Novoa, B., Figueras, A. 2000. Virus-like particle associated with mortalities of the carpet-shell clam *Ruditapes decussatus*. *Dis. Aquat. Org.* 39, 147-149.

[82] Carballal, M.J., Villalba, A., Iglesias, D., Hine, P.M. 2003. Virus-like particles associated with large foci of heavy hemocytic infiltration in cockles *Cerastoderma edule* from Galicia (NW Spain). *J. Invertebr. Pathol.* 84, 234-237.

[83] Oprandy, J.J., Chang, P.W., Pronovost, A.D., Cooper, K.R., Brown, R.S., Yates, V.J. 1981. Isolation of a viral agent causing hematopoietic neoplasia in the soft-shell clam, *Mya arenaria*. *J. Invertebr. Pathol.* 38, 45-51.

[84] Oprandy, J.J., Chang, P.W. 1983. 5-Bromodeoxyrudine induction of hematopoietic neoplasia and retrovirus activation in the soft-shell clam, *Mya arenaria*. *J. Invertebr. Pathol.* 42, 196-206.

[85] Romalde, J.L., Vilariño, M.L., Beaz, R., Rodríguez, J.M., Díaz, S., Villalba, A., Carballal, M.J. 2007. Evidence of retroviral etiology for disseminated neoplasia in cockles (*Cerastoderma edule*). *J. Invertebr. Pathol.* 94, 95-101.

[86] Manso, C.F., Día, S., Carballal, M.J., Villalba, A., Romalde, J.L. 2012. Detection of reverse transcriptase activity in golden carpet shell clams (*Venerupis aurea*) with disseminated neoplasia. *Bull. Eur. Ass. Fish Pathol.* 32, 56-63.

[87] Farley, C.A., Banfield, W.G., Kasnic Jr, G., Foster, W.S. 1972. Oyster herpes-type virus. *Science* 178, 759-760.

[88] Hine, P.M., Wesney, B., Hay, B.E. 1992. Herpes viruses associated with mortalities among hatchery-reared larval Pacific oysters, *Crassostrea gigas*. *Dis. Aquat. Org.* 12, 135-142.

[89] Nicolas, J.L., Comps, M., Cochennec, N. 1992. Herpes-like virus infecting Pacific oyster larvae, *Crassostrea gigas*. *Bull. Eur. Ass. Fish Pathol.* 12, 11-13.

[90] Hine, P.M., Thorne, T., 1997. Replication of herpes-like viruses in haemocytes of adult flat oysters *Ostrea angasi* (Sowerby 1871): an ultrastructural study. *Dis. Aquat. Org.* 29, 197-204.

[91] Hine, P.M., Wesney, B., Besant, P. 1998. Replication of herpes-like viruses in larvae of the flat oyster (*Tiostrea chilensis*) at ambient temperatures. *Dis. Aquat. Org.* 32, 161-171.

[92] Renault, T., Le Deuff, R.-M., Chollet, B., Cochennec, N., Gérard, A., 2000. Concomitant herpes-like virus infections in hatchery-reared larvae and nursery-cultured spat *Crassostrea gigas* and *Ostrea edulis*. *Dis. Aquat. Org.* 42, 173-183.

[93] Renault, T., Lipart, C., Arzul, I. 2001. A herpes-like virus infecting non-ostreid bivalve species: virus replication in *Ruditapes philippinarum* larvae. *Dis. Aquat. Org.* 45, 1-7.

[94] Renault, T., Lipart, C., Arzul, I. 2001. A herpes-like virus infecting *Crassostrea gigas* and *Ruditapes philippinarum* larvae in France. *J. Fish Dis.* 24, 369-376.

[95] Renault, T., Arzul, I. 2001. Herpes-like virus infections in hatchery-reared bivalve larvae in Europe: specific viral detection by PCR. *J. Fish Dis.* 24, 161-167.

[96] Arzul, I., Renault, T., Lipart, C., 2001. Experimental herpes-like viral infections in marine bivalves: demonstration of interspecies transmission. *Dis. Aquat. Org.* 46, 1-6.

[97] Arzul, I., Nicolas, J.-L., Davison, A.J., Renault, T. 2001. French scallops: a new host for ostreid herpesvirus-1. *Virology* 290, 342-349.

[98] Davison, A.J., Trus, B.L., Cheng, N., Steven, A.C., Watson, M.S., Cunningham, C., Le Deuff, R.M., Renault, T. 2005. A novel class of herpesvirus with bivalve hosts. *J. Gen. Virol.* 86, 41-53.

[99] Garcia, C., Thébault, A., Dégremont, L., Arzul, I., Miossec, L., Robert, M., Chollet, B., François, C., Joly, J.-P., Ferrand, S., Kerdudou, N., Renault, T. 2011. Ostreid herpesvirus 1 detection and relationship with *Crassostrea gigas* spat mortality in France between 1998 and 2006. *Vet. Res.* 42, 73.

[100] Segarra, A., Pépin, J.F., Arzul, I., Morga, B., Faury, N., Renault, T. 2010. Detection and description of a particular Ostreid herepesvirus 1 genotype associated with massive mortality outbreaks of Pacific oysters, *Crassostreea gigas*, in France in 2008. *Virus Res.* 153, 92-99.

[101] Davison, A.J., Eberle, R., Ehlers, B., Hayward, G.S., McGeoch, D.J., Minson, A.C., Pellet, P.E., Roizman, B., Studdert, M.J., Thiry, E. 2009. The order Herpesvirales. *Arch. Virol.* 154, 171-177.

[102] Le Deuff, R.M., Renault, T. 1999. Purification and partial genome characterization of a herpes-like virus infecting the Japanese oyster, *Crassostrea gigas*. *J. Gen. Virol.* 80, 1317-1322.

[103] Renault, T., Lipart, C. 1998. Diagnosis of herpes-like virus infections in oysters using moleculatr techniques. *Eur. Aquacult. Soc. Spec. Publ.* 26, 235-236.

[104] Renault, T., Le Deuff, R.M., Lipart, C., Delsert, C. 2000. Development of a PCR procedure for the detection of a herpes-like virus infecting oysters in France. *J. Virol. Methods* 88, 41-50.

[105] Lipart, C. Renault, T. 2002. Herpes-like virus detection in infected *Crassostrea gigas* spat using DIG-labelled probes. *J. Virol. Methods* 101, 1-10.

[106] Friedman, C.S., Estes, R.M., Stokes, N.A., Burge, C.A., Hargove, J.S., Barber, B.J., Elston, R.A., Burreson, E.M., Reece, K.S. 2005. Herpes virus in juvenile Pacific oysters *Crassostrea gigas* from Tomales Bay, California, coincides with summer mortality episodes. *Dis. Aquat. Org.* 63, 33-41.

[107] Moss, J.A., Burreson, E.M., Cordes, J.F., Cungan, C.F., Brown, G.D., Wang, A., Wu, X., Reece, K.S. 2007. Pathogens in *Crassostrea ariakensis* and other Asian oyster species: implications for non-native oyster introduction in Chesapeake Bay. *Dis. Aquat. Org.* 77, 207-233.

[108] Martenot, C., Oden, E., Trvaillé, E., Malas J.-P., Houssin, M. 2011. Detection of different variants of ostreid herpesvirus 1 in the Pacific oyster, *Crassostrea gigas* between 2008 and 2010. *Virus Res.* 160, 25-31.

[109] Peeler, E.J., Reese, R.A., Cheslett, D.L., Geoghegan, F., Power, A., Thrush, M.A. 2012. Investigation of mortality in Pacific oysters associated with *Ostreid herpesvirus*-1 μVar in the Republic of Ireland in 2009. *Prev. Vet. Med.* 105, 136-143.

[110] Roque, A., Carrasco, N., Andree, K.B., Lacuesta, B., Elandaloussi, L., Gairin, I., Rodgers, C.J., Furones, M.D. 2012. First report of OsHV-1 microvar in Pacific oyster (*Crassostrea gigas*) cultures in Spain. *Aquaculture* 324-325, 303-306.

[111] OIE. 2010. Immediate notification Oyster herpesvirus (OsHV-1 Var), United Kingdom 21/07/2010. WAHID Interface-OIE World Animal Health Information Database. http://www.oie.int/wahis_2/public/wahid.php/reviewreport/Review?page_refer=MapEventSummary&reportid=9527.

[112] Dundon, W.G., Arzul, I., Omnes, R., Robert, M., Magnabosco, C., Zambon, M., Gennari, L., Toffan, A., Terregino, C., Capua, I, Arcangeli, G. 2011. Detection of type 1 ostreid herpes variant (OsHV-1 μVar) with no associated mortality in French-origin Pacific cupped oyster *Crassostrea gigas* farmed in Italy. *Aquaculture* 314, 49-52.

[113] Engelsma, M.Y. 2012. *Verslag monitoring oester herpesvirus 2011: Resultaten van onderzoek naar de aanwezigheid van oester herpesvirus in Japanese oester van het Grevelingenmeer en the Oosterschelde* (Oyster herpes virus monitoring report 2011: Results of testing for the presence of oyster herpesvirus in Japanese oyster in the Grevelingen and the Oosterschelde). *CVI-rapport 12/CVI0098.* Centraal Veterinari Instituut, Lelystad.

[114] Gittenberger, A., Voorbergen-Laatman, M.A., Emgelsma, M.Y. 2016. Ostreid herpesvirus OsHV-1 μVar in Pacific oysters *Crassostrea gigas* (Thunberg 1793) of the Wadden Sea, a UNESCO world heritage site. *J. Fish Dis.* 39, 105-109.

[115] Jenkins, C., Hick, P., gabor, M., Spiers, Z., Fell, S.A., Gu, X., Read, A., Go, J., Dove, M., O'Connor, W., Kirkland, P.D., Frances, J. 2013. Identification and characterisation of an ostreid herpesvirus-1 microvariant (OsHV-1 μ-var) in *Crassostrea gigas* (Pacific oysters) in Australia. *Dis. Aquat. Org.* 105, 109-126.

[116] Keeling, S.E., Brosnahan, C.L., Williams, R., Gias, E., Hannah, M., Bueno, R., McDonald, W.L., Johnston, C. 2014. New Zealand juvenile oyster mortality associated with ostreid herpesvirus 1 – an opportunistic longitudinal study. *Dis. Aquat. Org.* 109, 231-239.

[117] Hwang, J.Y., Park, J.J., Yu, H.J., Hur, Y.B., Arzul, I., Couraleau, Y., Park, M.A. 2013. Ostreid herpesvirus 1 infection in farmed Pacific oyster larvae *Crassostrea gigas* (Thunberg) in Korea. *J. Fish Dis.* 36, 969-972.

[118] Schikorski, D., Faury, N., Pepin, J.F., Sailnier, D., Tourbiez, D., Renault, T. 2011. Experimental ostreid herpesvirus 1 infection of the Pacific oyster *Crassostrea gigas*: kinetics of virus DNA detection by qPCR in seawater and in oyster samples. *Virus Res.* 155, 28-34.

[119] Oden, E., Martenot, C., Berthaux, M., Travaillé, E., Malas, J.P., Houssin, M. 2011. Quantification of ostreid herpesvirus 1 (OsHV-1) in *Crassostrea gigas* by real-time PCR: determination of a viral load threshold to prevent summer mortalities. *Aquaculture* 317, 27-31.

[120] Barbosa-Solomieu, V., Miossec, L., Vázquez-Juárez, R., Ascencio-Valle, F., Renault, T. 2004. Diagnosis of ostreid herpesvirus 1 in fixed paraffin-embedded archival samples using PCR and in situ hybridization. *J. Virol. Methods* 119, 65-72.

[121] Glude, J.B. 1975. A summary report of the Pacific Coast oyster mortality investigations 1965-1972. Proceedings of the Third US-Japan Meeting on Aquaculture at Tokyo, Japan, p. 28.

[122] Perdue, J.A., Beattie, J.H. and Chew, K.K. 1981. Some relationships between gametogenic cycle and summer mortality phenomenon in the Pacific oyster (*Crassostrea gigas*) in Washington state. *J. Shellfish Res.* 1, 9-16.

[123] Hershberger, W.K., Perdue, J.A., Beattie, J.H. 1984. Genetic selection and systematic breeding in Pacific oyster culture. *Aquaculture* 39, 237-245.

[124] Dégremont, L., Berdier, E., Soletchnik, P., Ropert, M., Huvet, Moal, J., Samain, J.-F., Boudry, P. 2005. Relative importance of family, site, and field placement timing on survival, grwoth, and yield of hatchery-produced Pacific oyster spat (*Crassostrea gigas*). *Aquaculture* 249, 41-53.

[125] Huvet, A., Herpin, A., Dégremont, L., Labreuche, Y., Samain, J.-F., Cunningham, C. 2004. The identification of genes from the oyster *Crassostrea gigas* that are differentially expressed in progeny exhibiting opposed susceptibility to summer mortality. *Gene* 343, 211-220.

[126] Labreuche, Y., Lambert, C., Soudant, P., Boulo, V., Huvet, A., Nicolas, J.-L. 2006. Cellular and molecular hemocyte responses of the Pacific oyster, *Crassostrea gigas*, following bacterial infection with *Vibrio aestuarianus* strain 01/32. *Microb. Infect.* 8, 2715-2724.

[127] Sauvage, C., Pépin, J.F., Lapègue, S., Boudry, P., Renault, T. 2009. Ostreid herpesvirus 1 infection in families of the Pacific oyster, *Crassostrea gigas*, during a summer mortality outbreak: differences in viral DNA detection and quantification using real-time PCR. *Virus Res.* 142, 18-187.

[128] Sauvage, C., Boudry, P., de Koning D.-J., Haley, C.S., Heurtebise, S., Lapègue, S. 2010. QTL for resistance to summer mortality and OsHV-1 load in the Pacific oyster (*Crassostrea gigas*). *Anim Genet* 41, 390-399.

[129] Lang, R.P., Bayne, C.J., Camara, M.D., Cunningham, C., Jenny, M.J., Langdon, C.J. 2009. Transcriptome profiling of selectively bred Pacific oyster *Crassostrea gigas* families that differ in tolerance to heat shock. *Mar. Biotechnol.* 11, 650-668.

[130] Fleury, E., Moal, J., Boulo, V., Daniel, J.-Y., Mazauris, D., Hénaut, A., Corporeau, C., Boudry, P., Favrel, P., Huvet, A. 2010. Microarray-based identification of gonad transcripts differentially expressed between lines of Pacific oyster selected to be resistant or susceptible to summer mortality. *Mar. Biotechnol.* 12, 326-339.

[131] de Lorgeril, J., Zenagui, R., Rosa, R.D., Piquemal, D., Bachère, E. 2011. Whole transcriptome profiling of successful immune response to *Vibrio* infections in the oyster *Crassostrea gigas* by digital gene expression analysis. *PLoS One* 6, e23142.

[132] Fleury, E., Huvet, A. 2012. Microarray analysis highlights immune response of Pacific oysters as a determinant of resistance to summer mortality. *Mar. Biotechnol.* 14, 203-217.

[133] Rosa, R.D., de Lorgeril, J., Tailliez, P., Bruno, R., Piquemal, D., Bachère, E. 2012. A hemocyte gene expresion signature correlated with predictive capacity of oysters to survive *Vibrio* infections. *BMC Genomics* 13, 252.

[134] Green, T.J., Montagnani, C., Benkendorff, K., Robinson, N., Speck, P. 2014. Ontogeny and water temperature influences the antiviral response of the Pacific oyster, *Crassostrea gigas*. *Fish Shellfish Immunol.* 36, 151-157.

[135] Segarra, A., Baillon, L., Tourbiez, D., Benabdelmoura, A., Faury, N., Bourgougnon, N., Renault, T. 2014. Ostreid herpesvirus type 1 replication and host response in adult Pacific oysters, *Crassostrea gigas*. *Vet. Res.* 45, 103.

[136] Normand, J., Li, R., Quillien, V., Nicolas, J.-F., Boudry, P., Pernet, F., Huvet, A. 2014. Contrasted survival under field or controlled conditions displays associations between mRNA levels of candidate genes and response to OsHV-1 infection in the Pacific oyster *Crassostrea gigas*. *Mar. Genom.* 15, 95-102.

[137] Rosani, U., Varotto, L., Domeneghetti, S., Arcangeli, G., Pallavicini, A., Venier, P. 2015. Dual analysis of host and pathogen transcriptomes in ostreid herpesvirus 1-positive *Crassostrea gigas*. *Environ. Microbiol.* 17, 4200-4212.

[138] Pezzati, E., Canesi, L., Damonte, G., Salis, A., Marsano, F., Grande, C., Vezzulli, L., Pruzzo, C. 2015. Susceptibility of *Vibrio aestuarianus* 01/032 to the antibacterial activity of *Mytilus* haemolymph: identification of a serum opsonin involved in mannose-sensitive interactions. *Environ. Microbiol.* 17, 4271-4279.

[139] Petton, B., Bruto, M., James, A., Labreuche, Y., Alunno-Bruscia, M., Le Roux F. 2015. *Crassostrea gigas* mortality in France: the usual suspect, a herpes virus, may not be the killer in this polymicrobial opportunistic disease. *Front Microbiol.* 6, 686.

About the Editor

Jesús L. Romalde
Professor of Microbiology
Department of Microbiology and Parasitology,
Faculty of Biology and Institute of Aquaculture,
Universidade de Santiago de Compostela, Spain
Email: jesus.romalde@usc.es

Prof. Jesús L. Romalde has been involved in research in microbiology of aquatic organisms for more than 25 years, including different aspects of bacterial and viral pathogens. He was appointed Full Professor of Microbiology in 2010, being a world leader in bacterial taxonomy of fish and shellfish microorganisms. Since 1987, he has participated in 38 founded R+D projects from public calls, being the principal investigator in 15 of them. In addition, he has been involved in more than 20 industry-funded projects, being the scientist in charge in 10 of them. He has co-authored 211 publications (171 research papers and 40 book chapters and monographies) and has presented 25 invited talks. He has directed 12 Ph.D. thesis and 20 Master research projects. Between 2012 and 2014, he holded the position of Vice-rector for Degrees and Teaching and Research Staff of the University of Santiago de Compostela (Spain).

INDEX

C

D

E

F

G

H

I

J

L

M

N

O

P

R

S